Progress in Probability
Volume 23

Sums, Trimmed Sums and Extremes

Marjorie G. Hahn
David M. Mason
Daniel C. Weiner
Editors

1991

Birkhäuser
Boston · Basel · Berlin

Marjorie G. Hahn
Department of Mathematics
Tufts University
Medford, MA 02155

David M. Mason
Department of
  Mathematical Sciences
University of Delware
Newark, DE 19716

Daniel C. Weiner
Department of Mathematics
Boston University
Boston, MA 02215

Library of Congress Cataloging-in-Publication Data
Sums, Trimmed sums and extremes / Marjorie G. Hahn, Daniel C. Weiner,
  David M. Mason, editors.
      p.   cm. — (Progress in probability ; 23)
      Includes bibliographical references.
      ISBN-13: 978-1-4684-6795-6   e-ISBN-13: 978-1-4684-6793-2
      DOI: 1 0.1007/978-1-4684-6793-2

      1. Summability theory. 2. Stochastic sequences. 3. Asymptotic
expansions. I. Hahn, Marjorie G. II. Weiner, Daniel C.
III. Mason, David M. IV. Series.
QA292.S85   1991
515'.243—dc20   90-49892
                                                          CIP

Printed on acid-free paper.

Camera-ready copy provided by the editors.

9 8 7 6 5 4 3 2 1

ISBN-13: 978-1-4684-6795-6

PREFACE

The past decade has seen a resurgence of interest in the study of the asymptotic behavior of sums formed from an independent sequence of random variables. In particular, recent attention has focused on the interaction of the extreme summands with, and their influence upon, the sum. As observed by many authors, the limit theory for sums can be meaningfully expanded far beyond the scope of the classical theory if an "intermediate" portion (i.e., an unbounded number but a vanishingly small proportion) of the extreme summands in the sum are deleted or otherwise modified ("trimmed"). The role of the normal law is magnified in these intermediate trimmed theories in that most or all of the resulting limit laws involve variance-mixtures of normals.

The objective of this volume is to present the main approaches to this study of intermediate trimmed sums which have been developed so far, and to illustrate the methods with a variety of new results.

The presentation has been divided into two parts. Part I explores the approaches which have evolved from classical analytical techniques (conditioning, Fourier methods, symmetrization, triangular array theory). Part II is based on the quantile transform technique and utilizes weak and strong approximations to uniform empirical process.

The analytic approaches of Part I are represented by five articles involving two groups of authors. The first two articles expose the basic approaches, outline the problems considered, survey known results, and provide some new results together with a discussion of open problems. The next two articles illustrate and apply the techniques discussed earlier, with some emphasis on their flexibility. The final article in Part I rounds out the discussion of the analytic approaches to trimming by exploring possible techniques for multidimensional data and their implications. It may be noted that the first, second and final articles in Part I each presents its own views, motivation, and approach to trimming without attempting to be comprehensive in covering the subject.

Part II is devoted to the approach based on the quantile-transform and uniform empirical process. In many respects, this approach can be viewed as more probabilistically based and intuitively motivated. The first article in Part II provides a comprehensive survey which explains the basic approach, describes the range of problems considered, summarizes the results already obtained, and discusses new and open problems. The second article provides what is currently the shortest and most accessible proof of the basic weighted approximation theorem on which the quantile approach relies. The remaining papers in Part II illustrate the wide range of applicability of these methods.

This volume is written for both the non-specialist and specialist. Anyone who has had a solid year's course in measure theoretic probability should be able to understand the survey articles in each chapter. These survey articles provide a unified and complete introduction to the approaches taken in this subject. The remaining articles illustrate the frontiers of current research as well as the range of applications.

Each approach has its advantages and disadvantages. The quantile transform technique, which originally stems from asymptotic statistics, is purely one-dimensional. However, as can be seen from the variety of subjects addressed in the articles, it is very useful. A major advantage of the quantile approach is that the quantities considered often suggest appropriate new conjectures. On the other hand, the "analytic approach" is applicable to both multi-dimensional and non-identically distributed random variables. A disadvantage is that some of the analytic proofs are not particularly intuitive. In any case, the assets and debits of each approach become clear as this volume brings the reader from the basics to the frontiers of research in this area.

Marjorie G. Hahn
David M. Mason
Daniel C. Weiner

May, 1990

CONTENTS

Part I

Approaches to Trimming
and Self-normalization
Based on Analytic Methods

ASYMPTOTIC BEHAVIOR OF PARTIAL SUMS: A MORE ROBUST

APPROACH VIA TRIMMING AND SELF-NORMALIZATION

Marjorie G. Hahn*, Jim Kuelbs**, and Daniel C. Weiner***

1. Introduction. If $X_1, X_2, X_3, \ldots$, are independent, identically distributed (i.i.d.) random variables and $S_n = \sum_{i=1}^n X_i$ denotes the nth partial sum, then limit theorems such as the law of large numbers (LLN), the central limit theorem (CLT), and the law of the iterated logarithm (LIL) all involve a strong interplay between the maximal terms of the sample $\{|X_1|, \ldots, |X_n|\}$ and the asymptotic behavior of the partial sum S_n. Indeed, what is shown in the proofs of the classical formulation of each of these results is that the maximal or extreme terms of the sample are negligible in a sense required for the corresponding theorem. Furthermore, since the assumptions sufficient to prove the classical version of each of these results are also necessary, we see that extensions of these limit theorems will likely require methods that nullify or at least limit the effect of the extreme terms.

The possibility that extreme terms influence the limiting behavior of S_n was noticed as early as the 1930's by Lévy in connection with the CLT (see, for example, Lévy (1937)). In addition, Feller (1968) improved the LIL and Mori (1976, 1977) extended the LLN by examining the effect of extremes in each respective setting. Of course, there are many other papers that could be mentioned as well, but these

---

   * Supported in part by NSF grant DMS-87-02878.
  ** Supported in part by NSF grant DMS-85-21586.
*** Supported in part by NSF grants DMS-88-96217 and DMS-87-02878.

suffice to justify the claim that in order to broaden the scope of limit theorems for S_n, one must also understand the influence of the extreme terms.

As a concrete example, recall that if $\{a_n\}$ and $\{b_n\}$ are non-random sequences such that $\left\{\frac{S_n - b_n}{a_n}\right\}$ converges to a non-degenerate limit law, then Lévy already knew that

$$\mathcal{L}\left(\frac{S_n - b_n}{a_n}\right) \longrightarrow N(0,1) \qquad (1.1)$$

if and only if

$$\max_{1 \le j \le n} |X_j|/a_n \xrightarrow{p} 0. \qquad (1.2)$$

Hence, in order to extend the CLT with Gaussian limit law, this suggests that one must trim away or modify the maximal terms in such a manner that the remaining terms are all asymptotically negligible. Another possibility is to use random normalizations which would nullify the effect of the maximal terms in order to achieve overall asymptotic normality. We will examine both of these situations as well as point out a more general framework in which they represent special subcases. Most of the results we describe are in connection with the CLT, but some LIL results are presented later in the paper.

Throughout the paper we write $a_n \sim b_n$ if $\lim_{n \to \infty} a_n/b_n = 1$ and $a_n \approx b_n$ if $0 < \liminf_{n \to \infty} a_n/b_n \le \limsup_{n \to \infty} a_n/b_n < \infty$. HKW will refer to papers authored by Hahn, Kuelbs, and Weiner.

2. A Framework for Trimming and Self-normalization.

Let $X, X_1, X_2, \ldots$ be independent, identically distributed (i.i.d.) random variables with common distribution F and partial sums $S_n = \sum_{i=1}^n X_i$. If $E(X^2) < \infty$, $\mu = E(X)$, and

$\sigma^2 = E(X^2) - \mu^2 > 0$, then the classical CLT asserts

$$\mathcal{L}\left(\frac{S_n - n\mu}{\sigma\sqrt{n}}\right) \longrightarrow N(0,1). \tag{2.1}$$

In attempting to extend the convergence in law of S_n suitably normalized and centered when $E(X^2) = \infty$, one needs scale parameters σ_n to replace $\sigma\sqrt{n}$ and centrality parameters μ_n to replace μ so that

$$\mathcal{L}\left(\frac{S_n - n\mu_n}{\sigma_n}\right) \longrightarrow \beta \quad \text{(nondegenerate)}. \tag{2.2}$$

If β exists, it is necessarily a stable probability measure, $\beta = \mathcal{L}(S_\alpha)$, where S_α is a stable random variable of some index α, $0 < \alpha \le 2$. The distributions F for which μ_n and σ_n exist so that the above convergence holds are said to be in the domain of attraction of a stable law of index α, written $F \in DA(\alpha)$. Necessary and sufficient conditions for $\mathcal{L}(X) = F \in DA(\alpha)$ are known, e.g. Feller (1971):

(i)

$$\lim_{t \to \infty} \frac{t^2 P(|X| > t)}{E(X^2 \wedge t^2)} = \frac{2 - \alpha}{2}, \tag{2.3}$$

and if $0 < \alpha < 2$,

(ii)

$$\lim_{t \to \infty} \frac{P(X > t)}{P(|X| > t)} = \delta$$

for some $0 \le \delta \le 1$.

Naturally any $\tau_n \sim \sigma_n$ can be used to replace σ_n in (2.2). When $0 < \alpha < 2$, there are three well-known choices for suitable τ_n. They are determined by

(A) (the tail equation)

$$\tau_n = \sup\{t \ge 0 : nP(|X| > t) \ge 1\}$$

(B) (the truncated second moment equation)

$$\tau_n = \sup\{t \geq 0 : nE(X^2 \ I(|X| \leq t)) \geq t^2\} \qquad (2.4)$$

(C) (the censored second moment equation)

$$\tau_n = \sup\{t \geq 0 : \ nE(X^2 \wedge t^2) \geq t^2\}.$$

The term censoring refers to the replacement of X by the "censored" quantity $(|X| \wedge t)sgn(X)$ as opposed to the truncated quantity $XI(|X| \leq t)$. If τ_n is defined via (B) or (C), for all but the first few n, τ_n actually makes the relevant inequality used in its definition an equality. These two solutions extend to the case $\alpha = 2$ as well.

Given, $F \in DA(\alpha)$, $0 < \alpha \leq 2$, the scale parameters τ_n obtained above then allow us to define centering constants μ_n. For example, consider τ_n arising from the censoring situation so that for all but the first few n,

$$nE\left(\frac{X^2 \wedge \tau_n^2}{\tau_n^2}\right) = 1. \qquad (2.5)$$

Appropriate centering constants arising from this choice of n are given by

$$\mu_n \equiv E((|X| \wedge \tau_n)sgn(X)). \qquad (2.6)$$

One reason that τ_n is a suitable scale parameter and plays the role of the standard deviation is that when $E(X^2) = \infty$,

$$(E(|X| \wedge t))^2 = o(E(X^2 \wedge t^2)) \quad \text{as } t \to \infty. \qquad (2.7)$$

When τ_n is chosen via (A) or (B), then suitable centerings are given by

$$\mu_n = E(XI(|X| \leq \tau_n)). \tag{2.8}$$

In summary, if $F \in DA(\alpha)$, $0 < \alpha \leq 2$, then by using the scalings and center-ings described above, we obtain (2.2), thereby extending the classical CLT with the only possible limit laws being the corresponding stable laws. There are two immedi-ate weaknesses in this situation. First, the stable laws of index $0 < \alpha < 2$ are much less familiar than the normal law ($\alpha = 2$). Second–a more statistical consideration–when F and α are unknown, we are unable to even empirically construct the scalings τ_n and subsequent centerings μ_n.

In dealing with the first problem one attempts to keep the Gaussian law, or "nice" functions of it, as the limit law. In view of the equivalence of (1.1) and (1.2), this suggests the previously mentioned possibility of trimming away or modifying some of the largest values in the sample to improve the situation. The second problem can possibly be resolved by estimating τ_n empirically, yielding random scalings or self-normalizaitons $\hat{\tau}_n$ to replace τ_n. Then $\hat{\tau}_n$ could be used analogously to (2.6) or (2.8) to define "empirical centerings" which would in fact serve as the "trimmed" or "modified" sums just mentioned. In addition, $\hat{\tau}_n$ could serve as a random normalization of these same sums, as suggested at the end of Section 1. That is, perhaps scaling by $\hat{\tau}_n$ would mitigate the effect of the extreme terms and allow for asymptotic normality for the modified sums.

The approach we describe is based on determining τ_n empirically as well as deterministically. To do this, it is natural to examine normalizers (empirical scale constants) $\hat{\tau}_n$ obtained from the empirical version of (A), (B), or (C) in (2.4) when the distribution F is replaced by the empirical distribution $F_n = \frac{1}{n}\sum_{j=1}^{n}\delta_{X_j}$. If the censoring equation (C) is used, then the empirical version of τ_n is

$$\hat{\tau}_n \equiv \sup\left\{t \geq 0 : n\int(x^2 \wedge t^2)dF_n(x) \geq t^2\right\} = \left(\sum_{j=1}^{n}X_j^2\right)^{\frac{1}{2}} \qquad (2.9)$$

and the corresponding empirical version of the centering μ_n can be defined by

$$\hat{\mu}_n \equiv \int(|X|\wedge\hat{\tau}_n)\,sgn(x)\,dF_n(x) = \frac{1}{n}\sum_{j=1}^{n}(|X_j|\wedge\hat{\tau}_n)\,sgn(X_j). \qquad (2.10)$$

Since, clearly, $\hat{\tau}_n \geq \max_{1 \leq j \leq n}|X_j|$, notice that

$$n\hat{\mu}_n = \sum_{j=1}^{n}(|X_j|\wedge\hat{\tau}_n)sgn(X_j) = \sum_{j=1}^{n}X_j = S_n.$$

Thus the paper by Logan, Mallows, Rice and Shepp (1973) relates to the limiting behavior of the self-normalized sums

$$\frac{S_n - n\mu_n}{\hat{\tau}_n} = \frac{n(\hat{\mu}_n - \mu_n)}{\hat{\tau}_n}. \qquad (2.11)$$

Their results are for $F \in DA(\alpha)$, and if $1 < \alpha \leq 2$, they assume $E(X) = 0$ and set all centerings μ_n equal to zero. When $0 < \alpha \leq 1$, they again use $\mu_n = 0$ along with the assumption that X is strictly stable then $\alpha = 1$. With these centering assumptons in place, they then prove that all limit laws have a subgaussian tail which, unfortunately, depends in a complicated way on the parameter α. Furthermore, the limit laws have densities which have infinite discontinuities at ± 1, so they are far less familiar then the classical normal density. Finally, their results are only for distributions in the domain of attraction of some stable law, although suitable modifications of (2.9) and (2.10) would, in fact, be appropriate in a much wider setting.

To see how to modify the above $\hat{\tau}_n$, recall that $\hat{\tau}_n \geq \max_{1 \leq j \leq n} |X_j|$. Hence all large terms remain in both the numerator (cf. (2.10)) and denominator of (2.11). This latter fact accounts for the subgaussian tail in the limit law, but the persistence of the maximal terms still does not allow pure Gaussian limits when $0 < \alpha < 2$.

One way of improving this situation is to reduce the magnitude of the empirical scalings $\hat{\tau}_n$ in (2.9) (as they also appear as empirical censoring levels in $\hat{\mu}_n$), and similarly for their probabilistic analogues τ_n obtained via (2.4) (C). This can be achieved by considering the scale quantities a_n determined as in (2.4)(C) but satisfying

$$nE\left(\frac{X^2 \wedge a_n^2}{a_n^2}\right) = r_n \tag{2.12}$$

with $r_n > 1$ instead of the equation resulting from (2.4)(C). Once the normalizations are determined, they are used to determine centerings γ_n given by

$$\gamma_n = E((|X| \wedge a_n)sgn(X)). \tag{2.13}$$

Our study then concentrates on the corresponding empirical versions, $\hat{a}_n$ and $\hat{\gamma}_n$, of a_n and γ_n defined through the empirical analogues of (2.12)-(2.13) and satisfying

$$\sum_{j=1}^{n} \frac{X_j^2 \wedge \hat{a}_n^2}{\hat{a}_n^2} = r_n, \quad \hat{\gamma}_n = \frac{1}{n}\sum_{j=1}^{n}(|X_j| \wedge \hat{a}_n)sgn(X_j). \tag{2.14}$$

Of course, since $\hat{\gamma}_n$ is defined through $\hat{a}_n$, a detailed study of $\hat{\gamma}_n$ entails an understanding of the joint behavior of $(\hat{\gamma}_n, \hat{a}_n)$. Furthermore, we mention that $\hat{a}_n$ is well-defined by the first equation in (2.14) on the set where $\#\{j \leq n : |X_j| > 0\} > r_n$. For definiteness, define $\hat{a}_n = 0$ elsewhere. Then, as shown by (3.4) in HKW (1990a), we have $\hat{a}_n \to \infty$ a.s.

To interpret $r_n \geq 1$ in the above, observe that on the event $[\hat{a}_n > 0]$ we have

$$\#\{j \leq n : |X_j| \geq \hat{a}_n\} \leq r_n. \tag{2.15}$$

If not, then since

$$\hat{a}_n = \sup \left\{ t \geq 0 : \sum_{j=1}^{n} \frac{(X_j^2 \wedge t^2)}{t^2} \geq r_n \right\},$$

we would have a contradiction. Hence r_n represents an upper bound on the number of terms modified in forming $\hat{\gamma}_n$ as in (2.14).

If $\{r_n\}$ is bounded (light modification), there is typically no improvement with respect to asymptotic normality. (Compare Maller (1982) and Mori (1984).) Hence we consider the case $r_n \to \infty$. Furthermore, in order to study scale and centrality quantities which in the limit involve the entire distribution, it is necessary that $\lim_n r_n/n = 0$ (*intermediate modification*). Now the case $\lim_n \frac{r_n}{n} = c > 0$ (*heavy modification*) a la Stigler (1973) is also of interest, but here we always assume, unless specifically noted otherwise, that $\{r_n\}$ is a sequence of *integers* such that

$$1 \leq r_n \to \infty \text{ and } \frac{r_n}{n} \to 0. \tag{2.16}$$

It is worth noting that the tail equation or truncated second moment equation analogues of (2.12) would serve as the basis of two other approaches. That is, determining the truncation level and scale as in equation (2.4) (A), (B), or (C), in deterministic fashion or the empirical fashion using $r_n = 1$ or $r_n > 1$, represents the framework we alluded to at the start of this secton.

For example, if F is continuous and

$$\hat{b}_n = \sup\{t \geq 0 : \ nG_n(t) \geq r_n\}, \tag{2.17}$$

where

$$G_n(t) = \frac{1}{n} \#\{j \leq n : \ |X_j| > t\},$$

then by the empirical analogue of (2.8),

$$n\hat{\gamma}_n = \sum_{j=1}^{n} X_j I(|X_j| \le \hat{b}_n) = S_n - \sum_{j=1}^{r_n} X_j^{(n)} \tag{2.18}$$

where

$$|X_1^{(n)}| > |X_2^{(n)}| > \ldots > |X_n^{(n)}|$$

is the arrangement of $\{X_1, \ldots, X_n\}$ in decreasing order of magnitude. Since F is assumed continuous, ties form a null event, so in the above situation $n\hat{\gamma}_n$ is the classical trimmed sum obtained via magnitude trimming. This viewpoint was put to use in HKW (1990b) and we present some futher details and comparisons later.

Using the approach of (2.4)(B) with $r_n > 1$ relates to the material in Hahn and Kuelbs (1988), and again we present further details below. We now turn to a summary of the results of HKW (1990a) where the censoring equation (2.4)(C) was employed. The approach based on this equation has received in-depth attention mainly because its solutions in the deterministic and empirical cases are analytically well-enough behaved to allow for a complete set of self-normalized results.

3. Summary of Intermediate Censoring Results for Totally Censored Sums.
Besides the definitions already made in Section 1, we recall the following definitions, for $t > 0$ and integers $p \ge 1$:

$$G(t) = P(|X| > t), \quad G^+(t) = P(X > t), \quad G^-(t) = P(X < -t)$$

$$G_n^+(t) = \frac{1}{n}\#\{j \le n: \ X_j > t\}, \quad G_n^-(t) = \frac{1}{n}\#\{j \le n: \ X_j < -t\}$$

$$M(p,t) = E((|X| \wedge t)sgn(X))^p), \quad \tilde{M}(p,t) = E(X^p I(|X| \le t)) \tag{3.1}$$

$$m(p,t) = t^{-p}E((|X| \wedge t)sgn(X))^p, \quad \tilde{m}(p,t) = t^{-p}E(X^pI(|X| \le t))$$

$$m_n(p,t) = t^{-p}\frac{1}{n}\sum_{j=1}^{n}((|X_j| \wedge t)sgn(X_j))^p, \quad \tilde{m}_n(p,t) = t^{-p}\frac{1}{n}\sum_{j=1}^{n}X_j^pI(|X| \le t).$$

In particular,

$$M(p,t) = t^p m(p,t) \quad \text{and} \quad \tilde{M}(p,t) = t^p \tilde{m}(p,t).$$

Let $\{r_n\}$ be a sequence of integers satisfying (2.16). In the intermediate modification setting, universal asymptotic normality holds for the totally **censored** sums normalized by constants obtained from the censored second moment equation. The foundation of all the empirical results is the following result using deterministic normalizers:

Theorem 3.2. *(HKW (1990a), Proposition 4.3.) Assume $\{r_n\}$ satisfies (2.16). Then*

(i) For any nondegenerate F,

$$\mathcal{L}\left(\frac{n}{\sqrt{r_n}}(m_n(1,a_n) - m(1,a_n))\right) \equiv \mathcal{L}\left(\frac{\sum_{j=1}^{n}(|X_j| \wedge a_n)sgn(X_j) - n\gamma_n}{a_n\sqrt{r_n}}\right)$$
$$\to \mathcal{N}(0, a^2(X))$$

(3.3)

where

$$a^2(X) = \begin{cases} Var(X)/EX^2 & \text{if } EX^2 < \infty \\ 1 & \text{if } EX^2 = \infty. \end{cases}$$

(3.4)

(ii) The sequence $\{\frac{n}{\sqrt{r_n}}(m_n(1,a_n) - m(1,a_n), m_n(2,a_n) - m(2,a_n))\}$ is tight in $\mathbf{R}^2$ and every subsequential limit is mean zero bivariate normal with (possibly degenerate) covariance matrix of the form

$$\Sigma = \begin{pmatrix} a^2(X) & b \\ b & c^2 \end{pmatrix},$$

(3.5)

where

$$\begin{cases} 0 < a^2(X) \le 1 \ \ \text{and} \ \ b = c^2 = 0 & \text{if } EX^2 < \infty \\ a^2(X) = 1, \ 0 \le c^2 \le 1, \ \ \text{and} \ \ b^2 \le c^2 & \text{if } EX^2 = \infty. \end{cases}$$

Convergence in distribution occurs along subsequences for which both

$$\frac{n}{r_n} m(3, a_n) \to b \ \text{ and}$$
$$\frac{n}{r_n} m(4, a_n) \to c^2. \tag{3.6}$$

Finally, if X is asymptotically symmetric, i.e.,

$$\lim_{t \to \infty} \frac{G^+(t) - G^-(t)}{G(t)} = 0, \tag{3.7}$$

then for every subsequential limit law, $b = 0$.

The universal asymptotic normality of this result makes normality a fundamental part of all that follows. All limits are consequently normals or "nice" functions of them.

A self-normalized version of the asymptotic normality is valid for all *symmetric* distributions.

Theorem 3.8. *Assume $\{r_n\}$ satisfies (2.16). If F is nondegenerate and symmetric about the origin,*

$$\mathcal{L}\left(\frac{n\hat{\gamma}_n}{\sqrt{r_n}\hat{a}_n}\right) = \mathcal{L}\left(\frac{\sum_{j=1}^{n}(|X_j| \wedge \hat{a}_n)sgn(X_j)}{\sqrt{\sum_{j=1}^{n}(X_j^2 \wedge \hat{a}_n^2)}}\right) \to \mathcal{N}(0,1). \tag{3.9}$$

A procedure for attempting to relax the symmetry condition becomes evident from considering the *fundamental equation* relating the behavior of quantities censored at the level $\hat{a}_n$ to those censored at a_n as in Theorem 3.2:

$$\frac{n}{\hat{a}_n\sqrt{r_n}}(\hat{\gamma}_n - \gamma_n) = \frac{a_n}{\hat{a}_n}\frac{n}{\sqrt{r_n}}(m_n(1,a_n) - m(1,a_n)) + \frac{n}{\hat{a}_n\sqrt{r_n}}\int_{a_n}^{\hat{a}_n}\{G_n^+(s) - G_n^-(s)\}\,ds$$

$$\sim \frac{a_n}{\hat{a}_n}(Z_n + W_n)$$

$$(3.10)$$

where

$$W_n \equiv \int_0^{\sqrt{r_n}(\frac{\hat{a}_n}{a_n}-1)} \frac{n}{r_n}\{G^+(a_n(1 + \frac{u}{\sqrt{r_n}})) - G^-(a_n(1 + \frac{u}{\sqrt{r_n}}))\}\,du$$

and

$$Z_n \xrightarrow{D} \mathcal{N}(0, a^2(X))$$

where $a^2(X)$ is given by (3.4).

Now, from the fundamental equation, one can see that asymptotic normality would be achieved in the empirical result if both

(i) $\frac{\hat{a}_n}{a_n} \to 1$ in probability ("consistency")

and

(ii) $\{\frac{a_n}{\hat{a}_n}W_n\} \to 0$ in probability.

On the other hand, tractable limits are likely if both (i) holds and

(ii$'$) $\{\frac{a_n}{\hat{a}_n}W_n\}$ is tight.

As observed in HKW (1990a):

I. *Tightness of* $\{\sqrt{r_n}\left(\frac{\hat{a}_n}{a_n}-1\right)\} \Rightarrow$ *tightness of* $\{\frac{a_n}{\hat{a}_n}W_n\}$:

The basic reason is that given $\rho_{n_k} \to \infty$ with $\rho_{n_k}/\sqrt{r_{n_k}} \to 0$, the integrand for W_n is eventually dominated uniformly in $|u| \le \rho_{n_k}$, by

$$2\frac{n_k}{r_{n_k}}P(|X| > a_{n_k}(1+\frac{u}{\sqrt{r_{n_k}}})) \le 2\frac{n_k}{r_{n_k}}P(|X| > a_{n_k}(1-\frac{\rho_{n_k}}{\sqrt{r_{n_k}}}))$$

$$\le 2\frac{n_k}{r_{n_k}}E\left(\frac{X^2 \wedge a_{n_k}^2(1-\frac{\rho_{n_k}}{\sqrt{r_{n_k}}})^2}{a_{n_k}^2(1-\frac{\rho_{n_k}}{\sqrt{r_{n_k}}})^2}\right) \sim 2.$$

Since by tightness, $P(|\sqrt{r_{n_k}}\left(\frac{\hat{a}_{n_k}}{a_{n_k}}-1\right)| > \rho_{n_k}) \to 0$, we obtain $P(|W_{n_k}| > \rho_{n_k}) \to 0$.

II. *Consistency:* (HKW (1990a), Theorem 4.7)

$\frac{\hat{a}_{n_k}}{a_{n_k}} \to 1$ in probability $\iff \exists\ \rho_{n_k} \to \infty$ with $\frac{\rho_{n_k}}{\sqrt{r_{n_k}}} \to 0$ such that the following condition holds:

$$\lim_{k\to\infty}\frac{n_k}{\sqrt{r_{n_k}}}(m(2, a_{n_k}(1 - \rho_{n_k}/\sqrt{r_{n_k}})) - m(2, a_{n_k})) \equiv$$

$$\lim_{k\to\infty}\frac{n_k}{\sqrt{r_{n_k}}}\left(E\left(\left(\frac{X}{a_{n_k}(1 - \rho_{n_k}/\sqrt{r_{n_k}})}\right)^2 \wedge 1\right) - E\left(\left(\frac{X}{a_{n_k}}\right)^2 \wedge 1\right)\right)$$

$$= \infty.$$

$$(3.11)$$

This *Tail-controlling condition*, prevents too much heaviness in the tail of the distribution.

Remarks.

A. Consistency holds for all X in the **Feller Class**, i.e. with

$$\limsup_{t\to\infty}\frac{t^2 P(|X| > t)}{E(X^2 \wedge t^2)} < 1.$$

This is precisely the class of random variables with normalizable partial sums which are *stochastically compact*, i.e. tight with no degenerate limits.

B. The Feller Class is precisely the setting in which condition (3.11) holds *for every* $\{r_n\}$ satisfying (2.16). (HKW (1990a), Corollary 4.29)

C. Consistency is not universal. In fact, it can fail for X with a slowly varying tail (HKW (1990a), Example 6.14).

III. $\left\{\sqrt{r_{n_k}}\left(\frac{\tilde{a}_{n_k}}{a_{n_k}} - 1\right)\right\}$ is tight $\iff$ $\forall\,\rho_{n_k} \to \infty$ with $\frac{\rho_{n_k}}{\sqrt{r_{n_k}}} \to 0$, condition (3.11) holds. (HKW (1990a), Theorem 4.7)

IV. The sequence $\left\{\sqrt{r_{n_k}}\left(\frac{\tilde{a}_{n_k}}{a_{n_k}} - 1\right)\right\}$ is tight with only mean zero normal or degenerate limit laws if and only if both

$$
\begin{aligned}
&\limsup_{k\to\infty} \frac{n_k}{r_{n_k}} G(a_{n_k}) < 1, \quad \text{and} \\
&\forall x: \ \lim_{k\to\infty} \frac{n_k}{r_{n_k}} \left\{ G\left(a_{n_k}\left(1 + \frac{x}{\sqrt{r_{n_k}}}\right)\right) - G(a_{n_k}) \right\} = 0
\end{aligned}
\tag{3.12}
$$

hold. (HKW (1990a) Proposition 4.35)

V. Stochastic compactness of $\left\{\sqrt{r_{n_k}}\left(\frac{\tilde{a}_{n_k}}{a_{n_k}} - 1\right)\right\}$ with only mean zero normal subsequential limits is equivalent to both (3.12) and

$$
\liminf_{k\to\infty} \frac{n_k}{r_{n_k}} m(4, a_{n_k}) > 0.
\tag{3.13}
$$

holding. Convergence to $\mathcal{N}(0, \kappa^2)$, $\kappa^2 \geq 0$, occurs along subsequences where (3.12) holds and

$$
\frac{r_n m(4, a_n)}{n\tilde{m}(2, a_n)^2} \to 4\kappa^2.
$$

(HKW (1990a) Proposition 4.35)

Since $\hat{\gamma}_n$ is defined through $\hat{a}_n$, it is evident that the theory developed in HKW (1990a) must contain results on the distributional behaviors of both $\hat{a}_n$ and $(\hat{\gamma}_n, \hat{a}_n)$. Tightness of the scales is actually equivalent to tightness of the joint centers and scales. Moreover, all subsequential limits are convex functions of the normal, as the following main joint self-normalization theorem indicates:

Theorem 3.14. *(HKW (1990a), Theorem 5.1)*

(I.) The sequence

$$\left\{ \mathcal{L} \left(\frac{n_k}{\hat{a}_{n_k} \sqrt{r_{n_k}}} (\hat{\gamma}_{n_k} - \gamma_{n_k}), \sqrt{r_{n_k}} \left(1 - \frac{a_{n_k}}{\hat{a}_{n_k}} \right) \right) \right\} \tag{3.15}$$

is tight in $\mathbf{R}^2$ if and only if (3.11) holds for every $\rho_{n_k} \to \infty$ such that $\rho_{n_k}/\sqrt{r_{n_k}} \to 0$. Every subsequential limit is of the form

$$\mathcal{L} \left(Z_1 + \Phi(\Psi^{-1}(Z_2)), \Psi^{-1}(Z_2) \right), \tag{3.16}$$

where (Z_1, Z_2) is mean zero bivariate normal with covariance

$$\Sigma = \begin{pmatrix} a^2(X) & b \\ b & c^2 \end{pmatrix} \tag{3.17}$$

with $0 \leq c^2 \leq 1$, $b^2 \leq c^2$ and $c^2 = 0$ when $a^2(X) < 1$ and $a^2(X)$ is as in (3.4), Ψ is a strictly increasing convex function with range $(-\infty, \infty)$ such that $\Psi(0) = 0$ and $2(1 - c^2) \leq \Psi' \leq 2$, and $\Phi = \Phi^+ - \Phi^-$, where $\Phi^{\pm}$ are nondecreasing concave functions with $\Phi^{\pm}(0) = 0$ and $(\Phi^+)' + (\Phi^-)' = 1 - \frac{1}{2}\Psi'$. A limit law represented by (3.16) and (3.17) is supported on all of $\mathbf{R}^2$ when $\det\Sigma \neq 0$ and is supported on a curve in $\mathbf{R}^2$ when $\det\Sigma = 0$.

(II.) The sequence (3.15) will have subsequential limits of the form (3.16) whose support is not all of $\mathbf{R}^2$, if and only if along some subsequence $\{n'\}$ of $\{n_k\}$

$$\lim_{n' \to \infty} \frac{n'}{r_{n'}} G^+(x a_{n'}) = \alpha I_{(0,c)}(x)$$

$$\lim_{n' \to \infty} \frac{n'}{r_{n'}} G^-(x a_{n'}) = \beta I_{(0,c)}(x)$$

(3.18)

where $\alpha\beta = 0$ and either

(i) $b^2 = c^2 = 0$ and $\alpha = \beta = 0$,

or

(ii) $b^2 = c^2 > 0$ and $\alpha + \beta = c^{-2}$.

In either case, the corresponding limit law for (3.15) is concentrated on a curve in $\mathbf{R}^2$.

Theorem 5.50 of HKW (1990a) provides necessary and sufficient conditions for subsequential convergence to each possible subsequential limit. Although these criteria form one of the main theorems, it is unnecessary for the applications in the current paper and will therefore be omitted due to its slightly involved statement.

Bivariate normality of the subsequential limits can also be characterized.

Theorem 3.19. *(HKW (1990a), Corollary (5.59)) The sequence*

$$\left\{ \mathcal{L} \left(\frac{n_k}{\hat{a}_{n_k} \sqrt{r_{n_k}}} (\hat{\gamma}_{n_k} - \gamma_{n_k}), \sqrt{r_{n_k}} \left(1 - \frac{a_{n_k}}{\hat{a}_{n_k}} \right) \right) \right\}$$

(3.20)

is tight with only mean zero bivariate normal subsequential limits if and only if (3.12) holds. The sequence (3.20) is stochastically compact in $\mathbf{R}^2$ with only (nondegenerate) mean zero bivariate normal subsequential limits if and only if (3.12) holds and for no subsequence of $\{n_k\}$ do all of the following hold: (3.18), $\frac{n}{r_n}m(3, a_n) \to b$, and $\frac{n}{r_n}m(4, a_n) \to c^2$ with b and c^2 as in (3.6).

Section 6.9 of HKW (1990a) constructs a relatively large class of examples generating joint asymptotic normality with nonsingular limiting covariance for the quantities $\hat{\gamma}_n$ and $\hat{a}_n$. The class consists of random variables with densities which mimick the Feller property and are computationally tractable. More specifically, suppose that $|X|$ has Lebesgue density f satisfying

(i) for some $T > 0$ either f is continuous on (T, ∞) or the function $g(t) = tf(t)$ is nonincreasing on (T, ∞)

and

(ii) $0 < \liminf_{\xi \to \infty} \dfrac{\xi^3 f(\xi)}{\int_0^\xi 2s^2 f(s)\, ds} \leq \limsup_{\xi \to \infty} \dfrac{\xi^3 f(\xi)}{\int_0^\xi 2s^2 f(s)\, ds} < 1.$

Then it can be easily verified that X is in the Feller class but outside the domain of partial attraction of the normal law and condition (3.12) holds. Consequently, Theorem 3.19 also applies.

Finally, under the asymptotic normality condition (3.12) for the full sequence, it is possible to empirically normalize $\hat{\gamma}_n - \gamma_n$ to obtain a standard normal limit. Additionally, if F is outside the domain of partial attraction of the normal, the vectors $(\hat{\gamma}_n - \gamma, \hat{a}_n - a_n)$ can be empirically matrix normalized to obtain a standard bivariate normal limit.

More specifically, let b, θ, μ and c^2 be subsequential limits of, respectively,

$$b_n = \frac{n}{r_n} m(3, a_n)$$
$$\theta_n = \frac{n}{r_n}(G^+(a_n) - G^-(a_n))$$
$$\mu_n = \frac{n}{r_n} \tilde{m}(2, a_n) \tag{3.21}$$
$$c_n^2 = \frac{n}{r_n} m(4, a_n).$$

The quantities appearing in (3.21) can be consistently estimated by the obvious empirical estimators, i.e.

(i) $\hat{b}_n - b_n \equiv \frac{n}{r_n} m_n(3, \hat{a}_n) - \frac{n}{r_n} m(3, a_n) \xrightarrow{P} 0$

(ii) $\hat{\theta}_n - \theta_n \equiv \frac{n}{r_n}(G_n^+(\hat{a}_n) - G_n^-(\hat{a}_n)) - \frac{n}{r_n}(G^+(a_n) - G^-(a_n)) \xrightarrow{P} 0$ (3.22)

(iii) $\hat{\mu}_n - \mu_n \equiv \frac{n}{r_n} \tilde{m}_n(2, \hat{a}_n) - m_n(2, \hat{a}_n) \xrightarrow{P} 0$

(iv) $\hat{c}_n^2 - c_n^2 \equiv \frac{n}{r_n} m_n(4, \hat{a}_n) - \frac{n}{r_n} m(4, a_n) \xrightarrow{P} 0.$

Let

$$\hat{A}_n^2 = 1 - \frac{n\hat{\gamma}_n^2}{r_n \hat{a}_n^2} = 1 - \frac{\hat{\gamma}_n^2}{M_n(2, \hat{a}_n)} \tag{3.23}$$

which is the sample variance for the sample $\{(|X_j| \wedge \hat{a}_n)sgn(X_j)/(\sqrt{r_n}\hat{a}_n) : j \le n\}$. Finally, define

$$\hat{\Pi}_n = \begin{pmatrix} 1 & \hat{\theta}_n/\hat{\mu}_n \\ 0 & 1/\hat{\mu}_n \end{pmatrix} \begin{pmatrix} \hat{A}_n^2 & \hat{b}_n \\ \hat{b}_n & \hat{c}_n^2 \end{pmatrix} \begin{pmatrix} 1 & 0 \\ \hat{\theta}_n/\hat{\mu}_n & 1/\hat{\mu}_n \end{pmatrix} \tag{3.24}$$

and take $\hat{\Pi}_n^{-\frac{1}{2}} = I$ if $\hat{\Pi}_n$ is singular. Then the following theorem provides an analogue to the classical results which hold under finite second or fourth moments.

Theorem 3.25. *(HKW (1990a), Theorem 5.81) Let $n_k = k$ and assume (3.12). Then*

$$\mathcal{L}\left(\left(\hat{A}_n^2 + 2\hat{\theta}_n\hat{b}_n/\hat{\mu}_n + \hat{\theta}_n^2\,\hat{c}_n^2/\hat{\mu}_n^2\right)^{-\frac{1}{2}}\frac{n}{\hat{a}_n\sqrt{r_n}}\left(\hat{\gamma}_n - \gamma_n\right)\right) \longrightarrow N(0,1). \qquad (3.26)$$

In addition, if $\liminf_{t\to\infty} G(t)/m(2,t) > 0$ holds, then

$$\mathcal{L}\left(\hat{\Pi}_n^{-\frac{1}{2}}\left(\frac{n(\hat{\gamma}_n - \gamma_n)}{\hat{a}_n\sqrt{r_n}}, \sqrt{r_n}\left(1 - \frac{a_n}{\hat{a}_n}\right)\right)^t\right) \longrightarrow N(0,I), \qquad (3.27)$$

and in particular $\hat{\Pi}_n$ is nonsingular with probability tending to one.

4. Application to Domains of Attraction.

Again fix a sequence $\{r_n\}$ as in (2.16). The results of the previous section simplify considerably for $F \in DA(\alpha)$. In particular, the relation between the heaviness of the tails of F and the mutually dependent estimabilities of the center and scale parameters by their empirically censored analogues can be clearly and quantitatively exposed. Finally, *full sequential joint asymptotic normality for the estimators $\hat{a}_n$ and $\hat{\gamma}_n$ of a_n and γ_n respectively is available throughout the entire class $DA(\alpha)$, $0 < \alpha \le 2$.*

Necessary and sufficient conditions for $F \in DA(\alpha)$ were quoted in (2.3). As usual, the cases $0 < \alpha < 2$ and $\alpha = 2$ will be considered separately.

(4.1) The case $0 < \alpha < 2$.

By (2.3),

$$0 < \lim_{t\to\infty}\frac{G(t)}{m(2,t)} = \frac{2-\alpha}{2} < 1. \qquad (4.2)$$

In particular, X is in the Feller class but outside the domain of partial attraction of the normal, and $a^2(X) = 1$. For some $p \in [0,1]$ and all $s > 0$,

$$\frac{n}{r_n}\,G^+(sa_n) \longrightarrow ps^{-\alpha}\frac{2-\alpha}{2} \quad \text{and} \quad \frac{n}{r_n}\,G^-(sa_n) \longrightarrow (1-p)s^{-\alpha}\frac{2-\alpha}{2}. \qquad (4.3)$$

Considering that $1 + \frac{x}{\sqrt{r_n}} = 1 + o(1)$ and that the quantities on the right sides in (4.3) are continuous in $s > 0$, it is easy to see that both conditions in (3.12) hold along the full sequence $\{n_k = k\}$.

Thus, Theorem 3.25 applies to give standard normal convergence for empirically normalized versions of $\hat{\gamma}_n - \gamma_n$ and $(\hat{\gamma}_n - \gamma_n, \hat{a}_n - a_n)$, respectively. But it is revealing to compute directly and note that the quantities appearing in (3.21) all tend to unique limits as $n \to \infty$, so that the estimator $\hat{\Pi}_n$ in (3.24) obeys $\hat{\Pi}_n \xrightarrow{P} \Pi_0$, where Π_0 is given by

$$\Pi_0 = \begin{pmatrix} 1 & \theta/\mu \\ 0 & 1/\mu \end{pmatrix} \begin{pmatrix} a^2(X) & b \\ b & c^2 \end{pmatrix} \begin{pmatrix} 1 & 0 \\ \theta/\mu & 1/\mu \end{pmatrix}.$$

Using (4.2), (4.3), and the regular variation of G, it is easy to compute the full sequential limits for the quantities in (3.21):

$$b_n = \frac{n}{r_n} m(3, a_n) \longrightarrow \int_0^1 3s^2 \frac{2-\alpha}{2}(2p-1)s^{-\alpha}ds = \frac{3(2-\alpha)(2p-1)}{2(3-\alpha)} = b$$

$$\theta_n = \frac{n}{r_n}\left(G^+(a_n) - G^-(a_n)\right) \longrightarrow \frac{(2p-1)(2-\alpha)}{2} = \theta$$

$$\mu_n = \frac{n}{r_n}\tilde{m}(2, a_n) \longrightarrow 1 - \frac{2-\alpha}{2} = \frac{\alpha}{2} = \mu$$

$$c_n^2 = \frac{n}{r_n} m(4, a_n) \longrightarrow \int_0^1 4s^3 \frac{2-\alpha}{2} s^{-\alpha}\, ds = \frac{2(2-\alpha)}{4-\alpha} = c^2. \qquad (4.4)$$

Thus, Theorem 3.19 guarantees the full sequential convergence

$$\mathcal{L}\left(\frac{n(\hat{\gamma}_n - \gamma_n)}{\hat{a}_n \sqrt{r_n}}, \sqrt{r_n}\left(1 - \frac{a_n}{\hat{a}_n}\right)\right) \longrightarrow N(0, \Pi_0), \qquad (4.5)$$

where

$$\Pi_0 = \begin{pmatrix} 1 & \frac{(2p-1)(2-\alpha)}{2/\alpha} \\ 0 & 2/\alpha \end{pmatrix} \begin{pmatrix} 1 & \frac{3(2-\alpha)(2p-1)}{4-2\alpha} \\ \frac{3(2-\alpha)(2p-1)}{2(3-\alpha)} & \frac{4-2\alpha}{4-\alpha} \end{pmatrix} \begin{pmatrix} 1 & 0 \\ \frac{(2p-1)(2-\alpha)}{\alpha} & 2/\alpha \end{pmatrix}. \qquad (4.6)$$

Specialize to the first marginal in (4.5). The asymptotic normal variance of the reduced centrality estimator $\frac{n}{\hat{a}_n\sqrt{r_n}}(\hat{\gamma}_n - \gamma_n)$ is

$$V(\alpha, p) = 1 + 2b\theta/\mu + \theta^2 c^2/\mu^2$$
$$= 1 + (2p - 1)^2(2 - \alpha)^2 \left\{ \frac{3}{\alpha(3 - \alpha)} + \frac{2(2 - \alpha)^2}{\alpha^2(4 - \alpha)} \right\}. \qquad (4.7)$$

Now (4.7) quantifies the balance between near asymptotic symmetry (i.e., p near $\frac{1}{2}$) and heaviness of tails (i.e., α near 0) in moderating the size of $V(\alpha, p)$.

Consider the extremely asymmetric cases, $p = 1$ or $p = 0$. As $\alpha \downarrow 0$, write

$$V(\alpha) = V(\alpha, 1) = V(\alpha, 0) = 1 + (1 + o(1))\alpha^{-2}. \qquad (4.8)$$

In (4.8), the unit represents the "natural" variance contribution of the first term in (3.10), $\frac{n}{\sqrt{r_n}}(m_n(1, a_n) - m(1, a_n))$, which is asymptotically standard normal due to the universal result in Theorem 3.2 (i). However, the term of order α^{-2} in (4.8) must be regarded as "noise" relative to the centrality estimation. This is essentially due to the relative inestimability of scale as $\alpha \downarrow 0$. In this special situation, equation (4.8) quantifies the general need, exposed in the fundamental equation, for considering the joint asymptotic behavior of $\hat{\gamma}_n$ and $\hat{a}_n$ in order to determine the estimability of γ_n alone. Moreover, when seeking general results on the asymptotic behavior of $\{\frac{n}{\hat{a}_n\sqrt{r_n}}(\hat{\gamma}_n - \gamma_n)\}$ for possibly asymmetric F, we have seen the need for restricting attention to distributions satisfying (3.11), which is essentially a "heavy-tails controlling" condition. Here, in a readily computable situation, (4.8) quantifies the domination of the "noise" with respect to increasingly heavy tails.

The importance of "noise" in the variance quantity $V(\alpha) = V(\alpha, 1)$ in (4.8) is nonnegligible even for moderate values of α. Indeed, for the extremely asymmetric

Cauchy case ($\alpha = p = 1$), the noise contribution is already $\frac{13}{6}$, or over two-thirds of the total!

(4.9) The Case $\alpha = 2$.

When $\alpha = 2$, $\lim_{t \to \infty} \frac{G(t)}{m(2,t)} = 0$. Since for every $s > 0$, $\frac{n}{r_n} G(sa_n) \to 0$ (using (2.3) (i) with $\alpha = 2$ and the subsequent slow variation of $\tilde{M}(2, \cdot)$), it follows that $\frac{n}{r_n} m(4, a_n) \to 0$ and that (3.12) holds with $n_k = k$. From this we see $c_n^2 \to 0$, $\theta_n \to 0$, $b_n \to 0$ and $\mu_n \to 1$ in (4.4). Now only the first part of Theorem 3.25 applies, since $\liminf_{t \to \infty} \frac{G(t)}{m(2,t)} = 0$, because the estimation of $\hat{a}_n$ is sharp:

$$\sqrt{r_n}\left(1 - \frac{a_n}{\hat{a}_n}\right) \xrightarrow{p} 0. \tag{4.10}$$

Consequently, estimation of γ_n is relatively unaffected by "noise" due to the estimation of a_n, in contrast to the situation for $0 < \alpha < 2$. Consequently,

$$\mathcal{L}\left(\frac{n}{\hat{a}_n \sqrt{r_n}}(\hat{\gamma}_n - \gamma_n)\right) \longrightarrow N(0, a^2(X)). \tag{4.11}$$

We summarize our results in

Theorem 4.12. *Suppose* $F \in DA(\alpha)$, *where* $0 < \alpha \leq 2$. *Then*

$$\mathcal{L}\left(\frac{n}{\hat{a}_n \sqrt{r_n}}(\hat{\gamma}_n - \gamma_n), \sqrt{r_n}\left(1 - \frac{a_n}{\hat{a}_n}\right)\right) \to \mathcal{N}(0, \Pi_0),$$

with Π_0 *given by (4.6). Moreover,*

$$\mathcal{L}\left((\hat{A}_n^2 + 2\hat{\theta}_n\hat{b}_n/\hat{\mu}_n + \hat{\theta}_n^2\hat{c}_n^2/\hat{\mu}_n^2)^{-1/2}\frac{n}{\hat{a}_n\sqrt{r_n}}(\hat{\gamma}_n - \gamma_n)\right) \to \mathcal{N}(0,1),$$

where $\hat{A}_n^2 = 1 - n\hat{\gamma}_n^2/(r_n\hat{a}_n^2)$ and $\hat{b}_n$, $\hat{c}_n$, $\hat{\mu}_n$, and $\hat{\theta}_n$ are given by (3.22). If $\alpha = 2$,

$$\frac{\hat{a}_n - a_n}{\hat{a}_n}\sqrt{r_n} \to 0 \text{ in probability.}$$

Finally, if $\alpha \neq 2$,

$$\mathcal{L}\left(\hat{\Pi}_n^{-1/2}(\frac{n}{\hat{a}_n\sqrt{r_n}}(\hat{\gamma}_n - \gamma_n), \sqrt{r_n}\left(1 - \frac{a_n}{\hat{a}_n}\right))\right) \to \mathcal{N}(0, I),$$

where $\hat{\Pi}_n$ is given by (3.24) and $\hat{\Pi}_n^{-1/2}$ is defined to be I when $\hat{\Pi}_n$ is singular.

Theorem 4.12 motivates a new approach to the problem of estimating the index α of regular variation for the tail of a distribution. Development of this approach, expanded to include joint estimation of α and the asymmetry parameter in the limiting stable law, appears in Hahn and Weiner (1990a), the next paper in this volume.

5. Interpretations of the center and scale parameters γ_n and a_n.

When estimating the mean $\mu = EX$ of a random variable with finite second moment $\tau^2 = EX^2 < \infty$ by the sample mean $\overline{X}_n \equiv \frac{1}{n}\sum_{j=1}^n X_j$, the interdependence of the estimates of center and scale are clear. Not only is $\overline{X}_n$ a consistent estimate of μ, but if $\sigma^2 = Var(X) = \tau^2 - \mu^2$, then the CLT establishes the desirable asymptotic normality of $\overline{X}_n$, i.e

$$\mathcal{L}\left(\frac{S_n - n\mu}{\sqrt{n}\sigma}\right) = \mathcal{L}\left(\frac{\sqrt{n}}{\sqrt{\tau^2 - \mu^2}}(\overline{X}_n - \mu)\right) \to \mathcal{N}(0, 1). \qquad (5.1)$$

Notice that the estimability of the mean parameter μ depends on another parameter τ^2. Typtically the parameter, $\sigma^2 = \tau^2 - \mu^2 = \int(x - \mu)^2 \, dF(x)$ is unknown. Conversion of the central limit theorem to a practical statistical tool is obtained through the self-normalized or "studentized" formulation obtained upon replacing μ^2 and τ^2 by their estimators:

$$\mathcal{L}\left(\frac{S_n - n\mu}{\sqrt{\sum_{j=1}^n X_j^2 - n\overline{X}_n^2}}\right) = \mathcal{L}\left(\frac{\sqrt{n}}{\sqrt{\frac{1}{n}\sum_{j=1}^n X_j^2 - \overline{X}_n^2}}(\overline{X}_n - \mu)\right) \to \mathcal{N}(0,1). \quad (5.2)$$

The actual interpretation of this asymptotic normality is elucidated by the reformulation

$$\overline{X}_n - \mu = \frac{\sqrt{\frac{1}{n}\sum_{j=1}^n X_j^2 - \overline{X}_n^2}}{\sqrt{n}} Z_n \quad (5.3)$$

$$\equiv R_n Z_n,$$

where Z_n is asymptotically standard normal in distribution. The quantity R_n denotes an empirically observable *margin of error*.

Since both the scale and centrality parameters are generally unknown, it is desirable to determine how the estimates of scale and center depend upon each other. A solution should allow the construction of joint confidence regions. Under classical assumptions the solution requires finite fourth moments. If $EX^4 < \infty$, then

$$\sqrt{n}A_n^{-\frac{1}{2}}\left(\frac{\overline{X}_n - \mu}{\frac{1}{n}\sum_{j=1}^n X_j^2 - \tau^2}\right) \to \mathcal{N}(0, I), \quad (5.4)$$

where

$$A_n \equiv \left(\begin{matrix} \frac{1}{n}\sum_{j=1}^n X_j^2 - \overline{X}_n^2 & \frac{1}{n}\sum_{j=1}^n X_j^3 - \frac{1}{n}\sum_{j=1}^n X_j^2\overline{X}_n \\ \frac{1}{n}\sum_{j=1}^n X_j^3 - \frac{1}{n}\sum_{j=1}^n X_j^2\overline{X}_n & \frac{1}{n}\sum_{j=1}^n X_j^4 - (\frac{1}{n}\sum_{j=1}^n X_j^2)^2 \end{matrix}\right)$$

and $\mathcal{N}(0, I)$ denotes the standard bivariate normal law. The proof uses the Cramer-Wold device and the fact that $\{uX_j + vX_j^2\}$ are i.i.d. with finite variance. This affine normalization of the joint errors in the estimation of center and scale, which uses a sample covariance matrix for the random vectors (X_j, X_j^2), is empirically observable. Note the role played by EX^3 in determining the asymptotic dependence of the centrality and scale estimators.

When $EX^2 = \infty$ and possibly $E|X| = \infty$ we have seen that there are a variety of ways to introduce centrality parameters μ_n to replace μ and scale parameters σ_n to replace $\sigma\sqrt{n}$. The main purpose of this section is to provide some interpretation of these parameters in the censored case.

For simplicity and convenience of discussion consider the two relatively large subclasses of the Feller class ($\limsup_{t\to\infty} G(t)/m(2,t) < 1$) governed by the conditions

$$(A) \quad \liminf_{t\to\infty} \frac{G(t)}{m(2,t)} > \frac{1}{2} \tag{5.5}$$

and

$$(B) \quad \limsup_{t\to\infty} \frac{G(t)}{m(2,t)} < \frac{1}{2}, \tag{5.6}$$

respectively. Notice that Class A contains all random variables $X \in DA(\alpha)$ with $0 < \alpha < 1$, while Class B contains all $X \in DA(\alpha)$ with $1 < \alpha \leq 2$.

When X belongs to Class A, $E|X| = \infty$ and $n/(r_n a_n) \to 0$, as is easily verified using standard regular/dominated variational techniques (e.g., Feller(1971)). If we wish to view γ_n as the centrality predictor for samples of size $[n/r_n]$, and a_n as measuring the scale of the error (or deviation) of the natural estimator $\overline{X}_{[n/r_n]}$ of γ_n, we are vindicated by the stochastic compactness (cf. Feller (1967)) of

$$\frac{1}{a_n}\left[\frac{n}{r_n}\right](\overline{X}_{[n/r_n]} - \gamma_n) = \frac{1}{a_n}\sum_{j=1}^{[n/r_n]}(X_j - \gamma_n) \tag{5.7}$$

but, in a sense, are contradicted by the *inconsistency* of $\overline{X}_{[n/r_n]}$, i.e., the failure of $(\overline{X}_{[n/r_n]} - \gamma_n) \to 0$ in probability, due to $n/(r_n a_n) \to 0$. In this situation, the content of Theorem 3.14 may profitably be viewed as concerning the relative (and dimensionless) centrality quantity $\delta_n \equiv n\gamma_n/r_n a_n$. It is not difficult to show that, along a subsequence where the sums in (5.7) converge to an infinitely divisible law with Levy measure μ,

$$\delta_n = \frac{n\gamma_n}{r_n a_n} = \frac{n}{r_n} m(1, a_n) \to \int (|x| \wedge 1) sgn(x) \, d\mu(x) \equiv \delta. \qquad (5.8)$$

The quality of the estimation of δ_n by the obvious estimator $\hat{\delta}_n = n\hat{\gamma}_n/(r_n \hat{a}_n)$ can be gauged, via Theorem 3.14, from

$$\begin{aligned}
\sqrt{r_n}(\hat{\delta}_n - \delta_n) &= \frac{n}{\sqrt{r_n}} \left(\frac{\hat{\gamma}_n}{\hat{a}_n} - \frac{\gamma_n}{a_n} \right) \\
&= \frac{n}{\sqrt{r_n}\hat{a}_n}(\hat{\gamma}_n - \gamma_n) - \frac{n\gamma_n}{r_n a_n}\sqrt{r_n}\left(1 - \frac{a_n}{\hat{a}_n}\right) \qquad (5.9) \\
&= \frac{n}{\sqrt{r_n}\hat{a}_n}(\hat{\gamma}_n - \gamma_n) - \delta\sqrt{r_n}\left(1 - \frac{a_n}{\hat{a}_n}\right) + o_p(1),
\end{aligned}$$

due to (5.8). The joint asymptotic behavior of $\hat{\delta}_n$ and $\hat{a}_n$ can similarly be derived. Of course, (5.9) cannot be regarded as leading to inference on the parameter δ on the right of (5.8), unless the rate of convergence in (5.8) is faster than $1/\sqrt{r_n}$. In general, δ_n can be estimated but estimation of δ suffers bias.

When X belongs to Class B, $E|X| < \infty$ and $n/(r_n a_n) \to \infty$. Thus the stochastic compactness of the sums in (5.7) yields the desired consistency $\overline{X}_{[n/r_n]} - \gamma_n \to 0$, and also $\overline{X}_{[n/r_n]} \to EX$ in probability (since $\gamma_n \to EX$). But of course, $r_n/n \to 0$ guarantees, by the Strong Law of Large Numbers, that $\overline{X}_{[n/r_n]} \to EX$ almost surely. We claim that $\hat{\gamma}_n \to EX$ almost surely as well.

As was established following (3.4) in HKW (1990a), $\hat{a}_n \to \infty$ a.s. For $C > 0$, on the event $[\hat{a}_n > C]$,

$$
\begin{aligned}
|\overline{X}_n - \hat{\gamma}_n| &= \frac{1}{n}|\sum_{j=1}^{n}(|X_j| - \hat{a}_n)I(|X_j| > \hat{a}_n)sgn(X_j)| \\
&\leq \frac{1}{n}\sum_{j=1}^{n}|X_j|I(|X_j| > \hat{a}_n) \\
&\leq \frac{1}{n}\sum_{j=1}^{n}|X_j|I(|X_j| > C).
\end{aligned}
$$

Letting $n \to \infty$ on the almost sure event $[\hat{a}_n > C$ for all but finitely many $n]$, we have, almost surely,

$$
\limsup_{n \to \infty} |\overline{X}_n - \hat{\gamma}_n| \leq E|X|I(|X| > C) \to 0,
$$

after letting $C \to \infty$. Since $\overline{X}_n \to EX$, almost surely $\hat{\gamma}_n \to .EX$.

Thus, although in Class B the analogue of (5.8) is unavailable, it is also unnecessary for an interpretation of γ_n. But we caution that even in this desirable situation, Theorem 3.14 *cannot* (in general) be viewed as a theorem on inference concerning EX, for although $\left\{\frac{n}{\sqrt{r_n}\hat{a}_n}(\hat{\gamma}_n - \gamma_n)\right\}$ is tight, it may be that $\left\{\frac{n}{\sqrt{r_n}\hat{a}_n}(\gamma_n - EX)\right\}$ is unbounded, because $n/(r_n a_n) \to \infty$. In such a case, the squared bias of the estimator $\hat{\gamma}_n$ of EX may be much larger than its variance rendering it useless. As an example, take $X > 0$ with $G(t) \sim t^{-\alpha}$, where $1 < \alpha < 2$. Then $a_n \sim c(n/r_n)^{1/\alpha}$

(where $0 < c < \infty$ is a constant), so that

$$\frac{n}{\sqrt{r_n a_n}}(EX - \gamma_n) = \frac{n}{\sqrt{r_n a_n}} \int_{a_n}^{\infty} G(t)\, dt$$

$$\sim (\text{constant})\sqrt{n}\left(\frac{n}{r_n}\right)^{\frac{1}{2}-\frac{1}{\alpha}} \int_{(n/r_n)^{\frac{1}{\alpha}}}^{\infty} t^{-\alpha}\, dt$$

$$= (\text{constant})\sqrt{n}\left(\frac{n}{r_n}\right)^{(\frac{1}{2}-\frac{1}{\alpha})\mp\frac{1}{\alpha}(1-\alpha)}$$

$$= (\text{constant})\sqrt{n}\left(\frac{n}{r_n}\right)^{-1/2} = (\text{constant})\sqrt{r_n} \to \infty.$$

6. Some connections with other results and further comments. This section consists, for the most part, of a number of subsections describing some recent results and how they fit into the framework or approach described in Section 2. As a result, it is hoped that the reader will appreciate that this approach is broad and applicable in a variety of situations. Furthermore, it should also be clear that the matrix of possibilities allowable has not been completely studied. Throughout this section, references to collaborative work involving subsets of the authors Griffin, Hahn, Kuelbs, Ledoux, Samur, and Weiner will be denoted using appropriate letters G, H, K, L, S, and W.

6.1 Some further comments on $\{r_n\}$ in the censored case. Recall from (2.15) that on the event $\{\hat{a}_n > 0\}$ we have

$$\#\{j \le n : |X_j| > \hat{a}_n\} = nG_n(\hat{a}_n) \le r_n.$$

Under the fairly weak conditions,

(i) $\hat{a}_n$ is consistent for a_n, i.e.

$\frac{\hat{a}_n}{a_n} \xrightarrow{p} 1$, and   (6.1.1)

(ii) X is not in the domain of partial attraction of a Gaussian law, i.e.

$\liminf_{t \to \infty} \frac{G(t)}{m(2,t)} = \lambda > 0,$

the number of terms modified can be shown to actually be at least $(\lambda - \varepsilon)r_n$ for

each $\varepsilon > 0$. To make this precise, notice that in the censored mean

$$\hat{\gamma}_n = \frac{1}{n}\sum_{j-1}^{n}(|X_j| \wedge \hat{a}_n)sgn(X_n)$$

the number of X_j actually altered by the censoring operation is $nG_n(\hat{a}_n)$. Let $\varepsilon > 0$.

Now

$$\begin{aligned} Var\left(\frac{n}{r_n}G_n(a_n(1+\varepsilon))\right) &\leq \frac{n}{r_n^2}G(a_n(1+\varepsilon)) \\ &\leq \frac{n}{r_n^2}m(2, a_n(1+\varepsilon)) \\ &\leq \frac{n}{r_n^2}m(2, a_n) \text{ since } m(2,t) \downarrow \text{ as } t \nearrow 0 \\ &= 1/r_n \longrightarrow 0, \end{aligned}$$

so that

$$\frac{n}{r_n}\left(G_n(a_n(1+\varepsilon)) - G(a_n(1+\varepsilon))\right) \xrightarrow{P} 0. \tag{6.1.2}$$

Hence, by consistency of $\hat{a}_n$, with probability tending to one

$$\begin{aligned} \frac{1}{r_n}\#\{j \leq n : |X_j| > \hat{a}_n\} &= \frac{n}{r_n}G_n(\hat{a}_n) \\ &\geq \frac{n}{r_n}G(a_n(1+\varepsilon)) + o_p(1) \text{ by (6.1.2)} \\ &= \frac{G(a_n(1+\varepsilon))}{m(2, a_n)} + o_p(1) \tag{6.1.3} \\ &= \frac{a_n^2 G(a_n(1+\varepsilon))}{E(X^2 \wedge a_n^2)} + o_p(1) \\ &\geq (1+\varepsilon)^{-2}\frac{t_n^2 G(t_n)}{E(X^2 \wedge t_n^2)} + o_p(1) \end{aligned}$$

where $t_n = a_n(1+\varepsilon)$, and we use the fact that $E(X^2 \wedge t^2) \uparrow$ as $t \uparrow$. Hence with

probability tending to one, as n gets large

$$\frac{1}{r_n}\#\{j \leq n : |X_j| > \hat{a}_n\} \geq \lambda(1 - \varepsilon)(1 + \varepsilon)^{-2}$$

Since $\varepsilon > 0$ was arbitrary, the claim is proved. In the Feller class, since $\lambda < 1$, it is possible to also show by similar arguments, that with probability tending to one, as n gets large

$$\frac{1}{r_n}\#\{j \leq n : |X_j| > \hat{a}_n\} < 1.$$

6.2. Magnitude trimming in the intermediate case using the tail equation. It was briefly mentioned in Section 2 that if F is continuous, then our framework encompasses magnitude trimming. Here we continue this discusssion and its relationship to the results for magnitude Winsorized sums in HKW (1990b) and those of Pruitt (1988). For further details, references, and relationships the reader might consider HKW (1990b), Pruitt (1988), and Griffin and Pruitt (1987).

The purpose of HKW (1990b) was to present a new approach to tightness and weak convergence for laws formed from constantly normalized partial sums, by first studying the problem for self-normalized partial sums with truncation levels and scalings determined by the empirical version of (2.4)(A) and $\{r_n\}$ as in (2.16). The specific situation studied involved magnitude Winsorized sums of symmetric random variables.

Let $X, X_1, X_2, \ldots$ be a sequence of nondegenerate symmetric independent random variables with common distribution function F. Arrange $\{X_1, \cdots, X_n\}$ in decreasing order of magnitude,

$$|X_1^{(n)}| \geq |X_2^{(n)}| \geq \cdots \geq |X_n^{(n)}|, \tag{6.2.1}$$

breaking ties in some manner, such as according to priority of index.

Given integers $\{r_n\}, 0 \le r_n \le n$, the magnitude Winsorized sums

$$W_n \equiv \sum_{j=r_n+1}^{n} X_j^{(n)} + \left|X_{r_n+1}^{(n)}\right| \sum_{j=1}^{r_n} sgn\left(X_j^{(n)}\right)$$

$$= \sum_{j=1}^{n} \left(|X_j| \wedge \left|X_{r_n+1}^{(n)}\right|\right) sgn(X_j)$$

(6.2.2))

modify the r_n observations which are largest in magnitude and retain the sign of these observations. They are closely related to the classical magnitude trimmed sums

$$T_n = \sum_{j=r_n+1}^{n} X_j^{(n)},$$

(6.2.3)

which, for continuous F, are equal to the randomly totally truncated sums

$$\sum_{j=1}^{n} X_j I(|X_j| \le \hat{b}_n),$$

(6.2.4)

truncated at any level $\hat{b}_n$ satisfying the equation

$$nG_n(\hat{b}_n) = r_n.$$

(6.2.5)

In particular, for continuous F one can take $\hat{b}_n = |X_{r_n+1}^{(n)}|$ in (6.2.5). The equation (6.2.5) is the empirical version of the equation

$$nG(b_n) = r_n$$

(6.2.6)

which can be used to determine the probabilistic scales in the Feller class outside the domain of partial attraction of the normal. Pruitt (1985) and Griffin and Pruitt (1987) have complete results for the asymptotic behavior of the properly deterministically normalized classically trimmed sums in (6.2.3) provided X is symmetric and $\{r_n\}$ satisfies (2.16). Effectively, they condition on $\hat{b}_n$, work with totally truncated sums at the conditioned level, and then uncondition. When asymptotic normality holds, an alternative proof could be based on comparing the totally empirically truncated sums in (6.2.4) with the totally deterministically truncated sums

$$\sum_{j=1}^{n} X_j I(|X_j| \le b_n). \tag{6.2.7}$$

whose asymptotic normality when properly normalized by constants is easy to establish. Asymptotic normality can fail for even slight departures from symmetry. (Griffin and Pruitt (1987) give an example of a random variable in the domain of normal attraction of a symmetric stable law for which the classically trimmed sums, deterministically normalized–i.e. normalized by constants determined from the distribution as opposed to empirical normalization by statistics determined from the sample–are not asymptotically normal.)

The self-normalized or "studentized" results for magnitude Winsorized sums of symmetric nondegenerate random variables are particularly easy to obtain and hold universally. In precise terms the study HKW (1990b) concerns the magnitude Winsorized sum W_n, which can be written

$$W_n = \sum_{j=1}^{n} \left(|X_j| \wedge \hat{b}_n \right) sgn(X_j) \tag{6.2.8}$$

where

$$\hat{b}_n = G_n^{-1}(r_n/n) \tag{6.2.9}$$

when $G_n(0) > r_n/n$, and $\hat{b}_n = 0$ otherwise. Here G_n^{-1} is the left-continuous generalized inverse of

$$G_n(t) \equiv \frac{1}{n} \#\{j \le n : |X_j| > t\}.$$

When $\hat{b}_n > 0$, then

$$G_n(\hat{b}_n-) \ge r_n/n \ge G_n(\hat{b}_n),$$

and when F is continuous, $\hat{b}_n$ is simply the $(r_n + 1)st$ largest of $\{|X_j| : j \leq n\}$, namely $X_{r_n+1}^{(n)}$.

Setting

$$V_n = \left(\sum_{j=1}^{n} \left(X_j^2 \wedge \hat{b}_n^2 \right) \right)^{1/2}, \tag{6.2.10}$$

the basic result about symmetric self-normalized magnitude-Winsorized sums is the following universal asymptotic normality result.

Theorem 6.2.11 (HKW (1990b)). *For symmetric nondegenerate X and $\{r_n\}$ satisfying (2.16),*

$$\mathcal{L}(W_n/V_n) \longrightarrow N(0,1). \tag{6.2.12}$$

The form of the possible limit laws of $\{W_n/c_n\}$ where $\{c_n\}$ is a sequence of constants is always a variance mixture of normals. The source of the mixing is due to the behavior of $\{V_n/c_n\}$ as can be seen in

Theorem 6.2.13 (HKW (1990b)). *A subsequence of $\{\mathcal{L}(W_n/c_n)\}$ has a (non-degenerate) weak limit iff the corresponding subsequence $\left\{ \frac{V_n}{c_n} \right\}$ has a weak limit (different from δ_0). Furthermore, if*

$$\mathcal{L}(V_{n_k}/c_{n_k}) \longrightarrow \mathcal{L}(V),$$

then

$$\mathcal{L}\left(\frac{W_{n_k}}{c_{n_k}} \right) \longrightarrow \mathcal{L}(ZV) \tag{6.2.14}$$

where Z is $N(0,1)$ and independent of $V \geq 0$, and every such law can actually arise as some subsequential limit law of the sequence $\{\mathcal{L}(W_n/c_n)\}$ provided X is a suitable universal law.

Applying similar techniques to classically trimmed sums, we find that self-normalized asymptotic normality is no longer universal in the symmetric case, but we do have

Theorem 6.2.15 (HKW (1990b), Theorem 13). *If $\mathcal{L}(X)$ is continuous and symmetric and*

$$\lim_{k \to \infty} n_k b_{n_k}^{-2}(\alpha) EX^2 I(|X| \leq b_{n_k}(\alpha)) = \infty \qquad (6.2.16)$$

uniformly on compact sets, then

$$\mathcal{L}\left(\frac{T_{n_k}}{\left(\sum_{j=r_{n_k}+1}^{n_k}(X_j^{(n_k)})^2\right)^{1/2}}\right) \to N(0,1). \qquad (6.2.17)$$

The precise criterion for asymptotic normality of the intermediate studentized magnitude trimmed sums of (6.2.17) is slightly weaker than (6.2.16), and was recently obtained by Griffin and Mason (private communication). Specifically, they show that (6.2.17) holds if and only if the convergence to infinity in (6.2.16) is in measure with respect to $N(0,1)$ rather than uniformly on compact sets. Moreover, Proposition 2.1 of HW (1989b) establishes that the intermediate studentized magnitude trimmed sums $\{T_n/(\sum_{j=r_n+1}^{n}(X_j^{(n)})^2)^{1/2}\}$ are always stochastically compact. Furthermore, whenever the quantity in (6.2.16) converges in measure to a constant $c \in [0, \infty)$ with respect to $N(0,1)$, the intermediate studentized magnitude trimmed sums of (6.2.17) converge along the same subsequence to a limit. The limit can be identified as the usual Rademacher random variable if $c = 0$ and for $0 < c < \infty$ can be expressed as ratios of random series whose terms involve independent Rademachers and powers of products of i.i.d. uniform (0,1) random variables. (See Theorems 5.3 and 5.32 of HW (1989b).)

As shown in HKW (1990b), Theorem 6.2.15 leads easily to

Theorem 6.2.18 (Pruitt (1988)). *For symmetric continuous $\mathcal{L}(X)$ in the Feller class,*

$$\mathcal{L}\left(\frac{T_n}{nE(X^2 I(|X| \le b_n))}\right) \longrightarrow N(0,1),$$

where

$$b_n = G^{-1}(r_n/n)$$

and G^{-1} is the left-continuous inverse of $G(t) \equiv P(|X| > t)$.

6.3 Magnitude trimming in the intermediate case using the truncated second moment equation. *In HK (1989) $\{r_n\}$ satisfies (2.16) and $\{a_n\}$ ($\{\hat{a}_n\}$) is determined through the deterministic (random) version of (2.4)(B) slightly modified. Throughout $X, X_1, X_2, \ldots$ are i.i.d., and we assume $E(X^2) = \infty$ to simplify some notation.*

Define

$$b = \inf\{t \ge 1: \ t^{-2}E(X^2 I(|X| \le t)) > 0\}, \tag{6.3.1}$$

and for $n \geq 1$ set

$$b_n = \inf_{1 \leq j \leq n} \max\{1, |X_j|\}. \tag{6.3.2}$$

Note that with probability one $b_n \sim b$ as $n \to \infty$. For $n \geq 1$, define scale parameters

$$a_n = \inf\{t \geq b+1: \quad nE(X^2 I(|X| \leq t)) \leq t^2 r_n\} \tag{6.3.3}$$

and their empirical version

$$\hat{a}_n = \inf\{t \geq b_n+1: \quad \sum_{j=1}^{n} X_j^2 I(|X_j| \leq t) \leq t^2 r_n\}. \tag{6.3.4}$$

Then, the sequence (a_n) is comprised of strictly positive numbers such that

$$\lim_n a_n = \infty, \tag{6.3.5}$$

and

$$a_n^{-2} E(X^2 I(|X| \leq a_n)) = \frac{r_n}{n} \tag{6.3.6}$$

for $a_n \geq b+2$. Similarly, the random sequence $(\hat{a}_n)$ is such that on a set of probability one

(i) $\hat{a}_n$ exists for $n \geq 1$

(ii) $\lim_n \hat{a}_n = \infty$ \hfill (6.3.7)

(iii) $(\hat{a}_n)^{-2} E(X^2 I(|X| \leq \hat{a}_n)) = r_n/n$ for $\hat{a}_n \geq b_n+2 \sim b+2$

Define centering constants

$$\gamma_n = E(XI(|X| \leq a_n)), \tag{6.3.8}$$

and for $\psi_n \geq 0$ with $n \geq \psi_n r_n$, let

$$^{(\psi_n r_n)} S_n = S_n - \sum_{j=1}^{[\psi_n r_n]} X_j^{(n)} I(|X_j^{(n)}| > a_n). \tag{6.3.9}$$

Here $[\cdot]$ denotes the greatest integer function and $X_1^{(n)}, \cdots, X_n^{(n)}$ are the order statistics (6.2.1). Hence $^{(\psi_n r_n)} S_n$ denotes S_n with the $[\psi_n r_n]$ largest terms of the sample $\{|X_1|, \cdots, |X_n|\}$ trimmed or deleted provided they exceed a_n in magnitude. Then $^{(\psi_n r_n)} S_n$ is termed the *conditionally trimmed sum*. We define similarly

$$^{(\psi_n r_n)} \hat{S}_n = S_n - \sum_{j=1}^{[\psi_n r_n]} X_j^{(n)} I(|X_j^{(n)}| > \hat{a}_n). \tag{6.3.10}$$

Using the deterministic scaling constants $\{a_n\}$ as in (6.3.3) and centerings $\{\gamma_n\}$ as in (6.3.8), the following universal asymptotic normality holds (X need not be symmetric).

Theorem 6.3.11 (HK (1988)). *If X is nondegenerate with $EX^2 = \infty$ and $\{r_n\}$ satisfies (2.16), then there exists a sequence $\{\psi_n\}$ such that*

$$\lim_n \frac{\psi_n r_n}{n} = 0 \tag{6.3.12}$$

and

$$\mathcal{L}\left(\frac{^{(\psi_n r_n)} S_n - n\gamma_n}{\sqrt{r_n} a_n} \right) \longrightarrow N(0,1). \tag{6.3.13}$$

One particular choice of $\{\psi_n\}$ which always works is $\psi_n = nP^\Delta(|X| > a_n)/r_n$ where $\Delta \in (0,1)$. If X is in the Feller class, i.e. see the remark following (3.11), the sequence $\{\psi_n\}$ can be replaced by a single number $\psi > 0$ suitably chosen.

Unlike for the censored equation approach, we have the empirical analogue of (6.3.13) only when X is symmetric. The difficulties in the non-symmetric case appear substantial and at this point have not been resolved. For symmetric X, however, we have the following.

Theorem 6.3.14 (HK (1988)). *If X is symmetric, $EX^2 = \infty$, and $\{r_n\}$ satisfies (2.16), then there exists $\{\hat{\psi}_n\}$ such that*

$$\lim_n \frac{\hat{\psi}_n r_n}{n} = 0 \qquad (6.3.15)$$

with probability one, and

$$\mathcal{L}\left(\frac{(\hat{\psi}_n r_n)\hat{S}_n}{\sqrt{r_n}\hat{a}_n}\right) \longrightarrow N(0,1). \qquad (6.3.16)$$

If $E|X| < \infty$, then any sequence of real numbers $\{\hat{\psi}_n\}$ such that

$$\lim_{n=\infty} \frac{(n/r_n)^{1/2}}{\hat{\psi}_n} = 0 \qquad (6.3.17)$$

will suffice for (6.3.16). Furthermore, it is possible to choose $\{\hat{\psi}_n\}$ such that both (6.3.15) and (6.3.17) hold if $E|X| < \infty$.

Remarks.

(1) If $\hat{\psi}_n = (n/r_n)\left(\sum_{j=1}^n I(|X_j| > \hat{a}_n)/n\right)^{\delta}$ for some $\delta \in (0,1)$, then (6.3.15) and (6.3.16) hold.

(2) In HKW(1989) a universal LIL was proved for conditionally deterministically trimmed real valued sums obtained as above and normalized by the constants $(r_n \, L_2 n)^{1/2} a_n$. For the LIL results, it is also required that $\{r_n\}$ is a real sequence such that $r_n \uparrow \infty, n/r_n \uparrow \infty$, and $\liminf_n r_n/L_2 n > 0$. It was also shown that the LIL holds universally for conditionally censored sums with constant normalizations $(r_n L_2 n)^{1/2} a_n$. Here a_n is determined through the censored second moment equation and $\{r_n\}$ is a real sequence as in the previous case.

6.4 Magnitude trimming for nonidentically distributed sequences. When $X_1, X_2, X_3, \ldots$ are independent but not necessarily identically distributed, magnitude trimming, which alters summands according to their distance from the fixed center 0, is less apt to lead to useful results outside the setting of symmetry for all the summands. Simply consider a sequence $\{X_j = Y_j - c_j\}$ where $\{Y_j\}$ is i.i.d. but $c_j \to \infty$ rapidly.

However, when each X_j is symmetric about the origin, some of the self-normalized magnitude censoring results discussed earlier do generalize nicely. Here we focus on two related "near-universal" asymptotic normality results in this setting, generalizing results of HKW (1990a, 1990b) and Weiner (1990b).

Let $\{X_j\}$ be random variables which are independent, nondegenerate and symmetric about the origin. Consider the self-normalized censored sums

$$S_n = \frac{\sum_{j=1}^{n}(|X_j| \wedge \hat{c}_n)sgn(X_j)}{\left\{\sum_{j=1}^{n}(X_j^2 \wedge \hat{c}_n^2)\right\}^{1/2}} \tag{6.4.1}$$

where $\hat{c}_n$ is determined empirically by $\{X_1, \ldots, X_n\}$ in one of two ways:

(A) (Winsorizing) For integers $r_n \to \infty$, $r_n/n \to 0$, let $\hat{c}_n$ be the (r+1)st largest element in $\{|X_j|: \ j \le n\}$. (Cf. HKW (1990b).)

(B) (Implicit Censoring) Fix a continuous function $h: \ (0, \infty) \to (0, \infty)$ such that $h(t)/t^2 \uparrow \infty$, and define $\hat{c}_n$ empirically by the equation

$$\sum_{j=1}^{n}(X_j^2 \wedge \hat{c}_n^2) = h(\hat{c}_n). \tag{6.4.2}$$

(Cf. HKW (1990a) and Weiner (1990b).)

It is clear, upon inspecting the similar proofs of the universal symmetric asymptotic normality results for S_n in HKW (1990a, 1990b) for these two versions of $\hat{c}_n$ in
the i.i.d. case, that the same proof will apply here provided only that $P(\hat{c}_n > 0) \to 1$
as $n \to \infty$. Naturally, in the non-i.i.d. case, this latter fact does not always hold
automatically.

Theorem 6.4.3. *Let* $X_1, X_2, X_3, \ldots$ *be independent, nondegenerate random variables which are symmetric about the origin.*

(a.) *Define* $\hat{c}_n$ *by method (A). If* $\{r_n\}$ *are integers such that* $r_n \to \infty$ *and*
$r_n/n \to 0$, *then*

$$S_n \xrightarrow{D} N(0,1)$$

provided

$$\eta \equiv \liminf_{n\to\infty} \frac{1}{r_n} \sum_{j=1}^{n} P(X_j \neq 0) > 1. \qquad (6.4.4)$$

(b.) *If* $h : (0,\infty) \to (0,\infty)$ *is continuous and such that* $h(t)/t^2 \uparrow \infty$, *then
with probability tending to one, (6.4.2) uniquely defines* $\hat{c}_n > 0$ *and* $S_n \xrightarrow{D}$
$N(0,1)$, *provided*

$$\exists \tau > 0 : \sum_{j=1}^{\infty} E(X_j^2 \wedge \tau^2) = \infty. \qquad (6.4.5)$$

Remarks. 1. It follows easily from results in Weiner (1990a) that the failure of
(6.4.5) is equivalent to the almost sure convergence of the random series $\sum_{j=1}^{\infty} X_j$;
thus, (6.4.5) is truly a minimal condition under which to consider self-normalized
asymptotic normality.

2. Part (b) of Theorem 6.4.3 can be regarded as an "empiricalization" (in the symmetric case) of Theorem 5.6 in Weiner (1990b), since there the "centering functions" γ_j all vanish identically due to symmetry.

Proof. As observed earlier, it suffices in each case to verify $P(\hat{c}_n > 0) \to 1$.

(a): Let $J_n = \#\{j \le n : X_j \neq 0\}$. Now $\hat{c}_n = 0 \iff J_n \le r_n$. Given $1 < \eta' < \eta$, (6.4.4) implies $EJ_n = \sum_{j=1}^{n} P(X_j \neq 0) > \eta' r_n \to \infty$, so that $Var(J_n) \le \sum_{j=1}^{n} P(X_j \neq 0) = EJ_n = o((EJ_n)^2)$. Thus, $Var(J_n/EJ_n) \to 0$, whence $J_n/EJ_n \xrightarrow{p} 1$. But then

$$P(\hat{c}_n = 0) \le P(J_n \le r_n) \le P(J_n/EJ_n \le 1/\eta') \to 0,$$

since $1/\eta' < 1$.

(b): In Weiner (1990a) it is shown that if (6.4.5) holds for some $\tau > 0$ then it holds for every $\tau > 0$. Let $Y_n(\tau) = \sum_{j=1}^{n}(X_j^2 \wedge \tau^2)$. Then (6.4.2) reads, $Y_n(\hat{c}_n) = h(\hat{c}_n)$. Now for each $\tau > 0$, $E\left(\frac{Y_n(\tau)}{\tau}\right) \to \infty$ by (6.4.5), while

$$Var\left(\frac{Y_n(\tau)}{\tau}\right) \le \sum_{j=1}^{n} E\left(\frac{X_j^4 \wedge \tau^4}{\tau^4}\right) \le \sum_{j=1}^{n} E\left(\frac{X_j^2 \wedge \tau^2}{\tau^2}\right)$$

$$= E\left(\frac{Y_n(\tau)}{\tau^2}\right) = o\left(E\left(\frac{Y_n(\tau)}{\tau}\right)^2\right).$$

Thus, as in part (a), for each $\tau > 0$ we have $Y_n(\tau)/EY_n(\tau) \xrightarrow{p} 1$. It follows that $Y_n(\tau)/h(\tau) \to \infty$ as $n \to \infty$, yet $Y_n(\tau)/h(\tau) \to 0$ as $\tau \to \infty$ due to $Y_n(\tau)/h(\tau) \le n\tau^2/h(\tau) \to 0$ as $\tau \to \infty$, by the assumptions on h. Now, arguing just as in Weiner (1990b) (where deterministic solutions of the deterministic analogues of (6.4.2) were being sought), we see that due to continuity and strict monotonicity of the functions $\tau \to Y_n(\tau)/h(\tau)$, the two conditions $Y_n(\tau)/h(\tau) \to \infty$ $(n \to \infty)$ and $Y_n(\tau)/h(\tau) \to 0$ $(\tau \to \infty)$ together guarantee that with probability tending to one, the equation $Y_n(\tau)/h(\tau) = 1$ has a unique positive solution $\tau = \hat{c}_n$. Thus, (6.4.2) holds with $\hat{c}_n > 0$ with probability tending to one. ∎

6.5 Magnitude trimming for the vector valued situation and some LIL results. In some earlier papers KL (1987a), KL (1987b), KL (1990), and HKS (1987) results on asymptotic normality and the LIL were obtained for trimmed and also conditionally trimmed vector valued sums normalized by constants. Here $X, X_1, X_2, \cdots$ have values in a Banach space B, and we let $\{X_1^{(n)} \cdots, X_n^{(n)}\}$ be the arrangement of $\{X_1, \cdots, X_n\}$ in decreasing order of magnitude with respect to the norm $\|\cdot\|$ on B, i.e.

$$\|X_1^{(n)}\| \geq \|X_2^{(n)}\| \geq \cdots \geq \|X_n^{(n)}\| \tag{6.5.1}$$

breaking ties by priority of index.

The papers KL (1987b) and HKS (1987) prove asymptotic normality results for conditionally trimmed sums normalized by constants. To be precise, given a continuous, nonnegative function $d(\cdot)$ increasing to ∞, define

$$^{(\xi r)}_{\tau} S_n = S_n - \sum_{j=1}^{[\xi r]} X_j^{(n)} I(\|X_j^{(n)}\| > \tau d([n/r])) \tag{6.5.2}$$

where $[\cdot]$ is the greatest integer function. The relevant centerings in this situation are

$$\gamma_n(\tau, r) = E(X I(\|X\| \leq \tau d([n/r]))). \tag{6.5.3}$$

In KL (1987) asymptotic normality is studied for the i.i.d. situation, and B any type 2 Banach space, where in HKS (1987) we only assume $X, X_1, X_2, \cdots$ strictly stationary and Φ-mixing, but restrict their values to a Hilbert space H.

As a sample of the type of results obtained we state a theorem from each paper.

Theorem 6.5.4 (KL (1987b)). *Let $X, X_1, X_2, \cdots$ be i.i.d. B-valued where B is a type 2 Banach space and assume:*

(i) *for the centerings $\{\overline{\delta}_n\}$ and positive normalizing constants $\{\overline{d}_n\}$ the sequence*

$$\{(S_n - \overline{\delta}_n)/\overline{d}_n\} \tag{6.5.5}$$

is tight, and

(ii) *for some continuous linear functional h the sequence*

$$\{h(S_n - \overline{\delta}_n)/\overline{d}_n\} \tag{6.5.6}$$

is stochastically compact.

Then, there is an increasing continuous function $d(t)$ defined in $[0, \infty)$ such that

$$d(n) \sim \overline{d}_n, \tag{6.5.7}$$

and for each positive sequence $\{r_n\}$ satisfying (2.16) and each $\tau > 0$ we have a number $\xi > 0$ such that

$$\left\{ \frac{\frac{(\xi r_n)}{\tau} S_n - n\gamma_n(\tau, r_n)}{\sqrt{r_n} d(n/r_n)} \right\} \tag{6.5.8}$$

is a tight sequence in B with only centered Gaussian limits. Further, there is a $\tau_0 > 0$ such that $\tau \geq \tau_0$ implies the limits of (6.5.8) are all nondegenerate Gaussian.

Remarks

(1) The function $d(t)$ in Theorem (6.5.4) is defined through the censored second moment

$$E\left(h^2(X) \wedge s^2\right) \tag{6.5.9}$$

analogous to (2.4)(C). Here h is the continuous linear functional in (6.5.6) and for all large t

$$d(t) = \inf\{s > 0 : \ s^{-2}E(h^2(X) \wedge s^2) = 1/t\}. \tag{6.5.10}$$

Since $s^{-2}E(h^2(X) \wedge s^2)$ is continuous and strictly decreasing on $[a, \infty)$ where $a = \inf\{x : \ P(|h(X)| \leq x) > 0\}$, one can use either a sup or an inf in (6.5.10) for all $t > 0$ sufficiently large.

(2) If $X \in DA(\alpha)$, $0 < \alpha \leq 2$, then the sequence in (6.5.8) converges weakly to a Gaussian measure $\mathcal{L}(G_\tau)$ where G_τ is a nondegenerate centered Gaussian random variable for each $\tau > 0$.

The asymptotic normality of conditionally trimmed sums of Φ-mixing random variables was obtained in Theorem 1 of HKS (1987). The main result was:

Theorem 6.5.11 (HKS (1987)). *Let* $X, X_1, X_2, \cdots$ *be a strictly stationary sequence with values in a separable Hilbert space H which is Φ-mixing which $\Phi(1) < 1$.*

Let $S_n = X_1 + \cdots + X_n$ and assume there are normalizing constants $\{d(n)\}$ such that $\lim_n d(n) = \infty$,

$$\{S_n/d(n)\} \text{ is tight,} \tag{6.5.12}$$

$$\{S_n/d(n)\} \text{ has only nondegenerate limits,} \tag{6.5.13}$$

and if $\{q_n\}$ is any sequence of integers such that

$$\lim_n q_n/n = 0, \text{ then } \mathcal{L}\left(S_{q_n}/d(n)\right) \xrightarrow{w} \delta_0. \tag{6.5.14}$$

Let $\{r_n\}$ satisfy (2.16) and be such that

$$\lim_n r_n \Phi^{1/2}(n/r_n) = 0. \tag{6.5.15}$$

Then, for each $\tau > 0$ and any sequence $\{\xi_n\}$ such that $\lim_n \xi_n = \infty$, the sequence

$$\left\{ \frac{\frac{(\xi_n r_n)}{\tau} S_n - n\gamma_n(\tau, r_n)}{\sqrt{r_n} d(n/r_n)} \right\} \tag{6.5.16}$$

is tight with only centered Gaussian limits. Furthermore, for all $\tau > 0$ sufficiently large, the limits are all nondegenerate.

Remarks.

(1) If $\mathcal{L}\left(S_n/d(n)\right) \xrightarrow{w} \mathcal{L}(Z)$ when Z is nondegenerate, then it is pointed out in HKS (1987), that assumptons (6.5.12), (6.5.13), and (6.5.14) all hold (see HKS (1987), Remarks 3, 5, and 6).

(2) Theorem 6.5.12 has an application to what is known as Ibragimov's conjecture: If $\{X_j\}$ is a strictly stationary real-valued Φ-mixing sequence with $EX_j = 0$, $EX_j^2 = 1$ and if $\sigma_n^2 = E(S_n^2) \to \infty$ as $n \to \infty$, than S_n can

be normalized with constants so as to converge weakly to a $N(0,1)$ limit. This application is contained in Theorem 2 of HKS (1987) and shows that once the maximal terms of the sample $\{|X_1|, \cdots, |X_n|\}$ are suitably conditionally trimmed, the only possible nondegenerate limits must be centered Gaussian provided $\Phi(1) < 1$. Corollaries 3 and 4 of HKS (1987) then apply more directly to Ibragimov's conjecture, but still leave much to be determined. The condition $\Phi(1) < 1$ can be relaxed as can be seen by Remarks 9 and 10 of HKS (1987).

The papers KL (1987a) and KL (1990) contain LIL results. The paper KL (1987a) contains results for classically trimmed partial sums normalized by constants provided $X \in DA(2)$ and X takes values in any Banach space. Indeed, KL (1987a) provides necessary and sufficient conditions for the compact LIL in B provided exactly r terms are trimmed and r is any integer greater than or equal to zero. It also contains a result which yields the compact LIL in B for $X \in DA(2)$ provided $r_n = [\xi_n L_2 n]$ terms are trimmed. Here $[\cdot]$ denotes the greatest integer function and $\xi_n \to 0$ depends on X.

The paper KL (1990) presents a variety of bounded LIL and compact LIL results in the B-valued setting. To give a brief description of the bounded LIL results we set $n_0 = 0$ and for each $\beta > 1$, let $n_k = [\beta^k]$, and $I(k) = (n_k, n_{k+1}]$ for $k \geq 0$ where $[\cdot]$ is the greatest integer function. Then, for $r > 0$, $n \geq r, \tau > 0$ and a positive function $d(t)$ defined on $[0, \infty)$, let

$$^{(r)}S_n(\beta, \tau) = S_n - \sum_{j=1}^{[r]} X_j^{(n)} I(\|X_j^{(n)}\| > \tau d\overline{\alpha}(n_k)) \qquad (6.5.17)$$

aprovided $n \in I(k)$, $\alpha(t) = t/L_2 t$, and $\overline{\alpha}(s) = [\alpha(s)]$. Hence $^{(r)}S_n(\beta, \tau)$ denotes the partial sum S_n with the $[r]$ largest terms of the sample $\{||X_1||, \cdots,, ||X_n||\}$ trimmed (conditionally) provided they exceed $\tau d\overline{\alpha}(n_k)$ in norm when $n \in I(k)$. The relevant centerings for $n \in I(k)$ are

$$\gamma_n(\beta, \tau) = E(XI(||X|| \le \tau d\overline{\alpha}(n_k))). \tag{6.5.18}$$

We then have

Theorem 6.5.19 (KL (1990)). *Let* $X, X_1, X_2, \cdots$ *be i.i.d.* B-*valued such that for centerings* $\{\overline{\delta}_n\}$ *and positive normalizing constants* $\{\overline{d}_n\}$

$$\{(S_n - \overline{\delta}_n)/\overline{d}_n\} \text{ is stochastically bounded,} \tag{6.5.20}$$

and for some continuous linear functional on B

$$\{h(S_n - \overline{\delta}_n)/\overline{d}_n\} \text{ is stochastically compact.} \tag{6.5.21}$$

Then there is an increasing continuous function $d(t)$ *defined on* $[0, \infty)$ *such that*

$$d(n) \approx \overline{d}_n. \tag{6.5.22}$$

Furthermore, for each $\tau > 0$ *there is a positive constant* ξ_0 *such that if* $r_n = \xi_n L_2 n$ *where* $\xi_n \ge \xi_0$ *and* $r_n \le n$, *then for all* $\beta \in (1, \Lambda]$

$$\overline{\lim}_n ||^{(r_n)}S_n(\beta, \tau) - n\gamma_n(\beta, \tau)||/(L_2 n \, d(n/L_2 n))$$
$$\le M(\tau)(\Lambda + 1) \tag{6.5.23}$$

where $M(\tau)$ *is a finite function of* $\tau > 0$ *which is independent of* $\beta > 1$ *and* $\Lambda > 1$. *In addition, there exists a* $\tau_0 > 0$ *such that if* $\tau \ge \max(\tau_0, 1)$ *and* $r_n = \xi_n L_2 n$ *as above, then*

$$\overline{\lim}_n ||^{(r_n)}S_n(\beta, \tau) - n\gamma_n(\beta, \tau)||/(L_2 n \, d(n/L_2 n)) > 0. \tag{6.5.24}$$

Remarks.

(1) It is possible to eliminate the block aspects of the definition of $^{(r)}S_n(\beta, \tau)$ if one is willing to accept almost sure boundedness results which are less precise than those in (6.5.23) and which exclude the nondegeneracy result in (6.5.24). More precisely as was pointed out in Remark III following Theorem 2 in KL (1990), let $r_n = \xi_0 \, L_2 n$ where ξ_0 is as in Theorem 6.5.19 and set:

$$^{(r_n)}S_n \div S_n = \sum_{j=1}^{[r_n]} X_j(n).$$

Then $^{(r_n)}S_n$ is the classically trimmed sum and for $\beta \in (1, \Lambda]$, $\tau > 0$

$$\overline{\lim}_n \|^{(r_n)}S_n - n\gamma_n(\beta, \tau)\| / (L_2 n \; d(n/L_2 n)) \leq M(\tau)(2\Lambda + 1) + \xi_0 \tau \quad (6.5.25)$$

with probability one.

(2) For real-valued X in the Feller class, a result similar to (6.5.25) combined with a nondegeneracy result was also obtained by Griffin (1988).

(3) In the recent paper GK (1988), LIL results were obtained for all self-normalized partial sums of i.i.d. real-valued random variables in the Feller class. The self-normalizations used were the classical

$$V_n^2 = \sum_{i=1}^{n} X_i^2 \qquad (6.5.26)$$

as in Logan, Mallows, Rice, and Shepp (1973). The most complete result obtained in GK (1988) is the following refinement of the Hartmann-Wintner LIL (also compare KL (1987a)).

Theorem 6.5.27 (GK (1988)). *If $X \in D(2)$ and $EX = 0$ if $EX^2 < \infty$, and V_n^2 is as in (6.5.26), then with probability one*

$$\left\{ \frac{S_n - nEX}{(2L_2n \ V_n^2)^{1/2}} \right\} \rightarrow\rightarrow [-1, 1]. \tag{6.5.28}$$

The notation $\{d_n\} \rightarrow\rightarrow D$ means the sequence converges and clusters throughout D.

(4) A subsequent paper GK (1989)) also studies the LIL. It improves, via a new proof, the classical Hartmann-Wintner Theorem, and it also contains a sharpened version of the Kolmogorov Erdos test. The approach throughout is the use of self-normalizers. Perhaps the overall message of GK (1988) and GK (1989) is that self-normalizations are superior to constant normalizations even in some classical problems. Of course as indicated previously, self-normalizations are also useful in the study of asymptotic normality outside the classical domain, because they allow an easy construction of normalizing constants and they also seem to improve the tail behavior of the limit laws.

6.6 Some general remarks on extensions. The overall theme of the results presented is that once the large values of the sample $\{X, X_2, \cdots, X_n\}$ are trimmed and suitably modified, then asymtotic normality (or some close relative) and LIL behavior are much more likely to hold for the related partial sums. This is not a new idea, but in recent years it has been pursued systematically via a variety of classical methods as well as through an approach based on the quantile transform and

Brownian Bridge approximations. We have presented results on trimmed sums, conditionally trimmed sums, and censored sums using the magnitude of the individual terms as the device to determine the "large values" of the sample. One important aspect of the proof of these results requires determining a scaling sequence (deterministic or random) for the partial sums which relates to the trimming or modification of the r_n largest terms, and then using the scaling sequence to determine the proper centerings. Once this is accomplished the proofs usually proceed via a comparison of the modified sum to something more classical to obtain the desired result. We will not go into further details here, but we do mention that modifications of data other than trimming, truncating or censoring are possible. The paper Ould-Rouis (1989) in this volume examines this problem and obtains invariance principles as well as asymptotic normality results for a variety of "influence functions" which interpolate between truncation and censoring. Two papers by Weiner (1990a, 1989b), the latter appearing in this volume, investigate how centers and scales can be determined simultaneously rather then as indicated above. These results were motivated by considering asymptotic normality results via censoring when $X_1, X_2, \cdots$ are independent, but not identically distributed. Here ranking the summands according to magnitudes may be inappropriate since it arbitrarily presumes zero as the unchanging center of concentration for the random variables. Hence the need to simultaneously select centers about which to censor as well as censoring levels. Even in the i.i.d. case, for asymmetric random variables, the magnitude based method emphasized in this survey suffers from a lack of location invariance. Some possibilities for overcoming this deficiency are suggested in Weiner (1990b).

Finally, although our methods and perhaps our point of view are distinct from that in Csörgő, Horvath and Mason (1986), some important motivations come from the results obtained there. Of course, debts to Feller (1968) and Mori (1976, 1977) were also obvious at the beginning.

References

Csörgő, S. Horvath, L. and Mason, D. M. (1986). What portion of the sample makes a partial sum asymptotically stable or normal? *Probab Th. Rel. Fields* **72**, 1-16.

Feller, W. (1967). On regular variation and local limit theorems. *Proc. Fifth Berkeley Symp. Math. Statist. Prob.*, vol. II, Part 1. University of California Press, Berkeley, California (1967), 373-388.

Feller, W. (1968). An extension of the law of the iterated logarithm to variables without variances. *J. Math. Mechan.* **18**, 343-355.

Feller, W. (1971). *An introduction to probability theory and its applications*, vol II. John Wiley, New York.

Griffin, P. S. (1988). The influence of extremes on the law of the iterated logarithm. *Probab. Th. Rel. Fields* **77**, 241-270.

Griffin, P. S. and Kuelbs, J. (1988). Self-normalized laws of the iterated logarithm. To appear in *Ann. Probab.*

Griffin, P. S. and Kuelbs, J. (1989). Some extensions of the LIL via self-normalizations. Preprint.

Griffin, P. S. and Pruitt, W. E. (1987). The central limit problem for trimmed sums. *Math. Proc. Camb. Phil. Soc.* **102**, 329-349.

Hahn, M. G. and Kuelbs, J. (1988). Universal asymptotic normality for conditionally trimmed sums. *Stat. Prob. Letters* **7**, 9-15.

Hahn, M. G., Kuelbs, J. and Samur, J. (1987). Asymptotic normality of trimmed sums of Φ-mixing random variables. *Ann. Probab.* **15**, 1395-1418.

Hahn, M. G., Kuelbs, J. and Weiner, D. C. (1989). A universal law of the iterated logarithm for trimmed and censored sums. In: *Proceedings of Probability Theory on Vector Spaces IV* (Cambanis and Weron, Eds.), Lancut, Poland. *Springer LNM* **1391**, 82-98.

Hahn, M. G., Kuelbs, J. and Weiner, D. C. (1990a). The asymptotic joint distribution of self-normalized censored sums and sums of squares. *Ann. Probab.* **18**, 1284-1341.

Hahn, M. G., Kuelbs, J. and Weiner, D. C. (1990b). The asymptotic distribution of magnitude-winsorized sums via self-normalization. *J. Theoret. Probab.* **3**, 137-168.

Hahn, M. G. and Weiner, D. C. (1990a). On estimating an exponent of regular variation for tail distributions. In this volume.

Hahn, M. G. and Weiner, D. C. (1990b). Asymptotic behavior of self-normalized trimmed sums: nonnormal limits. Preprint.

Kuelbs, J. and Ledoux, M. (1987a). Extreme values for the LIL. *Probab. Th. Rel. Fields* **74**, 319-340.

Kuelbs, J. and Ledoux, M. (1987b). Extreme values for vector valued random variables and a Gaussian cental limit theorem. *Probab. Th. Rel. Fields* **74**, 341-355.

Kuelbs, J. and Ledoux, M. (1990). Extreme values and LIL behavior. In: *Proceedings of the Sixth International Conference on Probability in Banach Spaces* (Eds.), .Birkhauser.

Lévy, P. (1937). *Théorie de l'Addition des Variables Aléatoires*. Gauthier-Villars, Paris.

Logan, B., Mallows, C., Rice, S. and Shepp, L. (1973). Limit distributions of self-normalized sums. *Ann. Probab.* **1**, 788-809.

Maller, R. (1982). Asymptotic normality of lightly trimmed means–a converse. *Math. Proc. Camb. Phil. Soc.* **92**, 535-545.

Maller, R. (1988). Asymptotic normality of trimmed means in higher dimensions. *Ann. Probab.* **16**, 1608-1622.

Mori, T. (1976). The strong law of large numbers when extreme values are excluded from sums. *Z. Wahrscheinlichkeitstheor. Verw. Geb.* **36**, 189-194.

Mori, T. (1977). Stability for sums of i.i.d. random variables when extreme terms are excluded. *Z. Wahrscheinlichkeitstheor. Verw. Geb.* **41**, 159-167.

Mori, T. (1984). On the limit distributions of lightly trimmed sums. *Math Proc. Camb. Phil. Soc.* **96**, 507-516.

Ould-Rouis, H. (1990). Invariance principles and self-normalizations for sums trimmed according to choice of influence function. In this volume.

Pruitt, W. (1985). Sums of independent random variables with the extreme terms excluded. In: *Probability and Statistics*. Essays in Honor of Franklin A. Graybill (J. N. Srivastava, Ed.), 201-216. Elsevier, Amsterdam. Regularly varying functions. Springer, Berlin.

Stigler, S. M. (1973). The asymptotic distribution of the trimmed mean. *Ann. Statist.* **1**, 472-477.

Weiner, D. C. (1990a). Center, scale and asymptotic normality for sums of independent random variables. In: *Proceedings of the Seventh International Conference on Probability in Banach Spaces* (Eberlein, Kuelbs, and Marcus, Eds.), 287-307. Birkhauser.

Weiner, D. C. (1990b). Centrality, scale and asymptotic normality for censored sums of independent random variables. In this volume.

Marjorie G. Hahn	Jim Kuelbs	Daniel C. Weiner
Depart. of Mathematics	Depart. of Mathematics	Depart. of Mathematics
Tufts University	University of Wisconsin	Boston University
Medford, MA 02155 USA	213 Van Vleck Hall	Boston, MA 02215 USA
	Madison, WI 53706 USA	

WEAK CONVERGENCE OF TRIMMED SUMS

Philip S. Griffin * & William E. Pruitt **

1. Introduction. Considerable progress has been made in the last few years on problems concerning the asymptotic distribution of trimmed sums. The aim of this paper is to describe some of these results, discuss some of the problems which remain open, and provide a solution to one of these in an important special case.

Let $X, X_1, X_2, \ldots$ be a sequence of nondegenerate i.i.d. random variables and $X_{n1}, \ldots, X_{nn}$ be the order statistics of $X_1, \ldots, X_n$; thus $X_{n1} \leq \ldots \leq X_{nn}$. For r_n and s_n sequences of integers define

$$(1.1) \qquad S_n(s_n, r_n) = X_{n,s_n+1} + \ldots + X_{n,n-r_n}.$$

This is what is usually referred to in the statistical literature as a trimmed sum. Several other forms of trimming have also been studied and here, by way of comparison, we will discuss one of these called the modulus trimmed sum. To define this let $^{(1)}X_n, \ldots, ^{(n)}X_n$ be an arrangement of $X_1, \ldots, X_n$ in decreasing order of magnitude, i.e. $|^{(1)}X_n| \geq \ldots \geq |^{(n)}X_n|$, and set

$$(1.2) \qquad ^{(r_n)}S_n = {}^{(r_n+1)}X_n + \ldots + {}^{(n)}X_n.$$

---
* Research supported in part by N.S.F. Grant DMS-8700928
** Research supported in part by N.S.F. Grant DMS-8902581

The question of how we break ties in the ordering is clearly irrelevant in (1.1) but in (1.2) it may not be, and so we break ties according to priority of index. That is, for $1 \leq j \leq n$, let $m_n(j)$ be the number of i for which either $|X_i| > |X_j|$ or $|X_i| = |X_j|$ and $i \leq j$. Then define $^{(r)}X_n = X_j$ if $m_n(j) = r$.

We should point out that there are reasons, other than comparison, for discussing both forms of trimming. The first is that both kinds of trimming are often encountered in the literature; $S_n(s_n, r_n)$ in the statistical literature dealing with estimation, and $^{(r_n)}S_n$ in the probabilistic literature dealing with problems where large values of $|X_i|$ play an important role, e.g. the law of the iterated logarithm - see Feller (1968) - and the strong law of large numbers - see Mori (1976). Secondly, as the reader will see, the results in the two cases have a similar flavor. In studying these sums we found that progress on a specific problem was often easier for one type of trimming than the other, and suggested an approach to the problem in the more difficult case.

The behavior of both trimmed sums depends on the sequences r_n and s_n. The most interesting case, and the one that we will concentrate on, is when

$$(1.3) \qquad r_n \to \infty, \ s_n \to \infty, \ r_n n^{-1} \to 0, \ s_n n^{-1} \to 0.$$

This is usually referred to as intermediate trimming. For the case of light trimming, i.e. when r_n and s_n are bounded, see for example Maller (1982), Maller (1987), Mori (1984) and Csörgő, Haeusler and Mason (1988a); for the case of heavy trimming, i.e. when r_n and s_n are proportional to n , see for example Maller (1988) and Stigler (1973).

The results which have been obtained for intermediate trimming include necessary and sufficient conditions for asymptotic normality and a description of the class of subsequential limit laws and their domains of partial attraction. These results will be described in the next section. In section 3 we discuss some of the problems which we think are interesting and which remain open. Finally section 4 contains a solution to an important special case of one of these problems - a description of the class of limit laws of the modulus trimmed sums along the entire sequence when r_n is monotone.

2. **Survey of Results.** For the remainder of the paper, unless otherwise stated, we will be assuming (1.3) holds. In order to describe the results we need to introduce some notation. For $x \in \mathbf{R}$, set

(2.1) $$F(x) = P(X \le x), \quad G(x) = P(|X| > x).$$

Let $\tilde{F}$ and $\tilde{G}$ denote the right continuous inverses of F and G respectively; thus

$$\tilde{F}(u) = \inf\{x : F(x) > u\} \qquad 0 \le u < 1,$$

$$\tilde{G}(u) = \sup\{x : G(x) > u\} \qquad 0 \le u < 1.$$

For completeness, we let $\tilde{F}(1) = \tilde{F}(1^-)$ and $\tilde{G}(1) = \tilde{G}(1^-)$. Now if U is uniform on $(0,1)$, then $X =^d \tilde{F}(U)$ and $|X| =^d \tilde{G}(U)$. Next for $\alpha, \beta \in \mathbf{R}$ let

$$u_n(\alpha) = 0 \vee (r_n - \alpha r_n^{1/2})n^{-1} \wedge 1, \quad v_n(\beta) = 0 \vee (s_n - \beta s_n^{1/2})n^{-1} \wedge 1,$$

and define

$$a_n(\alpha) = \tilde{F}(v_n(\alpha)), \quad b_n(\beta) = \tilde{F}(1 - u_n(\beta)), \quad c_n(\alpha) = \tilde{G}(u_n(\alpha)).$$

Finally let

(2.2) $$\sigma_n(\alpha, \beta) = (\text{Var}(\tilde{F}(v_n(\alpha) \vee (U \wedge (1 - u_n(\beta))))))^{1/2},$$

(2.3) $$\tau_n(\alpha) = (E(\tilde{G}(U)^2; U \geq u_n(\alpha)))^{1/2}.$$

Observe that (2.2) can be written more simply as

$$\sigma_n(\alpha, \beta) = (\text{Var}(a_n(\alpha) \vee (X \wedge b_n(\beta))))^{1/2},$$

and when G is continuous (2.3) can be rewritten as

$$\tau_n(\alpha) = (E(X^2; |X| \leq c_n(\alpha)))^{1/2}.$$

For later reference we point out a few elementary properties of τ_n: For any $\lambda > 0$, if n is sufficiently large

(2.4) τ_n is nonnegative, nondecreasing and continuous on $[-\lambda, \lambda]$

and

(2.5) τ_n^2 is convex on $[-\lambda, \lambda]$.

The first of these is immediate from the definition of τ_n, while (2.5) follows from (2.17) and (2.18) in Griffin and Pruitt (1987).

We are now ready to state the results on asymptotic normality. We let $N(0,1)$ denote a standard normal random variable.

THEOREM 2.1. *There exist γ_n, δ_n such that $(S_n(s_n, r_n) - \delta_n)\gamma_n^{-1} \to N(0,1)$ if and only if*

(2.6) $\dfrac{\sigma_n(\alpha, \beta)}{\sigma_n(0,0)} \to 1$ *for all α, β.*

In this case one may take $\delta_n = nE(\tilde{F}(U) : v_n(0) < U < 1 - u_n(0))$ *and then let*
$\gamma_n = (n\sigma_n^2(0,0))^{1/2}.$

THEOREM 2.1'. *Assume* X *is symmetric. Then there exists* γ_n *such that* $^{(r_n)} S_n \gamma_n^{-1} \to N(0,1)$ *if and only if*

(2.7) $$\frac{\tau_n(\alpha)}{\tau_n(0)} \to 1 \qquad \text{for all } \alpha.$$

In this case one may take $\gamma_n = (n\tau_n^2(0))^{1/2}.$

REMARKS 1. The formulations given above and in the remainder of the paper are taken from Griffin and Pruitt (1987), (1989). However, the notation used here is slightly different. For the quantile versions, see Csörgő, Haeusler and Mason (1988b) and (1990).

2. The restriction to symmetric distributions in Theorem 2.1' is not just a technicality. The problem in the case of asymmetric distributions is genuinely more difficult and remains open. Indeed this is the case for modulus trimming in all the problems that we will discuss.

3. There are subsequential versions of these results. For example in the case of Theorem 2.1, there exist γ_n, δ_n such that

$$(S_{n_k}(s_{n_k}, r_{n_k}) - \delta_{n_k})\gamma_{n_k}^{-1} \to N(0,1)$$

if and only if (2.6) holds along the subsequence n_k. Similarly for Theorem 2.1'.

EXAMPLES 1. If X is in the domain of attraction of a stable law which is not completely asymmetric, then (2.6) holds for all sequences r_n, s_n.

2.   If X is symmetric and in the Feller class, i.e.

$$\limsup_{x \to \infty} \frac{x^2 P(|X| > x)}{EX^2 1(|X| \le x)} < \infty,$$

then (2.7) holds for any r_n. In fact, a converse is also true: If X is symmetric and (2.7) holds for all r_n, then X is in the Feller class. However, (2.6) fails for some symmetric distributions in the Feller class even if $r_n = s_n$.

3.   If X is symmetric with distribution given by

$$P(|X| > x) = \frac{1}{(\log x)^\rho} \qquad x \ge e,$$

then (2.6) holds if and only if

$$r_n n^{-2/(2+\rho)} \to \infty, \qquad s_n n^{-2/(2+\rho)} \to \infty,$$

and (2.7) holds if and only if

$$r_n n^{-2/(2+\rho)} \to \infty.$$

In comparing the two types of trimming, it is interesting to note that in the symmetric case, if there exist γ_n, δ_n such that

$$(S_n(r_n, r_n) - \delta_n)\gamma_n^{-1} \to N(0, 1),$$

then there exists γ_n such that

$$^{(2r_n)}S_n \gamma_n^{-1} \to N(0, 1),$$

but the converse is false.

The next question that arises is: What is the class of possible limit laws for these normalized trimmed sums? Before addressing this question we will consider the problem of finding the class of all possible subsequential limit laws. Thus let

$$\mathcal{I}(\{r_n\}, \{s_n\}) = \{Z : \text{there exist } X, \gamma_n, \delta_n, n_k \text{ such that}$$

$$(S_{n_k}(s_{n_k}, r_{n_k}) - \delta_{n_k})\gamma_{n_k}^{-1} \to Z\},$$

$$\mathcal{I}'(\{r_n\}) = \{Z : \text{there exist } X \text{ symmetric}, \gamma_n, n_k \text{ such that}$$

$$^{(r_{n_k})}S_{n_k}\gamma_{n_k}^{-1} \to Z\}.$$

THEOREM 2.2. *Fix $\{r_n\}$ and $\{s_n\}$ satisfying (1.3). Then $\mathcal{I}(\{r_n\}, \{s_n\})$ coincides with the class of random variables of the form $\xi N_1 + f(N_2) - g(N_3) + \mu$ where N_i, $i = 1, 2, 3$ are i.i.d. $N(0,1)$ variables, $\xi \geq 0$, $\mu \in \mathbf{R}$, and f and g are arbitrary nondecreasing convex functions which vanish at the origin.*

THEOREM 2.2'. *Fix a sequence r_n satisfying (1.3). Then $\mathcal{I}'(\{r_n\})$ coincides with the class of random variables of the form $N_1\tau(N_2)$ where N_1, N_2 are i.i.d. $N(0,1)$ and τ is nonnegative, nondecreasing and τ^2 is convex.*

It is easy to show that τ in the above representation is unique. However in Theorem 2.2, ξ, f, g and μ need not be unique. This can be seen by considering the case where the subsequential limit law is $N(0,1)$. Then one can take $f(x) = ax$, $g(x) = bx$ and $\mu = 0$ provided $\xi^2 + a^2 + b^2 = 1$. When the limit law is not normal, we believe that the representation is unique, but have been unable to prove this. The question of uniqueness creates a problem when trying to characterize the domain of partial attraction of a given limit law as we will see below. It is also interesting to note that $\mathcal{I}(\{r_n\}, \{s_n\})$ and $\mathcal{I}'(\{r_n\})$ do not depend on the sequences

r_n and s_n provided they satisfy (1.3).

To describe the domain of partial attraction of a given limit law, we need to introduce some further notation. For $Z \in \mathcal{I}(\{r_n\}, \{s_n\})$ we write $Z \sim (\xi, f, g, \mu)$ to mean that $Z =^d \xi N_1 + f(N_2) - g(N_3) + \mu$ where ξ, f, g, μ satisfy the conditions in Theorem 2.2. We let

$$\mathcal{D}_p(\{r_n\}, \{s_n\}, Z) = \{X : \text{there exist } \gamma_n, \delta_n, n_k \text{ such that}$$

$$(S_{n_k}(s_{n_k}, r_{n_k}) - \delta_{n_k})\gamma_{n_k}^{-1} \to Z\}$$

be the domain of partial attraction of Z. Similarly for $Z \in \mathcal{I}'(\{r_n\})$ we write $Z \sim \tau$ if $Z =^d N_1 \tau(N_2)$ where τ satisfies the conditions of Theorem 2.2', and

$$\mathcal{D}'_p(\{r_n\}, Z) = \{X : \text{ there exist } \gamma_n, n_k \text{ such that } ^{(r_{n_k})}S_{n_k}\gamma_{n_k}^{-1} \to Z\}.$$

Let

$$X(n, \alpha, \beta) = \tilde{F}(U)1(v_n(\alpha) < U < 1 - u_n(\beta)),$$

$$f_n(\beta) = EX(n, 0, \beta) - EX(n, 0, 0),$$

$$g_n(\alpha) = EX(n, 0, 0) - EX(n, \alpha, 0),$$

$$\xi_n = (\operatorname{Var}X(n, 0, 0))^{1/2}.$$

THEOREM 2.3. *Assume $Z \in \mathcal{I}(\{r_n\}, \{s_n\})$ and $Z \sim (\xi, f, g, \mu)$ where Z is nondegenerate. Fix a subsequence n_k and assume there exist $\alpha_o, \beta_o \in \mathbf{R}$ such that $g'(\alpha_o)$ and $f'(\beta_o)$ exist, $\xi^2 + g'(\alpha_o)^2 + f'(\beta_o)^2 \neq 0$ and*

(i) $\xi_{n_k}\sigma_{n_k}^{-1}(\alpha_o, \beta_o) \to \xi(\xi^2 + g'(\alpha_o)^2 + f'(\beta_o)^2)^{-1/2}$,

(ii) $n_k^{1/2} f_{n_k}(\beta)\sigma_{n_k}^{-1}(\alpha_o, \beta_o) \to f(\beta)(\xi^2 + g'(\alpha_o)^2 + f'(\beta_o)^2)^{-1/2}$ *all $\beta \in \mathbf{R}$,*

(iii) $n_k^{1/2} g_{n_k}(\alpha)\sigma_{n_k}^{-1}(\alpha_o, \beta_o) \to g(\alpha)(\xi^2 + g'(\alpha_o)^2 + f'(\beta_o)^2)^{-1/2}$ *all $\alpha \in \mathbf{R}$.*

Then there exist γ_n, δ_n such that

(2.8) $$(S_{n_k}(s_{n_k}, r_{n_k}) - \delta_{n_k})\gamma_{n_k}^{-1} \to Z.$$

Conversely, if Z uniquely determines ξ, f, g and μ then the conditions (i)-(iii), *for all α_o, β_o such that $g'(\alpha_o), f'(\beta_o)$ exist and $\xi^2 + g'(\alpha_o)^2 + f'(\beta_o)^2 \neq 0$, are also necessary for* (2.8).

COROLLARY 2.4. *Assume $Z \in \mathcal{I}(\{r_n\}, \{s_n\})$ and $Z \sim (\xi, f, g, \mu)$ where Z is nondegenerate. Then $X \in \mathcal{D}_p(\{r_n\}, \{s_n\}, Z)$ if* (i)-(iii) *hold along some subsequence n_k. Conversely, if Z uniquely determines ξ, f, g and μ, then* (i)-(iii) *holding along some subsequence are necessary for $X \in \mathcal{D}_p(\{r_n\}, \{s_n\}, Z)$.*

THEOREM 2.3'. *Assume $Z \in \mathcal{I}'(\{r_n\})$ and $Z \sim \tau$ where Z is nondegenerate. Fix a subsequence n_k. Then*

$$(2.9) \qquad \text{there exists } \gamma_n \text{ such that } {}^{(r_{n_k})}S_{n_k}\gamma_{n_k}^{-1} \to Z$$

if and only if

$$(2.10) \qquad \text{for some (all) } \alpha_o \text{ with } \tau(\alpha_o) > 0, \quad \frac{\tau_{n_k}(\alpha)}{\tau_{n_k}(\alpha_o)} \to \frac{\tau(\alpha)}{\tau(\alpha_o)} \quad \text{all } \alpha \in \mathbf{R}.$$

COROLLARY 2.4'. *Assume $Z \in \mathcal{I}'(\{r_n\})$ and $Z \sim \tau$ where Z is nondegenerate. Then $X \in \mathcal{D}_p'(\{r_n\}, Z)$ if and only if* (2.10) *holds along some subsequence n_k.*

As we remarked earlier the question of uniqueness in the representation of Z in Theorem 2.3 creates a problem in characterizing $\mathcal{D}_p(\{r_n\}, \{s_n\}, Z)$. If Z is normally distributed, then the representation is not unique, but $\mathcal{D}_p(\{r_n\}, \{s_n\}, Z)$ is characterized in Theorem 2.1 in this case (see Remark 3 following it). If, as we believe, the representation is unique when Z is not normal, Theorem 2.3 does complete the characterization.

We now come to the analogous questions for limit laws along the entire sequence. Thus let

$$\mathcal{L}(\{r_n\}, \{s_n\}) = \{Y : \text{there exist } X, \gamma_n, \delta_n \text{ such that}$$

$$(S_n(s_n, r_n) - \delta_n)\gamma_n^{-1} \to Y\},$$

$$\mathcal{L}'(\{r_n\}) = \{Y : \text{there exist } X \text{ symmetric and } \gamma_n \text{ such that}$$

$$^{(r_n)}S_n\gamma_n^{-1} \to Y\}.$$

The problem then is to characterize $\mathcal{L}(\{r_n\}, \{s_n\})$ and $\mathcal{L}'(\{r_n\})$. (Note that, mod-
ulo the uniqueness question, a characterization of the domain of attraction of a
given limit law follows immediately from Theorems 2.3 and 2.3'.) These problems
appear harder than their subsequential analogues, not least because, unlike the subsequential case, the class of limit laws may depend on the sequences r_n and s_n. For
example, there exist sequences r_n and s_n such that $\mathcal{L}(\{r_n\}, \{s_n\}) = \mathcal{I}(\{r_n\}, \{s_n\})$
and $\mathcal{L}'(\{r_n\}) = \mathcal{I}'(\{r_n\})$. Thus the analogues of the infinitely divisible laws and
stable laws coincide. However, if we add the additional assumption

(2.11) r_n and s_n are nondecreasing,

then it is not hard to show that the inclusions $\mathcal{L}(\{r_n\}, \{s_n\}) \subset \mathcal{I}(\{r_n\}, \{s_n\})$ and
$\mathcal{L}'(\{r_n\}) \subset \mathcal{I}'(\{r_n\})$ are both strict. In section 4 we will prove that under (2.11)
there are only two possibilities for $\mathcal{L}'(\{r_n\})$: Either

(2.12) $\mathcal{L}'(\{r_n\}) = \{Y : Y \stackrel{d}{=} aN_1\exp(\lambda N_2) , \ a \geq 0 , \ \lambda \geq 0\}$

or

(2.13) $\mathcal{L}'(\{r_n\}) = \{Y : Y \stackrel{d}{=} aN_1 , \ a \geq 0\}.$

Further, given $\{r_n\}$, we can determine which situation we have. This is done by checking the following condition;

(S): For any sequences $m_k, n_k \to \infty$, if

$$\frac{r_{m_k}}{m_k} = \frac{r_{n_k}}{n_k} + O\left(\frac{r_{n_k}^{1/2}}{n_k}\right)$$

then

$$\frac{m_k}{n_k} \to 1.$$

If (S) holds then $\mathcal{L}'(\{r_n\})$ is given by (2.12), and if (S) fails then it is given by (2.13). As an indication of the type of sequences which satisfy (S), one can easily check that if

$$\frac{r_n}{n^{1-\epsilon}} \sim t_n$$

for some $\epsilon > 0$, where t_n is nonincreasing, then $\{r_n\}$ satisfies (S). But if we only know that $r_n n^{-1} \sim t_n$ where t_n is nonincreasing, then (S) may or may not hold and the quesiton of whether it does may be quite delicate. As an example,

$$r_n = [n/\log n] \text{ satisfies (S) but}$$

$$r_n = [n/[\log n]] \text{ does not satisfy (S)},$$

where [ ] is the greatest integer function. The latter sequence is not monotone but replacing it by $\tilde{r}_n = \max_{k \leq n} r_k$ makes it monotone and $\tilde{r}_n$ still fails to satisfy (S).

3. **Some Open Problems.** Perhaps the most annoying problem, and the one whose solution will round out the theory developed so far, is the question of uniqueness of the representation of Z.

1. If $Z \sim (\xi_1, f_1, g_1, \mu_1)$ and $Z \sim (\xi_2, f_2, g_2, \mu_2)$ and Z is not normally distributed, does it follow that $\xi_1 = \xi_2, f_1 = f_2, g_1 = g_2$ and $\mu_1 = \mu_2$?

Based on the results for modulus trimming one would expect to be able to make progress on

2. Characterize $\mathcal{L}(\{r_n\}, \{s_n\})$ where r_n, s_n are nondecreasing. The uniqueness question will create problems again here.

All of the results for modulus trimming are in the symmetric case. One would like to be able to solve the problems in general.

3. Solve any of the problems discussed so far in the non-symmetric case for modulus trimming. Perhaps the place to start is with Theorem 2.1'.

The remaining problems pertain to both types of trimming, but we will usually restrict the discussion to modulus trimming in the symmetric case, where it may be easiest to make some progress.

4. Characterize $\mathcal{L}'(\{r_n\})$ for r_n just satisfying (1.3). We have observed that there are at least three possibilities, namely (2.12), (2.13) or $\mathcal{L}'(\{r_n\}) = \mathcal{I}'(\{r_n\})$. In fact, there are other possibilities as well. We have made some progress on this problem but do not yet have a complete solution.

5. In keeping with the classical case (untrimmed sums) one would like to think of $\mathcal{I}'(\{r_n\})$ as the "infinitely divisible laws" of this theory. Thus, is $\mathcal{I}'(\{r_n\})$ the

class of limit laws of trimmed triangular arrays?

6. One curious fact is that not all laws in $\mathcal{I}'(\{r_n\})$ are infinitely divisible. However the classes described in (2.12) and (2.13) contain only infinitely divisible laws. Is there any intuitive explanation for this?

7. How does the rate of increase of r_n affect the rate of convergence to normality? Here we are thinking of analogues of the Berry-Esseen Theorem.

8. In addition to normalizing the trimmed sums by sequences of constants, it is also interesting to consider self-normalization, a common statistical proceedure. Let $M_n = (n - r_n - s_n)^{-1} S_n(s_n, r_n)$ and

$$W_n^2(r_n, s_n) = s_n(X_{n,s_n} - M_n)^2 + \sum_{i=s_n+1}^{n-r_n} (X_{n,i} - M_n)^2 + r_n(X_{n,n-r_n+1} - M_n)^2.$$

It follows from (7.1) in Griffin and Pruitt (1989) that if (2.6) holds, then there exists δ_n such that

$$\frac{S_n(s_n, r_n) - \delta_n}{W_n(s_n, r_n)} \to N(0, 1).$$

The question is whether the converse holds, i.e. is normalization by a sequence of constants equivalent to self-normalizing? This problem is still open even for untrimmed sums, although in the symmetric case they are known to be equivalent. In the case of modulus trimming the appropriate self-normalizer is

$$^{(r_n)}V_n^2 = \sum_{i=r_n+1}^{n} {}^{(i)}X_n^2.$$

Then for symmetric distributions it is known that (2.7) implies

(3.1) $$^{(r_n)}S_n \, {}^{(r_n)}V_n^{-1} \to N(0, 1),$$

but the converse is false. Necessary and sufficient conditions for (3.1) can be found in Griffin and Mason (1990); see also Hahn, Kuelbs and Weiner (1990).

Another interesting problem in this area is to characterize the possible limit laws of these self-normalized trimmed sums. Only partial results are available even in the untrimmed case; see Logan, Mallows, Rice, and Shepp (1973) and Hahn and Weiner (1989).

4. **Limit Laws.** The aim of this section is to prove the following result:

THEOREM 4.1. *Assume r_n is nondecreasing and satisfies (1.3). Then $\mathcal{L}'(\{r_n\})$ is given by (2.12) if (S) holds and by (2.13) if (S) fails.*

If r_n is not assumed to be nondecreasing, the situation is more complicated and there are other classes of limit laws. In general, $\mathcal{L}'(\{r_n\})$ depends on the structure of the set S given by

$$S = \left\{ (\beta, \rho) \in \mathbf{R} \times \mathbf{R}^+ : \text{there exist} \{m_k\}, \{n_k\} \text{ such that} \right.$$

$$\left. (\beta, \rho) = \lim \left(\frac{(u_{n_k}(0) - u_{m_k}(0))n_k}{r_{n_k}^{1/2}}, \left(\frac{n_k}{m_k}\right)^{1/2} \right) \right\},$$

where $\mathbf{R}^+ = [0, \infty)$. With no assumptions on r_n other than (1.3), S can be almost anything, but if r_n is nondecreasing then S must contain the line $\rho = 1$ and this simplifies matters. We will prove this assertion in the first lemma.

We start with some observations. Suppose $(\beta, \rho) \in S$ and $\{m_k\}, \{n_k\}$ are the corresponding subsequences. Then

$$\frac{u_{m_k}(0)}{u_{n_k}(0)} - 1 = \frac{(u_{m_k}(0) - u_{n_k}(0))n_k}{r_{n_k}^{1/2}} \frac{1}{r_{n_k}^{1/2}} \sim -\beta \frac{1}{r_{n_k}^{1/2}} \to 0$$

and so

$$(4.1) \qquad \frac{r_{m_k}^{1/2}}{m_k} = \left(\frac{r_{m_k}}{m_k}\right)^{1/2} \frac{1}{m_k^{1/2}} \sim \left(\frac{r_{n_k}}{n_k}\right)^{1/2} \frac{\rho}{n_k^{1/2}} = \rho \frac{r_{n_k}^{1/2}}{n_k}.$$

Thus by interchanging the roles of the sequences $\{m_k\}, \{n_k\}$ we see that if $\rho > 0$

$$(4.2) \qquad (\beta, \rho) \in \mathcal{S} \text{ implies } \left(-\frac{\beta}{\rho}, \frac{1}{\rho}\right) \in \mathcal{S}.$$

Note that this argument also shows that if there are sequences $\{m_k\}, \{n_k\}$ which give the limit (β, ∞), then $(0,0) \in \mathcal{S}$.

Fix $\gamma > 0$ and define

$$(4.3) \qquad m_n = m_n(\gamma) = \min\{m > n : u_m(0) < u_n(\gamma)\}.$$

Then we have

LEMMA 4.2. *Assume r_n is nondecreasing and satisfies (1.3). Then with m_n as in (4.3)*

$$(4.4) \qquad \lim_{n \to \infty} \frac{(u_n(0) - u_{m_n}(0))n}{r_n^{1/2}} = \gamma$$

and

$$(4.5) \qquad \liminf_{n \to \infty} \frac{m_n}{n} = 1.$$

PROOF. For (4.4), observe that since r_n is monotone,

$$u_{m_n}(0) < u_n(\gamma) \le u_{m_n - 1}(0) \le \frac{r_{m_n}}{m_n - 1} = u_{m_n}(0) + \frac{u_{m_n}(0)}{m_n - 1}$$

and then

$$u_n(0) - u_{m_n}(0) = u_n(\gamma) - u_{m_n}(0) + \gamma\frac{r_n^{1/2}}{n} = \gamma\frac{r_n^{1/2}}{n} + O\left(\frac{u_{m_n}(0)}{m_n}\right).$$

This is enough since

$$\frac{u_{m_n}(0)}{m_n} < \frac{u_n(0)}{n} = \left(\frac{r_n^{1/2}}{n}\right)^2 = o\left(\frac{r_n^{1/2}}{n}\right).$$

If (4.5) is false, then there exists a $\lambda > 1$ such that $m_n > \lambda n$ for all n sufficiently large. Hence $u_{[\lambda n]}(0) \geq u_n(\gamma)$ and consequently

(4.6) $$r_{[\lambda n]} \geq \frac{[\lambda n]}{n} r_n (1 - \gamma r_n^{-1/2}).$$

Thus for any $\xi < \lambda$ we have for $n \geq n_o$, say,

$$r_{[\lambda n]} \geq \xi r_n.$$

By iterating this inequality we obtain for all $k \geq 0$

$$r_{[\lambda^k n_o]} \geq \xi^k r_{n_o},$$

where we have used the monotonicity of r_n and the inequality $[\lambda x] \geq [\lambda[x]]$ for all $x \geq 0$. Hence

$$\liminf_{k\to\infty} \xi^{-k} r_{[\lambda^k n_o]} > 0.$$

Since $\xi < \lambda$ was arbitrary and r_n is nondecreasing, this implies

(4.7) $$\liminf_{n\to\infty} r_n n^{\epsilon-1} > 0$$

for all $\epsilon > 0$. With this initial bound, we now proceed to show that (4.7) holds with $\epsilon = 0$ which contradicts (1.3). Recall that (4.6) holds for all n sufficiently large, say

$n \geq n_1$. By iterating (4.6) we obtain for all $k \geq 0$

(4.8)
$$r_{[\lambda^k n_1]} \geq \prod_{j=0}^{k-1} \frac{[\lambda[\lambda^j n_1]]}{[\lambda^j n_1]} r_{n_1} \prod_{j=0}^{k-1} (1 - \gamma r_{[\lambda^j n_1]}^{-1/2})$$

$$\geq \lambda^k r_{n_1} \prod_{j=0}^{k-1} (1 - [\lambda^j n_1]^{-1}) \prod_{j=0}^{k-1} (1 - \gamma r_{[\lambda^j n_1]}^{-1/2}).$$

Both products converge: the first since $\sum \lambda^{-j} < \infty$ and the second since

$$\sum_j r_{[\lambda^j n_1]}^{-1/2} < \infty$$

by (4.7). Hence (4.8) implies that

$$\liminf_{k \to \infty} \frac{r_{[\lambda^k n_1]}}{\lambda^k} > 0.$$

Montonicity of r_n now shows that (4.7) holds for $\epsilon = 0$. ◊

Note that Lemma 4.2 implies that $(\beta, 1) \in S$ for all $\beta > 0$ and this then follows for $\beta < 0$ by (4.2). Taking $m_k = n_k = k$ gives the limit $(0,1)$. We will now use the fact that S contains the line $\rho = 1$ to show that the set of limit laws is contained in (2.12) when r_n is nondecreasing.

PROPOSITION 4.3. *Assume r_n is nondecreasing and satisfies (1.3). Then*

$$\mathcal{L}'(\{r_n\}) \subset \{Y : Y \overset{d}{=} a N_1 \exp(\lambda N_2), a \geq 0, \lambda \geq 0\}.$$

PROOF. Fix $(\beta, \rho) \in S$ and let $\{m_k\}, \{n_k\}$ be the corresponding sequences. Then by (4.1), as $k \to \infty$,

$$u_{m_k}(\alpha) = u_{m_k}(0) - \alpha \frac{r_{m_k}^{1/2}}{m_k} = u_{n_k}(0) + (-\beta + o(1)) \frac{r_{n_k}^{1/2}}{n_k} - \alpha(\rho + o(1)) \frac{r_{n_k}^{1/2}}{n_k}$$

$$= u_{n_k}(\alpha \rho + \beta + o(1)) \qquad \text{for all } \alpha \in \mathbf{R}.$$

This means that (see (2.3)) if $(\rho, \beta) \in \mathcal{S}$, then $\exists \{m_k\}, \{n_k\}$ such that

(4.9) $\tau_{m_k}(\alpha) = \tau_{n_k}(\alpha\rho + \beta + o(1))$ for all $\alpha \in \mathbf{R}$.

Now assume that $^{(r_n)}S_n\gamma_n^{-1} \to Y$. If Y is degenerate, take $a = 0$. Thus we may assume that Y is nondegenerate and then by Theorem 2.2′, $Y =^d N_1\tau(N_2)$ where τ is nonnegative, nondecreasing, τ^2 is convex and $\tau \not\equiv 0$. By Theorem 2.3′ we have for any α_o for which $\tau(\alpha_o) > 0$

(4.10) $\dfrac{\tau_n(\alpha)}{\tau_n(\alpha_o)} \to \dfrac{\tau(\alpha)}{\tau(\alpha_o)}$ for all $\alpha \in \mathbf{R}$

and the convergence is uniform on compact sets (u.c.). This latter fact is because τ_n^2 is convex on compacts for large n by (2.5) and any sequence of convex functions which converges necessarily converges u.c. Since $\mathcal{S}$ contains the line $\rho = 1$, we can now use (4.9) with $\rho = 1$ in (4.10) to obtain

(4.11) $\dfrac{\tau(\alpha)}{\tau(\alpha_o)} = \dfrac{\tau(\alpha + \beta)}{\tau(\alpha_o + \beta)}$ for all $\alpha, \beta \in \mathbf{R}$,

provided the denominators are positive. This is valid if $\tau(\alpha_o) > 0$ and $\beta \geq 0$ since then $\tau(\alpha_o + \beta) \geq \tau(\alpha_o) > 0$. Now, if $\tau(\alpha)$ is 0 for some α, then $\tau(\alpha + \beta) = 0$ for all $\beta \geq 0$ which would mean $\tau \equiv 0$ by monotonicity. Thus we must have that τ is never zero and (4.11) holds for all β and all α_o. Taking $\alpha_o = 0$ yields

$$\frac{\tau(\alpha + \beta)}{\tau(0)} = \frac{\tau(\alpha)}{\tau(0)}\frac{\tau(\beta)}{\tau(0)}, \qquad \alpha, \beta \in \mathbf{R},$$

and so $\tau(\alpha) = a \, \exp(\lambda\alpha)$. The fact that $a > 0$, $\lambda \geq 0$ follows from τ being positive and nondecreasing.   ◇

We now come to the proof of Theorem 4.1 which we separate into two parts.

THEOREM 4.4. *Assume r_n is nondecreasing and satisfies* (1.3). *If* (S) *fails, then*

$$\mathcal{L}'(\{r_n\}) = \{Y : Y \overset{\mathrm{d}}{=} aN_1, \ a \geq 0\}.$$

PROOF. Assume $Y \in \mathcal{L}'(\{r_n\})$ and is nondegenerate. If (S) fails, then there exist subsequences $\{m_k\}, \{n_k\}$ such that the limits defined in S are (β, ρ) with $\rho \neq 1$. (If $n_k/m_k \to \infty$, then $(0,0) \in S$ by the sentence following (4.2).) Thus we have at least one point in S not on the line $\rho = 1$. Now we proceed as in the proof of Proposition 4.3. Using (4.9) in (4.10) leads to

$$(4.12) \qquad \frac{\tau(\alpha)}{\tau(\alpha_o)} = \frac{\tau(\alpha\rho + \beta)}{\tau(\alpha_o\rho + \beta)} \qquad \text{for all } \alpha, \alpha_o \in \mathbf{R},$$

since we already know by Proposition 4.3 that $\tau(\alpha) = a\exp(\lambda\alpha)$ and $a > 0$ since Y is nondegenerate. Thus we may take $\alpha_o = 0$ in (4.12) to obtain

$$\frac{\tau(\alpha\rho + \beta)}{\tau(0)} = \frac{\tau(\alpha)}{\tau(0)}\frac{\tau(\beta)}{\tau(0)} \qquad \text{for all } \alpha \in \mathbf{R},$$

or

$$\exp(\lambda(\alpha\rho + \beta)) = \exp(\lambda\alpha)\exp(\lambda\beta) \qquad \text{for all } \alpha \in \mathbf{R}.$$

Hence, we must have $\lambda\alpha\rho = \lambda\alpha$ for all $\alpha \in \mathbf{R}$ and since $\rho \neq 1$, this means $\lambda = 0$. Therefore, the existence of the point in S not on the line $\rho = 1$ further restricts τ to be $\tau(\alpha) \equiv a$. Hence,

$$\mathcal{L}'(\{r_n\}) \subset \{Y : Y \overset{\mathrm{d}}{=} aN_1, a \geq 0\}.$$

The other inclusion follows from Example 2 (following Theorem 2.1'); normal limits can always be obtained for normalized modulus trimmed sums by taking X to have a symmetric distribution in the Feller class. $\diamond$

THEOREM 4.5. *Assume r_n is nondecreasing and satisfies* (1.3). *If* (S) *holds, then*

$$\mathcal{L}'(\{r_n\}) = \{Y : Y \overset{d}{=} aN_1\exp(\lambda N_2), \quad a \ge 0, \lambda \ge 0\}.$$

PROOF. We have already proved one inclusion in Proposition 4.3 and have also observed that the normal laws ($\lambda = 0$) are always in $\mathcal{L}'(\{r_n\})$. Of course, the degenerate law ($a = 0$) is also in $\mathcal{L}'(\{r_n\})$. Thus we may assume $a > 0$ and $\lambda > 0$. By Theorem 2.3′ it is sufficient to construct a symmetric distribution for X such that

(4.13) $$\frac{\tau_n(\alpha)}{\tau_n(0)} \to \exp(\lambda\alpha) \qquad \text{for all } \alpha \in \mathbf{R}.$$

We start by defining some sequences. Take

$$n_1 = \min\{n : u_n(0) \le 1/2\}$$

and for $k \ge 1$,

$$n_{k+1} = \min\{n > n_k : u_n(0) < u_{n_k}(1)\} = m_{n_k}(1)$$

where $m_n(\gamma)$ is defined in (4.3). Next let

$$\xi_k = u_{n_k}(0) = \frac{r_{n_k}}{n_k}, \quad \eta_k = \frac{r_{n_k}^{1/2}}{n_k};$$

note that

$$u_{n_k}(\alpha) = \xi_k - \alpha\eta_k.$$

We also have

$$\xi_{k+1} < u_{n_k}(1) = \xi_k - \eta_k.$$

By (4.4),

(4.14)
$$\frac{(\xi_k - \xi_{k+1})n_k}{r_{n_k}^{1/2}} \to 1$$

so that

(4.15)
$$\xi_{k+1} = \xi_k - (1 + o(1))\eta_k = u_{n_k}(1 + o(1)).$$

Thus by (S) with $m_k = n_{k+1}$, we must have

$$\frac{n_{k+1}}{n_k} \to 1,$$

and then by (4.1) and (4.14)

(4.16)
$$\frac{\eta_{k+1}}{\eta_k} \to 1.$$

Next we define $\delta_1 = \exp(-\lambda(1 - \xi_1)\eta_1^{-1})$ and

$$\delta_k = \exp(\lambda(\xi_k - \xi_{k-1})/2\eta_k), \quad k > 1,$$

$$\gamma_k = \exp(\lambda(\xi_k - \xi_{k+1})/2\eta_k), \quad k \geq 1.$$

By (4.14) and (4.16),

(4.17)
$$\delta_k \to e^{-\lambda/2}, \gamma_k \to e^{\lambda/2}.$$

We let $x_1 = 1$ and

$$x_k = \prod_{j=1}^{k-1} \frac{\gamma_j}{\delta_{j+1}}, \quad k > 1.$$

Finally, we define $G(x) = 1$ for $x < \delta_1 x_1$ and

$$G(x) = \xi_k - \frac{\eta_k}{\lambda}\log\frac{x}{x_k}, \quad \delta_k x_k \leq x < \gamma_k x_k, \quad k = 1, 2, \dots .$$

We note that $G(\delta_1 x_1) = 1$, $\gamma_k x_k = \delta_{k+1} x_{k+1}$ and that $G(\gamma_k x_k-) = G(\delta_{k+1} x_{k+1})$ so that G is well defined and continuous. Furthermore $G(x)$ decreases to 0 as $x \to \infty$ so we can construct a symmetric random variable X with $P(|X| > x) = G(x)$. We now need to show that (4.13) holds.

We start by observing that by the definition of G and (4.17)

$$E(X^2; \delta_k x_k < |X| \leq \gamma_k x_k) = \frac{\eta_k}{\lambda} \int_{\delta_k x_k}^{\gamma_k x_k} x^2 \frac{1}{x} dx = \frac{\eta_k x_k^2}{2\lambda}(\gamma_k^2 - \delta_k^2)$$

$$\sim \frac{\eta_k x_k^2}{2\lambda}(e^\lambda - e^{-\lambda}).$$

Furthermore, by (4.16) and (4.17)

$$\frac{\eta_k x_k^2}{\eta_{k-1} x_{k-1}^2} \sim \frac{x_k^2}{x_{k-1}^2} = \frac{\gamma_{k-1}^2}{\delta_k^2} \to e^{2\lambda},$$

so that we may sum these terms as if they were a geometric series to obtain

(4.18) $E(X^2; |X| \leq \gamma_k x_k) \sim \frac{\eta_k x_k^2}{2\lambda}(e^\lambda - e^{-\lambda})(1 - e^{-2\lambda})^{-1} = \frac{\eta_k x_k^2 e^\lambda}{2\lambda}.$

The next step is to estimate $c_{n_k}(\alpha)$. We want to solve

(4.19) $$G(c_{n_k}(\alpha)) = u_{n_k}(\alpha) = \xi_k - \alpha \eta_k.$$

We already have G written in a convenient form to solve this if $c_{n_k}(\alpha) \in [\delta_k x_k, \gamma_k x_k]$ but we need to express G in this form on adjacent intervals. If j is fixed and $\delta_{k+j} x_{k+j} \leq x < \gamma_{k+j} x_{k+j}$ we have by (4.15) and (4.16)

$$G(x) = \xi_{k+j} - \frac{\eta_{k+j}}{\lambda} \log \frac{x}{x_{k+j}} = \xi_k - (j + o(1))\eta_k - \frac{\eta_k}{\lambda}(1 + o(1))\log \frac{x}{x_{k+j}}$$

$$= \xi_k - \eta_k(j + \frac{1}{\lambda} \log \frac{x}{x_{k+j}} + o(1)).$$

Using this in (4.19) leads to

$$j + \frac{1}{\lambda}\log\frac{c_{n_k}(\alpha)}{x_{k+j}} + o(1) = \alpha$$

and then by (4.17)

$$c_{n_k}(\alpha) = x_{k+j}e^{\lambda(\alpha-j+o(1))} \sim x_k e^{\lambda\alpha}.$$

This will be valid for any fixed α since then $x_k e^{\lambda\alpha}$ will be in $[\delta_{k+j}x_{k+j}, \gamma_{k+j}x_{k+j})$ for some j. Now choose j so that $c_{n_k}(\alpha) \in [\delta_{k+j}x_{k+j}, \gamma_{k+j}x_{k+j})$. Then

$$E(X^2; \delta_{k+j}x_{k+j} < |X| \le c_{n_k}(\alpha)) = \frac{\eta_{k+j}}{\lambda}\int_{\delta_{k+j}x_{k+j}}^{c_{n_k}(\alpha)} x\,dx$$

$$= \frac{\eta_{k+j}}{2\lambda}(c_{n_k}^2(\alpha) - \delta_{k+j}^2 x_{k+j}^2)$$

$$\sim \frac{\eta_k}{2\lambda}x_k^2(e^{2\lambda\alpha} - e^{\lambda(2j-1)}).$$

In conjunction with (4.18), this yields

$$\tau_{n_k}^2(\alpha) = E(X^2; |X| \le c_{n_k}(\alpha)) \sim \frac{\eta_k x_k^2}{2\lambda}e^{2\lambda\alpha}$$

which gives (4.13) for the subsequence n_k. To see that it holds along the entire sequence, we will show that it is valid for an arbitrary subsequence m_j. We take k_j so that

$$\xi_{k_j+1} < u_{m_j}(0) \le \xi_{k_j}.$$

This is possible for large j since $\xi_k \searrow 0$. Then

$$0 \le \xi_{k_j} - u_{m_j}(0) < \xi_{k_j} - \xi_{k_j+1} = (1 + o(1))\eta_{k_j}$$

by (4.15), so for any subsequence of m_j we may take a further subsequence so that (we will not rename the subsequence)

$$\frac{\xi_{k_j} - u_{m_j}(0)}{\eta_{k_j}} \to \gamma \in [0, 1].$$

By (S) we must then have $m_j/n_{k_j} \to 1$ and then by (4.1)

$$u_{m_j}(\alpha) = u_{m_j}(0) - \alpha\frac{r_{m_j}^{1/2}}{m_j} = \xi_{k_j} - (\gamma + o(1))\eta_{k_j} - (\alpha + o(1))\eta_{k_j}$$

$$= u_{n_{k_j}}(\alpha + \gamma + o(1)).$$

This means that

$$\tau_{m_j}(\alpha) = \tau_{n_{k_j}}(\alpha + \gamma + o(1)),$$

and applying the known result (4.13) for the subsequence n_k it then follows that

$$\frac{\tau_{m_j}(\alpha)}{\tau_{m_j}(0)} \to \frac{e^{\lambda(\alpha+\gamma)}}{e^{\lambda\gamma}} = e^{\lambda\alpha}.$$

Since the limit is independent of the subsequence, this completes the proof.   ◊

References

[1] Csörgő, S., Haeusler, E., Mason, D.M. (1988a). *A probabilistic approach to the asymptotic distribution of sums of independent, identically distributed random variables.* Adv. Appl. Math. **9**, 259-333.

[2] Csörgő, S., Haeusler, E., Mason, D.M. (1988b). *The asymptotic distribution of trimmed sums.* Ann. Probab. **16**, 672-699.

[3] Csörgő, S., Haeusler, E., Mason, D.M. (1990). *The quantile-transform approach to the asymptotic distribution of modulus trimmed sums.* In this volume.

[4] Feller, W. (1968). *An extension of the law of the iterated logarithm to variables without variance.* J. Math. Mech. **18**, 343-355.

[5] Griffin, P.S., Mason, D.M. (1990). *On the asymptotic normality of self-normalized sums.* (Preprint).

[6] Griffin, P.S., Pruitt, W.E. (1987). *The central limit problem for trimmed sums.* Math. Proc. Camb. Phil. Soc. **102**, 329-349.

[7] Griffin, P.S., Pruitt, W.E. (1989). *Asymptotic normality and subsequential limits of trimmed sums.* Ann. Probab. **17**, 1186-1219.

[8] Hahn, M., Kuelbs, J., Weiner, D. (1990). *The asymptotic distribution of magnitude–Winsorized sums.* J. Theoret. Probab. **3**, 137-168.

[9] Hahn, M., Weiner, D. (1989). *Asymptotic behavior of self-normalized trimmed sums: nonnormal limits.* (Preprint).

[10] Logan, B.F., Mallows, C.L., Rice, S.O., Shepp, L.A. (1973). *Limit distributions of self-normalized sums.* Ann. Probab. **1**, 788-809.

[11] Maller, R.A. (1982). *Asymptotic normality of lightly trimmed means - a converse.* Math. Proc. Camb. Phil. Soc. **92**, 535-545.

[12] Maller, R.A. (1988). *Asymptotic normality of trimmed sums in higher dimensions.* Ann Probab. **16**, 1608-1622.

[13] Maller, R.A. (1987). *Some results on asymptotics of trimmed means in projection pursuit.* (Preprint).

[14] Mori, T. (1976). *The strong law of large numbers when extreme terms are excluded from sums.* Z. Wahrscheinlichkeitstheorie verw. Geb. **36**, 189-194.

[15] Mori, T. (1984). *On the limit distribution of lightly trimmed sums.* Math. Proc. Camb. Phil. Soc. **96**, 507-516.

[16] Stigler, S.M. (1973). *The asymptotic distribution of the trimmed mean.* Ann. Statist. **1**, 472-477.

Philip S. Griffin

Department of Mathematics

Syracuse University

Syracuse, NY 13244-1150

USA

William E. Pruitt

School of Mathematics

University of Minnesota

Minneapolis, MN 55455

USA

INVARIANCE PRINCIPLES AND SELF-NORMALIZATIONS
FOR SUMS TRIMMED ACCORDING TO CHOICE
OF INFLUENCE FUNCTION

Hamid Ould-Rouis

1. Introduction. Asymptotic normality for sums of independent random variables has numerous useful applications in probability and statistics. This paper considers ways of modifying the summands of sums of i.i.d. random variables in order to get invariance principles when the random variables fail to have second or even first moments. Asymptotic normality fails for sums of i.i.d. random variables principally because of terms of large magnitude (Lévy 1937). So conceivably central limit theorems might be obtainable if these terms of large magnitudes are moderated or discounted. The inspiration for this work comes from three papers by Hahn and Kuelbs (1988) and Hahn, Kuelbs and Weiner (1990a,b) which investigate universal asymptotic normality. Respectively the truncating function $T_t(x) = xI(|x| \leq t)$ and the censoring function $C_t(x) = (|x| \wedge t)\mathrm{sgn}(x)$ were applied to the sample to prove two universal type central limit theorems for totally modified and conditionally modified sums, and their empirical versions. Our goal is to obtain invariance principles for a large class of functions $H_t(x)$ which includes those for which $T_t(x) \leq H_t(x) \leq C_t(x)$. These functions H_t might be called influence functions because they determine what influence the large terms have on the partial sums. The approach here is necessarily different. However, the techniques used to

HAMID OULD-ROUIS

obtain the empirical results for symmetric random variables are similar to those in
the above three papers.

2. **Notation and Preliminaries.** Let $\{X_i\}_{i\geq 1}$ be a sequence of i.i.d. nondegenerate random variables. Let $\{r_n\}_{n\geq 1}$ be a sequence of positive real numbers such
that

$$r_n \to \infty \text{ and } \frac{r_n}{n} \to 0 \text{ as } n \to \infty. \tag{2.1}$$

Definition 2.2. Let

$$H_t(x) = k(x)I(|x| \leq t) + g_t(x)I(|x| > t)$$

where there exist $0 < C_1, C_2 < \infty$ such that

$$k(x) \neq 0 \quad \text{for} \quad x \neq 0,$$

$$|k(x)| \leq C_1|x| \quad \text{for } |x| \text{ large}$$

and   (2.3)

$$|g_t(x)| \leq C_2 t \quad \text{for } |x| > t.$$

Notationally,

$$u_n \sim v_n \quad \Longleftrightarrow \quad \lim_{n\to\infty} \frac{u_n}{v_n} = 1,$$

and

$$u_n \asymp v_n \quad \Longleftrightarrow \quad \exists \ m > 0 \text{ and } M < \infty \text{ such that } m v_n \leq u_n \leq M v_n \text{ for } n \text{ large}.$$

Many results of this paper rely on the following central limit theorem for
stochastic processes. Although it could be deduced from Kasahara and Watanabe's

(1986) more general results for point processes, a proof of this simplified version can be given directly and is included for completeness.

Theorem 2.4. *(Central Limit Theorem for Stochastic Processes). Let $B(t)$ be a standard Brownian motion and let f_n, for $n \in \mathbf{N}$, be measurable functions. Set*

$$M_n(s) \equiv \sum_{i \leq ns} (f_n(X_i) - E f_n(X_i)).$$

Suppose there exist positive constants b_n with $b_n \to 0$ as $n \to \infty$ such that

(A) $\sup_{x \in \mathbf{R}} |f_n(x)| \leq b_n$

and

(B) $\mathrm{Var} M_n(s) \longrightarrow s$ as $n \to \infty \quad \forall\, s \geq 0$.

Then

$$M_n(\cdot) \xrightarrow{D} B(\cdot) \quad \text{in } D[0,1] \text{ as } n \to \infty$$

in the usual Skorohod J_1-topology.

Proof. Convergence in $D[0,1]$ is equivalent to convergence of the finite-dimensional distributions and tightness.

Step 1. Convergence of the finite-dimensional distributions of $M_n(s)$ to those of $B(s)$:

Let

$$0 \leq s_0 < s_1 < \ldots < s_k \leq 1$$

and, for any j, define

$$M_n(s_j) = \sum_{i \leq n s_j} (f_n(X_i) - E f_n(X_i)) \equiv \sum_{i \leq n s_j} Y_{ni}.$$

(a) First, consider the one-dimensional distributions. To show that $M_n(s_j) \Longrightarrow$ $N(0, s_j)$, it suffices to verify the conditions of the Central Convergence Criterion for triangular arrays.

Since $|Y_{ni}| = |f_n(X_i) - E f_n(X_i)| \leq 2b_n \rightarrow 0$ as $n \rightarrow \infty$, the $\{Y_{ni}\}$ form an infinitesimal array and for every $\varepsilon > 0$, $nP(|Y_{ni}| > \varepsilon) \rightarrow 0$. Also, $\mathrm{Var} M_n(s_j) \rightarrow s_j$ by condition B. Therefore $M_n(s_j) \Longrightarrow N(0, s_j)$.

(b) The proof of convergence of the finite-dimensional distributions is merely a generalization of the following proof for the two-dimensional distributions. Exactly as for the one-dimensional distributions, for $j < l$,

$$M_n(s_l) - M_n(s_j) \Longrightarrow B(s_l) - B(s_j).$$

Then Theorem 3.2 of Billingsley, combined with the independence of $M_n(s_j)$ and $M_n(s_l) - M_n(s_j)$, yields

$$(M_n(s_j), \ M_n(s_l) - M_n(s_j)) \Longrightarrow (B(s_j), \ B(s_l) - B(s_j)).$$

Now let $h(x, y) = (x, y + x)$. Applying the Continuous Mapping Theorem,

$$(M_n(s_j), M_n(s_l)) = h(M_n(s_j), M_n(s_l) - M_n(s_j))$$

$$\Longrightarrow h(B(s_j), B(s_l) - B(s_j)) = (B(s_j), B(s_l)).$$

An easy generalization yields that the finite-dimensional distributions of $M_n(\cdot)$ converge to those of $B(\cdot)$.

Step 2. Tightness of $\{M_n(\cdot)\}$.

It is clear that $P(B(1) \neq B(1-)) = 0$. So, by Theorem 15.6 of Billingsley (1968), tightness will hold if there exist a nondecreasing continuous function F on $[0,1]$ and constants $\gamma \geq 0, \alpha \geq \frac{1}{2}$ such that

$$E\left\{|M_n(s) - M_n(s_1)|^\gamma \, |M_n(s_2) - M_n(s)|^\gamma\right\} \leq (F(s_2) - F(s_1))^{2\alpha} \qquad (2.5)$$

for $s_1 \leq s \leq s_2$ and $n \geq 1$.

By condition B, $\mathrm{Var}\, M_n(s) \to s$ as $n \to \infty$. Consequently,

$$s \longleftarrow \mathrm{Var}\, M_n(s) = [ns]EY_{ni}^2 = \frac{[ns]}{n} n EY_{ni}^2.$$

Since $\frac{[ns]}{n} \to s$, we may conclude that $nEY_{ni}^2 \to 1$ and therefore that there exists $C > 1$ for which

$$\left|nEY_{ni}^2\right| \leq C \quad \text{for all} \ \ n, i. \qquad (2.6)$$

Now, by independence,

$$E\left\{\left(M_n(s) - M_n(s_1)\right)^2 \left(M_n(s_2) - M_n(s)\right)^2\right\}$$
$$= E\left(M_n(s) - M_n(s_1)\right)^2 \cdot E\left(M_n(s_2) - M_n(s)\right)^2$$
$$= E\left(\sum_{[ns_1]<i\leq[ns]} Y_{ni}\right)^2 \cdot E\left(\sum_{[ns]<i\leq[ns_2]} Y_{ni}\right)^2 \qquad (2.7)$$
$$= ([ns] - [ns_1])([ns_2] - [ns])\left(EY_{ni}^2\right)^2$$

Two cases can arise:

Case 1: $s_2 - s_1 \leq \frac{1}{n}$.

In this situation, either s_1 and s lie in a subinterval of the form $\left(\frac{i-1}{n}, \frac{i}{n}\right)$ or else s and s_2 do. This means either $[ns_1] = [ns]$ or $[ns_2] = [ns]$. In either case, the right member of (2.7) is zero, and consequently

$$E\left\{(M_n(s) - M_n(s_1))^2 (M_n(s_2) - M_n(s))^2\right\} = 0.$$

Case 2: $s_2 - s_1 > \frac{1}{n}$.

In this situation, $[ns_2] - [ns_1] \leq (ns_2 - ns_1 + 1)$. Then there exists $C > 1$ such that

$$
\begin{aligned}
E\left\{(M_n(s) - M_n(s_1))^2 (M_n(s_2) - M_n(s))^2\right\} &\leq \left(\frac{[ns_2] - [ns_1]}{n}\right)^2 C^2 \text{ by (2.6)} \\
&\leq \left(\frac{ns_2 - ns_1 + 1}{n}\right)^2 C^2 \\
&= \left((s_2 - s_1) + \frac{1}{n}\right)^2 C^2 \\
&\leq (2(s_2 - s_1))^2 C^2 \\
&= 4C^2 (s_2 - s_1)^2.
\end{aligned}
$$

Combining the two cases, for all $s_1 \leq s \leq s_2$,

$$E\left\{(M_n(s) - M_n(s_1))^2 (M_n(s_2) - M(s))^2\right\} \leq 4C^2 (s_2 - s_1)^2.$$

Hence (2.5) holds with $F(s) = 2Cs$, $\gamma = 2$ and $\alpha = 1$. ∎

3. Intermediate Modifications: Invariance Principles for Totally and Partially Modified Sums.

The first invariance principle to be proved modifies all the terms. The level of modification is dependent on H_t and r_n where $r_n \to \infty$ but $\frac{r_n}{n} \to 0$.

Theorem 3.1. *Let H_t be defined as in (2.2) with the additional assumption that k is nondecreasing. Let $\beta = \frac{p}{q}$ where p and q are odd integers and $\alpha \in \mathbf{R}$ satisfies $0 < \beta \leq \alpha$. Given $\{r_n\}_{n \geq 1}$ satisfying (2.1), assume there exists a sequence $\{a_n\}$ such that $a_n \to \infty$ as $n \to \infty$ such that*

$$\frac{a_n^{2\alpha} r_n}{n} = EH_{a_n}^{2\beta}(X) \tag{3.2}$$

for n sufficiently large. If $E|k(X)|^\beta < \infty$, assume $E(k(X))^\beta = 0$. Let

$$M_n(s) = \sum_{i \leq ns} \frac{H_{a_n}^\beta(X_i) - EH_{a_n}^\beta(X)}{a_n^\alpha \sqrt{r_n}}.$$

Then

$$M_n(\cdot) \overset{D}{\longrightarrow} B(\cdot) \quad \text{in } D[0,1]$$

where B is a standard Brownian motion.

Prior to proving Theorem 3.1, we obtain a lemma which allows the replacement of $\text{Var} H_{a_n}^\beta(X)$ by $EH_{a_n}^{2\beta}(X)$ when $E(k(X))^{2\beta} = \infty$.

Lemma 3.3. *Let $0 \leq C_1, C_2 < \infty$. Define*

$$H_t(X) = k(X)I(|X| \leq t) + g_t(X)I(|X| > t)$$

where

$$|g_t(X)| \le C_2 t \quad \text{for} \quad |X| > t$$

and k is a nondecreasing function with $k(x) \ne 0$ for $x \ne 0$ and $|k(x)| \le C_1|x|$ for $|x|$ large. Let $\beta = \frac{p}{q}$ where p and q are odd integers and $\alpha \in \mathbf{R}$ satisfies $0 < \beta \le \alpha$. Assume $E(k(X))^{2\beta} = \infty$ and there exists a sequence $a_n \to \infty$ such that $\frac{a_n^{2\alpha} r_n}{n} = EH_{a_n}^{2\beta}(X)$ for n sufficiently large. Then $\left(EH_{a_n}^{\beta}(X)\right)^2 = o\left(EH_{a_n}^{2\beta}(X)\right)$.

Proof. Notice that $E(k(X))^{2\beta} = \infty$ implies $EH_{a_n}^{2\beta}(X) \to \infty$ as $n \to \infty$. Choose $t_n = \left(EH_{a_n}^{2\beta}(X)\right)^{1/p}$ where $p > 2\alpha$. It is obvious that $t_n \to \infty$ as $n \to \infty$. Also for n large,

$$\begin{aligned}
a_n &= \left(\frac{n}{r_n}\right)^{1/2\alpha} \left(EH_{a_n}^{2\beta}(X)\right)^{1/2\alpha} \\
&> \left(EH_{a_n}^{2\beta}(X)\right)^{1/2\alpha} \\
&> \left(EH_{a_n}^{2\beta}(X)\right)^{1/p} \\
&= t_n.
\end{aligned}$$

Thus, for n large,

$$\begin{aligned}
H_{a_n}(X)I(|X| \le t_n) &= k(X)I(|X| \le a_n)I(|X| \le t_n) \\
&\quad + g_{a_n}(X)I(|X| > a_n)I(|X| \le t_n) \\
&= k(X)I(|X| \le t_n).
\end{aligned}$$

Now

$$\begin{aligned}
EH_{a_n}^{\beta}(X) &= E\left(H_{a_n}^{\beta}(X)I\left(|X| \le t_n\right)\right) + E\left(H_{a_n}^{\beta}(X)I(|X| > t_n)\right) \\
&= E\left(k^{\beta}(X)I\left(|X| \le t_n\right)\right) + E\left(H_{a_n}^{\beta}(X)I(|X| > t_n)\right).
\end{aligned}$$

Combining this with $(a+b)^2 \le 2a^2 + 2b^2$ yields

$$\left(EH_{a_n}^{\beta}(X)\right)^2 \le 2\left(Ek^{\beta}(X)I(|X| \le t_n)\right)^2$$

$$+ 2\left(EH_{a_n}^{\beta}(X)I(|X| > t_n)\right)^2$$

$$\le 2k^{2\beta}(t_n) + 2EH_{a_n}^{2\beta}(X)EI(|X| > t_n)$$

by Cauchy Schwarz and the fact that

$k(x)$ is nondecreasing

$$= 2k^{2\beta}(t_n) + 2EH_{a_n}^{2\beta}(X)P(|X| > t_n).$$

For n large, $|k(t_n)| \le C_1 t_n$. Hence

$$\left(EH_{a_n}^{\beta}(X)\right)^2 \le 2C_1^{2\beta}t_n^{2\beta} + 2EH_{a_n}^{2\beta}(X)P(|X| > t_n).$$

Dividing both sides by $EH_{a_n}^{2\beta}(X)$,

$$\frac{\left(EH_{a_n}^{\beta}(X)\right)^2}{EH_{a_n}^{2\beta}(X)} \le \frac{2C_1^{2\beta}t_n^{2\beta}}{EH_{a_n}^{2\beta}(X)} + 2P(|X| > t_n)$$

$$= 2C_1^{2\beta}t_n^{2\beta-p} + 2P(|X| > t_n)$$

$$\longrightarrow 0$$

as $n \to \infty$ since $p > 2\alpha \ge 2\beta$ and $t_n \to \infty$. ∎

Proof of Theorem 3.1. Let $f_n(x) = \frac{H_{a_n}^{\beta}(x)}{a_n^{\alpha}\sqrt{r_n}}$. The two conditions of Theorem 2.4 will be verified.

Verification of Condition A:

$$\sup_{x \in \mathbb{R}} |f_n(x)| = \sup_{x \in \mathbb{R}} \left|\frac{H_{a_n}^{\beta}(x)}{a_n^{\alpha}\sqrt{r_n}}\right| \le \frac{(k(a_n) + C_2 a_n)^{\beta}}{a_n^{\alpha}\sqrt{r_n}}$$

$$\le \frac{(C_1 + C_2)^{\beta} a_n^{\beta}}{a_n^{\alpha}\sqrt{r_n}} \quad \text{for } n \text{ large}$$

$$\le \frac{(C_1 + C_2)^{\beta}}{\sqrt{r_n}} \longrightarrow 0 \text{ as } n \to \infty.$$

Verification of Condition B:

$$\mathrm{Var} M_n(s) = \mathrm{Var}\left(\sum_{i \leq ns} \frac{H^\beta_{a_n}(X_i) - EH^\beta_{a_n}(X)}{a_n^\alpha \sqrt{r_n}}\right)$$

$$= \sum_{i \leq ns} \mathrm{Var}\left(\frac{H^\beta_{a_n}(X_i) - EH^\beta_{a_n}(X)}{a_n^\alpha \sqrt{r_n}}\right)$$

$$= \frac{[ns]}{a_n^{2\alpha} r_n}\left(EH^{2\beta}_{a_n}(X) - (EH^\beta_{a_n}(X))^2\right).$$

If $E(k(X))^{2\beta} < \infty$, then $EH^{2\beta}_{a_n}(X) < \infty$ for all n and by assumption $E(k(X))^\beta = 0$.

Therefore,

$$\mathrm{Var} M_n(s) \simeq \frac{[ns]}{a_n^{2\alpha} r_n} EH^{2\beta}_{a_n}(X)$$

$$= \frac{[ns]}{n} \frac{n EH^{2\beta}_{a_n}(X)}{a_n^{2\alpha} r_n}$$

$$\longrightarrow s \text{ as } n \to \infty \text{ using } (3.2).$$

If $E(k(X))^{2\beta} = \infty$, then $EH^{2\beta}_{a_n}(X) \to \infty$ as $n \to \infty$ and

$$\mathrm{Var} M_n(s) = \frac{[ns]}{n} n \frac{EH^{2\beta}_{a_n}(X)}{a_n^{2\alpha} r_n}\left(1 - \frac{(EH^\beta_{a_n}(X))^2}{EH^{2\beta}_{a_n}(X)}\right) \longrightarrow s \text{ as } n \to \infty$$

upon again employing (3.2) and Lemma 3.3. Therefore

$$M_n(\cdot) \overset{D}{\longrightarrow} B(\cdot) \text{ in } D[0,1]. \quad \blacksquare$$

It is possible to modify by g_t only an asymptotically vanishing proportion of terms and still retain the validity of the invariance principle. Verification of this assertion is similar to the analogous trimmed central limit theorem in Hahn and Kuelbs (1988). Arrange the X_i's in descending order of magnitude $|X_1^{(n)}| \geq |X_2^{(n)}| \geq \ldots$.

Theorem 3.4. **Let H_t be defined as in (2.2) with the additional assumption that k is nondecreasing. Let $\beta = \frac{p}{q}$ where p and q are odd integers and $\alpha \in \mathbf{R}$ satisfies $0 < \beta \leq \alpha$. Given $\{r_n\}$ satisfying (2.1), assume there exists a sequence a_n converging to ∞ for which (3.2) is satisfied. Then there exists a sequence $\{\psi_n\}$ with $\frac{\psi_n r_n}{n} \to 0$ such that**

$$S_n(s) \equiv \frac{\sum_{i=[\psi_n r_n]+1}^{[ns]} k^\beta\left(X_i^{(n)}\right) + \sum_{i=1}^{[\psi_n r_n]} H_{a_n}^\beta\left(X_i^{(n)}\right) - [ns]EH_{a_n}^\beta(X)}{a_n^\alpha\sqrt{r_n}}$$

$$\xrightarrow{D} B(s) \quad \text{in } D[0,1].$$

Proof. Write

$$S_n(s) = \sum_{i=1}^{[ns]} \frac{H_{a_n}^\beta(X_i) - EH_{a_n}^\beta(X)}{a_n^\alpha\sqrt{r_n}}$$

$$+ \sum_{i=[\psi_n r_n]+1}^{[ns]} \frac{k^\beta\left(X_i^{(n)}\right) - H_{a_n}^\beta\left(X_i^{(n)}\right)}{a_n^\alpha\sqrt{r_n}}$$

$$\equiv M_n(s) + C_n(s).$$

By Theorem 3.1, $M_n(s)\xrightarrow{D} B(s)$ in $D[0,1]$. Thus it suffices to show that

$$\sup_{0\leq s\leq 1} |C_n(s)|\xrightarrow{P} 0.$$

We will actually establish the stronger statement

$$\sup_{0\leq s\leq 1} |A_n(s)| \equiv \sup_{0\leq s\leq 1} |a_n^\alpha\sqrt{r_n}C_n(s)|\xrightarrow{P} 0.$$

Observe that

$$\left\{ \sup_{0\leq s< \frac{[\psi_n r_n]+1}{n}} |A_n(s)| \neq 0 \right\}$$

$$= \left\{ \sup_{0\leq s< \frac{[\psi_n r_n]+1}{n}} \sum_{i=[\psi_n r_n]+1}^{[ns]} \left\{ k^\beta\left(X_i^{(n)}\right) - H_{a_n}^\beta\left(X_i^{(n)}\right)\right\} \neq 0 \right\}$$

$$= \emptyset.$$

Consequently,

$$P\left(\sup_{0\le s\le 1}|A_n(s)|\ne 0\right)=P\left(\sup_{\frac{[\psi_n r_n]+1}{n}\le s\le 1}|A_n(s)|\ne 0\right)$$

$$\le P\left\{\exists s\in\left[\frac{[\psi_n r_n]+1}{n},1\right]:\ \sum_{i=[\psi_n r_n]+1}^{[ns]}\left(k^\beta\left(X_i^{(n)}\right)-H_{a_n}^\beta\left(X_i^{(n)}\right)\right)\ne 0\right\}$$

$$\le P\left\{\exists i\ge[\psi_n r_n]+1:\ k^\beta\left(X_i^{(n)}\right)-H_{a_n}^\beta\left(X_i^{(n)}\right)\ne 0\right\}$$

$$\le P\left\{\exists i\ge[\psi_n r_n]+1:\ \left|X_i^{(n)}\right|>a_n\right\}$$

$$=P\{\text{more than }[\psi_n r_n]\text{ events }|X_i|>a_n\text{ occur, }i\le n\}$$

$$=P\left(T_n>\psi_n r_n\right)$$

where

$$T_n=\sum_{i=1}^n I\left(|X_i|>a_n\right).$$

Let

$$G(a_n)=P\left(|X_i|>a_n\right).$$

Then T_n is Binomial, $B(n,G(a_n))$.

Employing Chebyshev's inequality, and letting $\psi_n=\frac{n}{r_n}G^\delta(a_n)$ for some $0<\delta\le\frac12$,

$$P\left(T_n > \psi_n r_n\right) = P\left(T_n - ET_n > \psi_n r_n - ET_n\right)$$

$$\leq P\left(|T_n - ET_n| > \psi_n r_n - ET_n\right)$$

$$\leq \frac{\operatorname{Var} T_n}{\left(\psi_n r_n - ET_n\right)^2}$$

$$= \frac{nG(a_n)(1 - G(a_n))}{\left(\psi_n r_n - nG(a_n)\right)^2}$$

$$= \frac{G^{1-2\delta}(a_n)(1 - G(a_n))}{n(1 - G^{1-\delta}(a_n))^2}$$

$$\longrightarrow 0 \quad \text{as} \quad n \to \infty. \quad \blacksquare$$

Both Theorems 3.1 and 3.4 require condition (3.2). This condition is always satisfied whenever g_t has no downward jumps for t large. Furthermore, for all H_t such that $|k(x)| \geq C_3|x|$, there exists $n_0(C_3)$ such that the equations can be solved simultaneously for all $n \geq n_0(C_3)$. This latter fact will be required in Section 7.

Proposition 3.5. *Let* $\beta = \frac{p}{q}$ *where* p *and* q *are odd integers and* $\alpha \in \mathbf{R}$ *satisfies* $0 < \beta \leq \alpha$. *Let* X *be a nondegenerate random variable. Let* H_t *be as in* (2.2).

(i) If g_t has no downward jumps for t large, then there exists $a_n \to \infty$ such that the equation $\frac{a_n^{2\alpha} r_n}{n} = EH_{a_n}^{2\beta}(X)$ is satisfied for n sufficiently large.

(ii) Let $\tilde{H}_t(x) \equiv C_3 x I(|x| \leq t)$. Then there exists ·

$n_0 = n_0(C_3)$ such that $\frac{\tilde{a}_n^{2\alpha} r_n}{n} = E\tilde{H}_{\tilde{a}_n}^{2\beta}(X)$ for $n \geq n_0$. Additionally, if $|H_t| \geq |\tilde{H}_t|$

then $\frac{a_n^{2\alpha} r_n}{n} = EH_{a_n}^{2\beta}(X)$ and $a_n \geq \tilde{a}_n$ for $n \geq n_0$.

Proof. Define $h(t) = \frac{EH_t^{2\beta}(X)}{t^{2\alpha}}$. Let $C = \max\{C_1, C_2\}$ where C_1 and C_2 are as in (2.3). Then for t large,

$$h(t) \le E\left(\frac{H_t^{2\beta}(X)}{t^{2\beta}}\right) \le C^{2\beta} E\left(\frac{X^2 \wedge t^2}{t^2}\right)^\beta = C^{2\beta} E\left(\left(\frac{X}{t}\right)^2 \wedge 1\right)^\beta \longrightarrow 0$$

as $t \to \infty$. Choose t_1 such that $h(t_1) > 0$. For n sufficiently large, $\frac{r_n}{n} < h(t_1)$ (which is possible since $\frac{r_n}{n} \to 0$). The facts that g_t does not have downward jumps and $h(t) \to 0$ imply the existence of a_n such that $h(a_n) = \frac{r_n}{n}$. Moreover, since $0 < h(t) \to 0$ and $\frac{r_n}{n} \to 0$, $a_n \to \infty$. This verifies (i).

To prove (ii), assume $|H_t| \ge |\tilde{H}_t|$. Then

$$\frac{EH_t^{2\beta}(X)}{t^{2\alpha}} \ge \frac{E\tilde{H}_t^{2\beta}(X)}{t^{2\alpha}} \quad \text{for all } t. \tag{3.6}$$

Let $c = \sup_t \frac{E\tilde{H}_t^{2\beta}(X)}{t^{2\alpha}} > 0$ since X is nondegenerate . There exists $n_0 \equiv n_0(C_3)$ such that $\frac{r_n}{n} \le c$ for all $n \ge n_0$ since $\frac{r_n}{n} \to 0$ as $n \to \infty$. Fix $n \ge n_0$. By (3.6) there exists t_0 such that

$$\frac{EH_{t_0}^{2\beta}(X)}{t_0^{2\alpha}} \ge \frac{E\tilde{H}_{t_0}^{2\beta}(X)}{t_0^{2\alpha}} \ge \frac{r_n}{n}.$$

Since $\frac{EH_t^{2\beta}(X)}{t^{2\alpha}}$ and $\frac{E\tilde{H}_t^{2\beta}(X)}{t^{2\alpha}}$ both converge to zero and assume every value in $(0, \frac{E\tilde{H}_{t_1}^{2\beta}(X)}{t_1^{2\alpha}}]$, there exist $a_n \ge t_0$ and $\tilde{a}_n \ge t_0$ such that both $\frac{EH_{a_n}^{2\beta}(X)}{a_n^{2\alpha}} = \frac{r_n}{n}$ and $\frac{E\tilde{H}_{\tilde{a}_n}^{2\beta}(X)}{\tilde{a}_n^{2\alpha}} = \frac{r_n}{n}$ are satisfied. Moreover, (3.6) implies that $a_n \ge \tilde{a}_n$ for $n \ge n_0$. ∎

4. Intermediate Modifications: Empirical Central Limit Theorems for Totally and Partially Modified Sums.

If the X_i's are defined from an unknown underlying distribution, it is important to know whether use of empirical estimates of the modifying levels a_n still yields asymptotic normality. This will be verified for

Theorems 3.1 and 3.4 (using only $s = 1$) if $H_t(-x) = -H_t(x)$ and X is a symmetric random variable.

Let $F_n(x) = \frac{1}{n} \sum_{i=1}^{n} I(X_i \le x)$ be the empirical distribution function. E_n will denote expectation taken with respect to the empirical distribution. Let

$$b = \inf \left\{ t \ge 1 : \quad \frac{E\left(k(X)I(|X| \le t)\right)^{2\beta}}{t^{2\alpha}} > 0 \right\}$$

$$b_n = \inf_{1 \le i \le n} \{\max\{1, |X_i|\}\}$$

$$\hat{a}_n = \inf \left\{ t \ge b_n + 1 : \quad \frac{E_n H_t^{2\beta}(X_i)}{t^{2\alpha}} \le \frac{r_n}{n} \right\}.$$

Theorem 4.1. *Let H_t be as in (2.2) with the additional assumption that $H_t(-x) = -H_t(x)$. Suppose H_t is right continuous and g_t does not have downward jumps in t. Assume X is a nondegenerate symmetric random variable and $\{r_n\}$ is such that $r_n \to \infty$, $\frac{r_n}{n} \to 0$. Let $\beta = \frac{p}{q}$ where p and q are odd integers and $\alpha \in \mathbf{R}$ satisfies $0 < \beta \le \alpha$. Then*

$$\mathcal{L}\left(\sum_{j=1}^{n} \frac{H_{\hat{a}_n}^{\beta}(X_j)}{\hat{a}_n^{\alpha} \sqrt{r_n}} \right) \longrightarrow N(0,1).$$

Note that for β defined as above, the symmetry is maintained. The proof of Theorem 4.1 is analogous to that of Theorem 2 in Hahn and Kuelbs (1988). We first require a few facts about the empirical estimates $\hat{a}_n$.

Lemma 4.2. *Let H_t, α and β be as in Theorem 4.1. Suppose $E(k(X))^{2\beta} = \infty$. Then with probability one,*

(1) $\hat{a}_n$ exists for $n \geq 1$,

(2) $\lim_n \hat{a}_n = \infty$, and

(3) $\dfrac{E_n H_{\hat{a}_n}^{2\beta}(X)}{\hat{a}_n^{2\alpha}} = \dfrac{r_n}{n}$.

Proof. With probability one, both

$$\lim_{n \to \infty} b_n = b = \inf\{t \geq 1: \ t^{-2\alpha} E\left(k(X)I(|X| \leq t)\right)^{2\beta} > 0\} \geq 1$$

and

$$\lim_{t \to \infty} \frac{E_n H_t^{2\beta}(X)}{t^{2\alpha}} = 0 \ \text{ with } E_n H_t^{2\beta}(X) > 0 \quad \forall t \geq b_n.$$

Hence $\hat{a}_n$ exists for $n \geq 1$, thereby verifying (1).

To prove (2), assume there is a subsequence $\{n'\}$ such that $\lim_{n'} \hat{a}_{n'} = \gamma < \infty$ a.s. Set

$$\delta = EH_{b+\frac{1}{2}}^{2\beta}(X)/(b+\gamma+1)^{2\alpha}$$

$$= \frac{E\left(k(X)I\left(|X| \leq b+\frac{1}{2}\right) + \left|g_{b+\frac{1}{2}}(X)\right| \operatorname{sgn}(X)I\left(|X| > b+\frac{1}{2}\right)\right)^{2\beta}}{(b+\gamma+1)^{2\alpha}}$$

Then $\delta > 0$, and $b_{n'} + 1 \leq \hat{a}_{n'}$. So $\gamma \geq b+1$ a.s.

Now since downward jumps are not allowed,

$$\liminf_n \ \inf_{b_n+1 \leq t \leq b_n+\gamma+1} \sum_{i=1}^n \frac{\{k(X_i)I(|X_i| \leq t) + |g_t(X_i)| \operatorname{sgn}(X_i)I(|X_i| > t)\}^{2\beta}}{nt^{2\alpha}}$$

$$\geq \liminf_n \sum_{i=1}^n \frac{\left\{k(X_i)I(|X_i| \leq b+\frac{1}{2}) + |g_{b+\frac{1}{2}}(X_i)| \operatorname{sgn}(X_i)I(|X_i| > b+\frac{1}{2})\right\}^{2\beta}}{n(b_n+\gamma+1)^{2\alpha}}$$

$\geq \delta$ a.s. by the Strong Law of Large Numbers.

$$(4.3)$$

Since $\frac{r_n}{n} \to 0$, for n sufficiently large, $\frac{r_n}{n} < \delta$. Hence by the definition of $\hat{a}_n$, (4.3) implies that $\hat{a}_n \geq b + \gamma + 1$. This contradicts the fact that $\lim_n \hat{a}_{n'} = \gamma$. So (2) holds.

To prove (3), let

$$h_n(t) = \frac{E_n H_t^{2\beta}(X)}{t^{2\alpha}}.$$

Two cases can arise.

Case 1. Assume $\hat{a}_n$ is not a point of discontinuity. Then

$$\lim_{t \nearrow \hat{a}_n} h_n(t) = h_n(\hat{a}_n) = \frac{E_n H_{\hat{a}_n}^{2\beta}(X)}{\hat{a}_n^{2\alpha}} \geq \frac{r_n}{n}$$

and

$$h_n(\hat{a}_n) = \lim_{t \searrow \hat{a}_n} h_n(t) \leq \frac{r_n}{n}.$$

Thus $h_n(\hat{a}_n) = \frac{r_n}{n}$.

Case 2. Assume $\hat{a}_n$ is a point of discontinuity. Let $A_n = \{$all discontinuity points$\}$. If $\hat{a}_n \in A_n$,

$$\frac{r_n}{n} \geq \lim_{t \searrow \hat{a}_n} h_n(t) = h_n(\hat{a}_n) > \lim_{t \nearrow \hat{a}_n} h_n(t) \geq \frac{r_n}{n}.$$

Hence $\hat{a}_n \notin A_n$. This means that $\hat{a}_n$ is always a point of continuity. Therefore, $h_n(\hat{a}_n) = \frac{r_n}{n}$ and (3) is satisfied. ∎

Proof of Theorem 4.1. Let $\{X_i, i \geq 1\}$ be an i.i.d. sequence on $(\Omega, \mathcal{F}, P)$ and let $\{\varepsilon_i, i \geq 1\}$ be a sequence of independent Rademacher random variables on $(\Omega', \mathcal{F}', P')$, i.e. $P(\varepsilon_i = +1) = P(\varepsilon_i = -1) = \frac{1}{2}$. Define

$$\Omega'' = \Omega' \times \Omega, \quad \mathcal{F}'' = \mathcal{F}' \otimes \mathcal{F}, \quad P'' = P' \times P.$$

Since X and $H_t(X)$ are symmetric,

$$\mathcal{L}\left(\varepsilon_1 H_{\hat{a}_n}^\beta(X_1), \ldots, \varepsilon_n H_{\hat{a}_n}^\beta(X_n)\right) = \mathcal{L}\left(H_{\hat{a}_n}^\beta(X_1), \ldots, H_{\hat{a}_n}^\beta(X_n)\right).$$

If a_n^* is defined as $\hat{a}_n$ but in terms of $\{\varepsilon_i H_t(X_i), \ i = 1, \ldots, n\}$, then $\hat{a}_n = a_n^*$ a.s. since $(\varepsilon_i H_t(X_i))^{2\beta} = (H_t(X_i))^{2\beta}$.

Therefore, symmetry and $a_n^* = \hat{a}_n$ imply

$$\mathcal{L}\left(\varepsilon_1 H_{\hat{a}_n}^\beta(X_1), \ldots, \varepsilon_n H_{\hat{a}_n}^\beta(X_n), \hat{a}_n\right) = \mathcal{L}\left(H_{\hat{a}_n}^\beta(X_1), \ldots, H_{\hat{a}_n}^\beta(X_n), \hat{a}_n\right)$$

and further,

$$\mathcal{L}\left(\sum_{i=1}^n \frac{\varepsilon_i H_{\hat{a}_n}^\beta(X_i)}{\hat{a}_n^\alpha \sqrt{r_n}}\right) = \mathcal{L}\left(\sum_{i=1}^n \frac{H_{\hat{a}_n}^\beta(X_i)}{\hat{a}_n^\alpha \sqrt{r_n}}\right).$$

Let E_ε, E_X and E respectively denote expectations with respect to P', P and $P' \times P$.

Define

$$\lambda_n(\omega) \equiv E_\varepsilon\left(\exp\left\{it \sum_{j=1}^n \varepsilon_j(\omega') \frac{H_{\hat{a}_n}^\beta(X_j)(\omega)}{\hat{a}_n^\alpha(\omega)\sqrt{r_n}}\right\}\right)$$

$$= \prod_{j=1}^n E_\varepsilon\left(\exp it\left(\varepsilon_j(\omega') \frac{H_{\hat{a}_n}^\beta(X_j)}{\hat{a}_n^\alpha \sqrt{r_n}}\right)\right)$$

$$= \prod_{j=1}^n \left(\frac{1}{2}\exp\left\{it \frac{H_{\hat{a}_n}^\beta(X_j)}{\hat{a}_n^\alpha \sqrt{r_n}}\right\} + \frac{1}{2}\exp\left\{-it \frac{H_{\hat{a}_n}^\beta(X_j)}{\hat{a}_n^\alpha \sqrt{r_n}}\right\}\right)$$

$$= \prod_{j=1}^n \cos\left(\frac{t \, H_{\hat{a}_n}^\beta(X_j)}{\hat{a}_n^\alpha \sqrt{r_n}}\right).$$

Now

$$\cos\left(\frac{t H_{\hat{a}_n}^\beta(X_j)}{\hat{a}_n^\alpha \sqrt{r_n}}\right) = 1 - t^2 \frac{H_{\hat{a}_n}^{2\beta}(X_j)}{2\hat{a}_n^{2\alpha} r_n} + o\left(\frac{t^2 H_{\hat{a}_n}^{2\beta}(X_j)}{\hat{a}_n^{2\alpha} r_n}\right)$$

since $\dfrac{H^{\beta}_{\hat{a}_n}(X_j)}{\hat{a}^{\alpha}_n \sqrt{r_n}} \longrightarrow 0$ a.s. Hence

$$\log \lambda_n(\omega) = \sum_{j=1}^{n} \left(\log \left(1 - t^2 \frac{H^{2\beta}_{\hat{a}_n}(X_j)}{2\hat{a}^{2\alpha}_n r_n} + o \left(\frac{t^2 H^{2\beta}_{\hat{a}_n}(X_j)}{\hat{a}^{2\alpha}_n r_n} \right) \right) \right)$$

$$= \sum_{j=1}^{n} \left(\frac{-t^2 H^{2\beta}_{\hat{a}_n}(X_j)}{2\hat{a}^{2\alpha}_n r_n} + o \left(\frac{t^2 H^{2\beta}_{\hat{a}_n}(X_j)}{\hat{a}^{2\alpha}_n r_n} \right) \right).$$

By Lemma 4.2, $\sum_{j=1}^{n} \dfrac{H^{2\beta}_{\hat{a}_n}(X_j)}{\hat{a}^{2\alpha}_n r_n} = 1$ a.s. Thus $\log(\lambda_n(\omega)) \to \dfrac{-t^2}{2}$ which implies that $\lambda_n(\omega) \to e^{-t^2/2}$ as $n \to \infty$.

Now using Fubini's Theorem,

$$\lim_n E \left(\exp \left\{ it \sum_{j=1}^{n} \frac{\varepsilon_j H^{\beta}_{\hat{a}_n}(X_j)}{\hat{a}^{\alpha}_n \sqrt{r_n}} \right\} \right) = \lim_n E_X \left(E_\varepsilon \left(\exp \; it \sum_{j=1}^{n} \frac{\varepsilon_j H^{\beta}_{\hat{a}_n}(X_j)}{\hat{a}^{\alpha}_n \sqrt{r_n}} \right) \right)$$

$$= \lim_n E_X \left(\lambda_n(\omega) \right)$$

$$= E_X \left(\lim_n \lambda_n(\omega) \right) \text{ by the dominated convergence Theorem}$$

$$\text{since } |\lambda_n| \le 1 \text{ and } \lambda_n \to e^{-t^2/2}$$

$$= E_X e^{-t^2/2}$$

$$= e^{-t^2/2}.$$

Therefore,

$$\mathcal{L} \left(\sum_{j=1}^{n} \frac{\varepsilon_j H^{\beta}_{\hat{a}_n}(X_j)}{\hat{a}^{\alpha}_n \sqrt{r_n}} \right) \longrightarrow N(0,1)$$

which implies

$$\mathcal{L} \left(\sum_{j=1}^{n} \frac{H^{\beta}_{\hat{a}_n}(X_j)}{\hat{a}^{\alpha}_n \sqrt{r_n}} \right) \longrightarrow N(0,1). \quad \blacksquare$$

Theorem 4.4. *Assume the set-up and assumptions of Theorem 4.1. Then there exists a sequence of random variables $\{\hat{\psi}_n\}$ determined by $\{X_1, \cdots, X_n\}$ such that $\frac{\hat{\psi}_n r_n}{n} \to 0$ a.s. and*

$$\mathcal{L}\left(\frac{\sum_{j=[\hat{\psi}_n r_n]+1}^{n} k^\beta\left(X_j^{(n)}\right) + \sum_{j=1}^{[\hat{\psi}_n r_n]} H_{\hat{a}_n}^\beta\left(X_j^{(n)}\right)}{\hat{a}_n^\alpha \sqrt{r_n}}\right) \longrightarrow N(0,1).$$

Proof. Fix $0 < \delta < 1$ and let $\hat{\psi}_n$ satisfy

$$\frac{\hat{\psi}_n r_n}{n} = \left(\sum_{j=1}^{n} \frac{I(|X_j| > \hat{a}_n)}{n}\right)^\delta \qquad (4.5)$$

Write

$$\sum_{j=[\hat{\psi}_n r_n]+1}^{n} k^\beta\left(X_j^{(n)}\right) + \sum_{j=1}^{[\hat{\psi}_n r_n]} H_{\hat{a}_n}^\beta\left(X_j^{(n)}\right)$$

$$= \sum_{j=1}^{n} H_{\hat{a}_n}^\beta\left(X_j\right) + \sum_{j=[\hat{\psi}_n r_n]+1}^{n} \left(k^\beta\left(X_j^{(n)}\right) - H_{\hat{a}_n}^\beta\left(X_j^{(n)}\right)\right).$$

By Theorem 4.1,

$$\mathcal{L}\left(\sum_{j=1}^{n} \frac{H_{\hat{a}_n}^\beta(X_j)}{\hat{a}_n^\alpha \sqrt{r_n}}\right) \longrightarrow N(0,1).$$

Thus it remains to show that

$$\sum_{j=[\hat{\psi}_n r_n]+1}^{n} \frac{k^\beta\left(X_j^{(n)}\right) - H_{\hat{a}_n}^\beta\left(X_j^{(n)}\right)}{\hat{a}_n^\alpha \sqrt{r_n}} \xrightarrow{P} 0.$$

We will again prove the stronger statement

$$\sum_{j=[\hat{\psi}_n r_n]+1}^{n} \left(k^\beta\left(X_j^{(n)}\right) - H_{\hat{a}_n}^\beta\left(X_j^{(n)}\right)\right) \xrightarrow{P} 0.$$

Now

$$P\left\{\sum_{j=[\hat{\psi}_n r_n]+1}^{n}\left(k^\beta\left(X_j^{(n)}\right)-H_{\hat{a}_n}^\beta\left(X_j^{(n)}\right)\right)\neq 0\right\}$$

$$= P\left\{\sum_{j=[\hat{\psi}_n r_n]+1}^{n}\left\{k^\beta\left(X_j^{(n)}\right)-\left[k\left(X_j^{(n)}\right)I\left(\left|X_j^{(n)}\right|\leq \hat{a}_n\right)\right.\right.\right.$$

$$\left.\left.\left.+\left|g_{\hat{a}_n}\left(X_j^{(n)}\right)\right|\operatorname{sgn}\left(X_j^{(n)}\right)I\left(\left|X_j^{(n)}\right|>\hat{a}_n\right)\right]^\beta\right\}\neq 0\right\}$$

$$= P\left\{\#\left\{j\leq n:|X_j|>\hat{a}_n\right\}>\hat{\psi}_n r_n\right\}$$

$$= P\left(\sum_{j=1}^{n}I\left(|X_j|>\hat{a}_n\right)>\hat{\psi}_n r_n\right) \tag{4.6}$$

$$= EI\left(\sum_{j=1}^{n}I\left(|X_j|>\hat{a}_n\right)>\hat{\psi}_n r_n\right)$$

$$\leq E\left(\sum_{j=1}^{n}\frac{I\left(|X_j|>\hat{a}_n\right)}{\hat{\psi}_n r_n}\right)$$

$$= E\left(\sum_{j=1}^{n}\frac{I\left(|X_j|>\hat{a}_n\right)}{n}\right)^{1-\delta}$$

by the definition of $\hat{\psi}_n$ in (4.5). But

$$\lim_{n\to\infty} E\left(\sum_{j=1}^{n}\frac{I\left(|X_j|>\hat{a}_n\right)}{n}\right)^{1-\delta}=E\left(\lim_{n\to\infty}\sum_{j=1}^{n}\frac{I\left(|X_j|>\hat{a}_n\right)}{n}\right)^{1-\delta}$$

So for $M<\infty$, since $\hat{a}_n\to\infty$ a.s.,

$$\lim_{n}\sum_{j=1}^{n}\frac{I\left(|X_j|>\hat{a}_n\right)}{n}\leq \lim_{n}\sum_{j=1}^{n}\frac{I\left(|X_j|>M\right)}{n}=P(|X|>M)\quad\text{a.s.}$$

Since M is arbitrary and $\lim_{M\to\infty}P(|X|>M)=0$, we may conclude that

$$E\left(\sum_{j=1}^{n}\frac{I\left(|X_j|>\hat{a}_n\right)}{n}\right)^{1-\delta}\longrightarrow 0\quad\text{as}\quad n\to\infty.$$

Combining this with (4.6) yields

$$P\left(\sum_{j=[\dot\psi_n r_n]+1}^{n}\left(k^\beta\left(X_j^{(n)}\right)-H_{\hat a_n}^\beta\left(X_j^{(n)}\right)\right)\neq 0\right)\longrightarrow 0,$$

as desired. ∎

5. Rates of Convergence. We have seen that via the application of a variety of influence functions asymptotically normal limits may be obtained. A variety of factors have a bearing on the choice of influence function. Certainly if different influence functions yield different rates of convergence, it might be advantageous to use the one which provides the fastest rate of convergence. The objective of this section is to show that all functions H_t satisfying

$$H_t(x) = k(x)I(|x| \le t) + g_t(x)I(|x| > t)$$

with

$$H_t(-x) = -H_t(x),$$

$$C_3|x| \le |k(x)| \le C_1|x| \quad \text{for some} \quad 0 < C_1, C_3 < \infty, \tag{5.1}$$

and

$$|g_t(x)| \le C_2 t \quad \text{for} \quad |x| > t \text{ and some} \quad 0 \le C_2 < \infty$$

yield the same resultant rate of convergence in the case of a symmetric random variable X whose joint tail is regularly varying for t large,i.e.

$$P(|X| > t) = ct^{-\lambda} \quad \text{for} \quad t \ge t_0, \tag{5.2}$$

where $0 < \lambda < 2$. Note the additional constraint on H_t that $C_3|x| \le |k(x)| \le C_1|x|$ for all x. Condition (5.2) is precisely the condition for a symmetric random variable to be in the normal domain of attraction of a symmetric stable. The rate of convergence will be established by finding appropriate upper and lower bounds for the uniform error in the central limit theorem a la Hall (1982).

Theorem 5.3. *Let H_t be defined as in (2.2) with the additional assumptions in (5.1). Let $\{r_n\}$ be as in (2.1). Let $\{X_i\}_{i \geq 1}$ be a sequence of symmetric i.i.d. random variables with regularly varying joint tail satisfying (5.2). Let $\beta = \frac{m}{n}$ where $m \geq n$ are odd integers and $\alpha \in \mathbf{R}$ satisfies $\alpha \geq \beta \geq 1$. Assume g_t does not have downward jumps for t large. Then there exists $a_n \to \infty$ such that for n sufficiently large*

$$\frac{EH_{a_n}^{2\beta}(X)}{a_n^{2\alpha}} = \frac{r_n}{n} \tag{5.4}$$

and

$$\sup_{-\infty < x < \infty} \left| P\left(\sum_{j=1}^{n} \frac{H_{a_n}^{\beta}(X_j)}{a_n^{\alpha}\sqrt{r_n}} \leq x \right) - \Phi(x) \right| \asymp \frac{n a_n^{4\beta - 4\alpha - \lambda}}{r_n^2}. \tag{5.5}$$

Moreover, all such functions H_t yield the same rate of convergence.

The proof of Theorem 5.3 relies on the following theorem from Hall (1982) plus several lemmas. Hall's theorem provides general upper and lower bounds for the rate of convergence for triangular arrays of independent random variables.

Theorem 5.6. *(Hall (1982), Theorems 2.2 and 3.1). Let $\{Z_{nj}, 1 \leq j \leq n < \infty\}$ be a triangular array of rowwise independent random variables with $EZ_{nj} = 0$, $\sum_{j=1}^{\infty} EZ_{nj}^2 = 1$ and $\max_{1 \leq j \leq n} EZ_{nj}^2 \to 0$ as $n \to \infty$. Assume that $\forall \varepsilon > 0$, $\sum_{j=1}^{n} EZ_{nj}^2 I(|Z_{nj}| > \varepsilon) \to 0$ as $n \to \infty$. Then, for large n, there exist $0 < C_4, C_5 <$*

∞ , such that

$$C_4 \delta_{n1} \le \sup_{-\infty < x < \infty} \left| P\left(\sum_{j=1}^{n} Z_{nj} \le x \right) - \Phi(x) \right| \le C_5 \delta_n \qquad (5.7)$$

where

$$\delta_n = \sum_{j=1}^{n} E\{Z_{nj}^2 I(|Z_{nj}| > 1)\} + \sum_{j=1}^{n} E\{Z_{nj}^4 I(|Z_{nj}| \le 1)\}$$
$$+ \left| \sum_{j=1}^{n} E\{Z_{nj}^3 I(|Z_{nj}| \le 1)\} \right| \qquad (5.8)$$

and

$$\delta_{n1} = \sum_{j=1}^{n} P(|Z_{nj}| > 1) + \sum_{j=1}^{n} E\{Z_{nj}^4 I(|Z_{nj}| \le 1)\} + \left| \sum_{j=1}^{n} E\{Z_{nj}^3 I(|Z_{nj} \le 1)\} \right|. \quad (5.9)$$

Lemma 5.10. *Let X be a symmetric random variable with regularly varying joint tail satisfying (5.2). Let p be an even positive integer. Given p, β, λ choose $m = m(p, \beta, \lambda)$ such that $t \ge m t_0$ implies $\frac{p\beta}{p\beta - \lambda}(1 - (\frac{1}{m})^{p\beta - \lambda} > 1$. Then*

$$EX^{p\beta} I(|X| \le t) \asymp t^{p\beta - \lambda}, \quad \text{for} \quad t \ge m t_0.$$

Proof. Since $p\beta$ is even ,

$$EX^{p\beta} I(|X| \le t) = -t^{p\beta} P(|X| > t) + \int_0^t p\beta y^{p\beta - 1} P(|X| > y) \, dy$$
$$= -t^{p\beta} P(|X| > t) + \int_0^{t_0} p\beta y^{p\beta - 1} P(|X| > y) \, dy$$
$$+ \int_{t_0}^t p\beta y^{p\beta - 1} P(|X| > y) \, dy.$$

But if $t \geq t_0$,

$$\int_{t_0}^{t} p\beta y^{p\beta-1} P(|X| > y) \, dy = c \int_{t_0}^{t} p\beta y^{p\beta-\lambda-1} \, dy$$

$$= \frac{cp\beta}{p\beta - \lambda}(t^{p\beta-\lambda} - t_0^{p\beta-\lambda}).$$

Let $\varepsilon = \frac{p\beta}{p\beta-\lambda}(1 - (\frac{1}{m})^{p\beta-\lambda} - 1$. Hence

$$\frac{cp\beta}{p\beta - \lambda}(t^{p\beta-\lambda} - t_0^{p\beta-\lambda}) \leq \frac{cp\beta}{p\beta - \lambda}t^{p\beta-\lambda} \quad \text{if } t \geq t_0$$

and

$$\frac{cp\beta}{p\beta - \lambda}(t^{p\beta-\lambda} - t_0^{p\beta-\lambda}) \geq \frac{cp\beta}{p\beta - \lambda}t^{p\beta-\lambda}(1 - (\frac{1}{m})^{p\beta-\lambda}) \quad \text{if } t \geq mt_0.$$

Also

$$-t^{p\beta}P(|X| > t) = -ct^{p\beta-\lambda} \quad \text{for } t \geq t_0.$$

Therefore,

$$\varepsilon ct^{p\beta-\lambda} \leq EX^{p\beta}I(|X| \leq t) \leq \frac{cp\beta}{p\beta - \lambda}t^{p\beta-\lambda} \quad \text{for } t \geq mt_0. \quad \blacksquare$$

Proof of Theorem 5.3. Let $\tilde{H}_t(x) = C_3 x I(|x| \leq t)$. By Proposition 3.5, there

exist $n_0 = n_0(C_3)$, a_n and $\tilde{a}_n$ such that for $n \geq n_0$

$$a_n^{2\alpha} = \frac{n}{r_n}EH_{a_n}^{2\beta}(X) \quad \text{with} \quad a_n \to \infty,$$

$$\tilde{a}_n^{2\alpha} = \frac{n}{r_n}E\tilde{H}_{\tilde{a}_n}^{2\beta}(X) \quad \text{with} \quad \tilde{a}_n \to \infty$$

and

$$a_n \geq \tilde{a}_n.$$

This proves (5.4).

Let $Z_{nj} = \frac{H_{a_n}^{\beta}(X_j)}{a_n^{\alpha}\sqrt{r_n}}$. Notice that $H_{a_n}^{\beta}(-x) = -H_{a_n}^{\beta}(x)$ and X is symmetric. Therefore

$$EZ_{nj}^r = 0 \quad \text{for } r \text{ odd}. \tag{5.11}$$

Also, for $n \geq n_0$

$$|Z_{nj}| = \left| \frac{H_{a_n}^{\beta}(X_j)}{a_n^{\alpha}\sqrt{r_n}} \right| \leq \frac{1}{a_n^{\alpha-\beta}\sqrt{r_n}} \leq \frac{1}{\tilde{a}_n^{\alpha-\beta}\sqrt{r_n}} \to 0 \quad \text{as} \quad n \to \infty. \tag{5.12}$$

In particular,

$$EZ_{nj}^2 = E\left(\frac{H_{a_n}^{2\beta}(X_j)}{a_n^{2\alpha}r_n} \right) \leq \frac{1}{a_n^{2\alpha-2\beta}r_n} \leq \frac{1}{\tilde{a}_n^{2\alpha-2\beta}r_n} \to 0 \quad \text{as} \quad n \to \infty, \tag{5.13}$$

so that

$$\max_{1 \leq j \leq n} EZ_{nj}^2 \to 0 \quad \text{as} \quad n \to \infty. \tag{5.14}$$

Finally, for $n \geq n_0$

$$\sum_{j=1}^{n} EZ_{nj}^2 = \sum_{j=1}^{n} E\left(\frac{H_{a_n}^{2\beta}(X_j)}{a_n^{2\alpha}r_n} \right) = n\frac{EH_{a_n}^{2\beta}(X_j)}{a_n^{2\alpha}r_n} = 1. \tag{5.15}$$

Step 1. There exists n_1 such that for $n \geq n_1$,

$$\delta_{n1} = \delta_n = \frac{nEH_{a_n}^{4\beta}(X)}{a_n^{4\alpha}r_n^2}.$$

Proof. By (5.12), there exists n_2 such that for $n \geq n_2$, $|Z_{nj}| < 1$. Hence, combining this fact with (5.11), for $n \geq n_1 \equiv \max\{n_0, n_2\}$,

$$\delta_n = \delta_{n1} = \sum_{j=1}^{n} \frac{EH_{a_n}^{4\beta}(X_j)}{a_n^{4\alpha}r_n^2} = \frac{nEH_{a_n}^{4\beta}(X)}{r_n^2 a_n^{4\alpha}}. \quad \blacksquare$$

Step 2. For n large,

$$\sup_{-\infty < x < \infty} \left| P\left(\sum_{j=1}^{n} \frac{H_{a_n}^{\beta}(X_j)}{a_n^{\alpha}\sqrt{r_n}} \leq x \right) - \Phi(x) \right| \asymp \delta_n.$$

Proof. Conditions (5.11), (5.12), (5.14) and (5.15) show that the conditions of Theorem 5.6 are satisfied. Consequently, for n large, since $\delta_{n1} = \delta_n$ by step 1,

$$\sup_{-\infty < x < \infty} \left| P\left(\sum_{j=1}^{n} \frac{H_{a_n}^{\beta}(X_j)}{a_n^{\alpha}\sqrt{r_n}} \leq x \right) - \Phi(x) \right| \asymp \delta_n. \quad \blacksquare$$

Step 3. Let $\tilde{n} \geq n_0$ be such that $\tilde{a}_n \geq mt_0$. Then $\delta_n \asymp \frac{na_n^{4\beta - 4\alpha - \lambda}}{r_n^2}$ for $n \geq \tilde{n}$.

Proof.

(a) By the definition of H_t,

$$EH_{a_n}^{4\beta}(X) \leq C_1^{4\beta} E\{X^{4\beta} I(|X| \leq a_n)\} + C_2^{4\beta} a_n^{4\beta} P(|X| > a_n)$$

$$\asymp C_1^{4\beta} a_n^{4\beta - \lambda} + C_2^{4\beta} a_n^{4\beta - \lambda} \quad \text{for} \quad n \geq \tilde{n}.$$

using Lemma 5.10 and (5.2). Hence, there exists $\tilde{C}_1 < \infty$ such that for $n \geq \tilde{n}$,

$$\delta_n = \frac{n E H_{a_n}^{4\beta}(X)}{r_n^2 a_n^{4\alpha}}$$

$$\leq \tilde{C}_1 \frac{na_n^{4\beta - 4\alpha - \lambda}}{r_n^2}. \tag{5.16}$$

(b) Also for $n \geq \tilde{n}$,

$$EH_{a_n}^{4\beta}(X) \geq C_3^{4\beta} EX^{4\beta} I(|X| \leq a_n) \asymp C_3^{4\beta} a_n^{4\beta - \lambda},$$

again using Lemma 5.10. Hence, there exists $\tilde{C}_2 > 0$ such that for $n \geq \tilde{n}$,

$$\delta_n = \frac{nEH_{a_n}^{4\beta}(X)}{r_n^2 a_n^{4\alpha}} \geq \tilde{C}_2 \frac{na_n^{4\beta-4\alpha-\lambda}}{r_n^2}. \tag{5.17}$$

Combining (5.16) and (5.17) yields for $n \geq \tilde{n}$,

$$\delta_n \asymp \frac{na_n^{4\beta-4\alpha-\lambda}}{r_n^2}. \quad \blacksquare$$

Steps 2 and 3 together yield (5.5).

Step 4. All such functions H_t yield the same rate of convergence when applied to random variables with tails as in (5.2).

Proof. We will compare the rate of convergence induced by a general $\dot{H}_t$ to the rate induced by $\tilde{H}_t(x) \equiv C_3 x I(|x| \leq t)$. For $n \geq n_0$, let

$$\delta_n' = \frac{nE(\tilde{H}_{\tilde{a}_n}^{4\beta}(X))}{r_n^2 \tilde{a}_n^{4\alpha}} \quad \text{with} \quad \tilde{a}_n^{2\alpha} = \frac{n}{r_n} E(\tilde{H}_{\tilde{a}_n}^{2\beta}(X)). \tag{5.18}$$

Then for $n \geq \tilde{n} \geq n_0$ (where recall $\tilde{a}_{\tilde{n}} \geq mt_0$),

$$\begin{aligned}
\delta_n' &= \frac{n}{r_n^2 \tilde{a}_n^{4\alpha}} E(C_3^{4\beta} X^{4\beta} I(|X| \leq \tilde{a}_n)) \\
&\asymp \frac{n}{r_n^2 \tilde{a}_n^{4\alpha}} C_3^{4\beta} \tilde{a}_n^{4\beta-\lambda} \quad \text{by Lemma 5.10.} \\
&= C_3^{4\beta} \frac{n\tilde{a}_n^{4\beta-4\alpha-\lambda}}{r_n^2}.
\end{aligned}$$

Therefore, for $n \geq \tilde{n}$,

$$\frac{\delta_n'}{\delta_n} \asymp \left(\frac{\tilde{a}_n}{a_n}\right)^{4\beta-4\alpha-\lambda}$$

It remains to evaluate a_n and $\tilde{a}_n$. First consider $a_n \geq m t_0$.

$$
\begin{aligned}
a_n^{2\alpha} &= \frac{n}{r_n} E H_{a_n}^{2\beta}(X) \\
&= \frac{n}{r_n} \left\{ E(k^{2\beta}(X) I(|X| \leq a_n)) + E(g_{a_n}^{2\beta}(X) I(|X| > a_n)) \right\} \\
&\leq \frac{n}{r_n} \left\{ C_1^{2\beta} E(X^{2\beta} I(|X| \leq a_n)) + C_2^{2\beta} a_n^{2\beta} P(|X| > a_n) \right\} \\
&\asymp \frac{n}{r_n} \left\{ C_1^{2\beta} a_n^{2\beta-\lambda} + C_2^{2\beta} a_n^{2\beta-\lambda} \right\}.
\end{aligned}
$$

Thus, there exists $\tilde{C}_3 < \infty$ such that for all $n \geq \tilde{n}$,

$$
a_n^{2\alpha-2\beta+\lambda} \leq \frac{n}{r_n} \tilde{C}_3 (C_1^{2\beta} + C_2^{2\beta}). \tag{5.19}
$$

Similarly, there exists $\tilde{C}_4 > 0$ such that for all $n \geq \tilde{n}$,

$$
a_n^{2\alpha-2\beta+\lambda} \geq \frac{n}{r_n} \tilde{C}_4 C_3^{2\beta}. \tag{5.20}
$$

Together (5.19) and (5.20) imply that for all $n \geq \tilde{n}$,

$$
a_n^{2\alpha-2\beta+\lambda} \asymp \frac{n}{r_n}.
$$

Turning to $\tilde{a}_n$, for all $n \geq \tilde{n}$,

$$
\begin{aligned}
\tilde{a}_n^{2\alpha} &= \frac{n}{r_n} E(\tilde{H}_{\tilde{a}_n}^{2\beta}(X)) \\
&\asymp \frac{n}{r_n} \left(C_3^{2\beta} \tilde{a}_n^{2\beta-\lambda} \right)
\end{aligned}
$$

or equivalently,

$$
\tilde{a}_n^{2\alpha-2\beta+\lambda} \asymp \frac{n}{r_n}.
$$

Therefore for $n \geq \tilde{n}$, $\frac{\tilde{a}_n}{a_n} \asymp 1$ which implies that δ_n and δ_n' have the same order. This means all such functions H_t yield the same rate of convergence. ∎

Acknowledgement. I wish to express my gratitude to my adviser, Marjorie Hahn, for her insightful help and support throughout this paper. I would like to thank Ravi Chari for teaching me about point processes and Dan Weiner for many helpful discussions.

References

[1] Billingsley, P. (1968). *Convergence of Probability Measures.* John Wiley, New York.

[2] Feller, W. (1971). *An introduction to probability theory and its applications,* vol II. John Wiley, New York.

[3] Hahn, M. G. and Kuelbs, J. (1988). Universal asymptotic normality for conditionally trimmed sums. *Stat. Prob. Letters* **7**, 9-15.

[4] Hahn, M. G., Kuelbs, J. and Weiner, D. C. (1990a). The asymptotic distribution of magnitude-winsorized sums via self-normalization. *J. Theoret. Probab.* **3**, 137-168.

[5] Hahn, M. G., Kuelbs, J. and Weiner, D. C. (1990b). The Asymptotic Joint Distribution of Self-Normalized Censored Sums and Sums-of-Squares. To appear in *Ann. Probab.*

[6] Hall, P. (1988). On the effect of random norming on the rate of convergence in the central limit theorem. *Ann. Prob.* **16**, 1265-1280.

[7] Hall, P. (1980). Characterizing the rate of convergence in the central limit theorem. *Ann. Prob.*; **8**, 1037-1048.

[8] Hall, P. (1982). *Rates of Convergence in the central limit theorem.* Pitman Advanced Publishing Program.

[9] Kasahara, Y. and Watanabe, S. (1986). Limit theorems for point processes and their functionals. *J. Math. Soc. Japan,* **38**, 543-574

[10] Lévy, P. (1937). *Théorie de l'Addition des Variables Aléatoire.* Gauthier-Villars, Paris.

Hamid Ould-Rouis
1 Rue du 24 Fevrier 1956
Blida (09000), Algeria

ON JOINT ESTIMATION OF AN EXPONENT OF REGULAR VARIATION AND AN ASYMMETRY PARAMETER FOR TAIL DISTRIBUTIONS

Marjorie G. Hahn* and Daniel C. Weiner **

1. Introduction. A random variable X is said to have a joint tail distribution which is regularly varying of index $-\alpha$ if for each $c > 0$,

$$\lim_{t \to \infty} \frac{P(|X| > ct)}{P(|X| > t)} = c^{-\alpha}.$$

Regular variation plays a fundamental role in the probabilistic limit theory for sums and maxima of i.i.d. random variables. Consequently, it is not surprising that numerous methods have been proposed for estimating the index of regular variation. (See, e.g., the bibliography of Csorgo, Deheuvels, and Mason (1985).) Some of these estimates are motivated by Karamata's theorem characterizing regular variation, as is the one to be proposed here.

In addition to regular variation of the joint tail distribution, the classical limit theory for partial sums requires that when $0 < \alpha \leq 2$, the one sided tails be balanced in the sense that

$$\lim_{t \to \infty} \frac{P(X > t)}{P(|X| > t)} = \delta$$

* Supported in part by NSF grant DMS-87-02878.
** Supported in part by NSF grants DMS-87-02878 and DMS-88-96217.

for some $\delta \in [0,1]$, which in turn implies that

$$\lim_{t \to \infty} \frac{P(X > t) - P(X < -t)}{P(|X| > t)} = \delta - (1 - \delta) = 2\delta - 1 \equiv \beta$$

for some $\beta \in [-1,1]$. The current literature seems to lack joint estimates of (α, β), the index of stability and the asymmetry parameter, even when $0 < \alpha < 2$. (But see Davis and Resnick (1984).)

The purpose of this paper is to provide joint estimates of (α, β) which are *always* consistent and asymptotically normal for $0 < \alpha < \infty$ and $|\beta| \neq 1$. The proposed estimators are easily analyzable and the estimate of α can be shown to be at least as good (in terms of rates of convergence) as the "best" possible estimator for some well-known submodels. This will be illustrated in the case of the model considered in Hall and Welsh (1984).

The proposed estimates and proofs in this paper are an outgrowth of the results and techniques in Hahn, Kuelbs, and Weiner (1990a). However, the most specific motivation for the estimates comes from the special result of Theorem 6 of Hahn, Kuelbs, and Weiner (1990b), in this volume, which establishes the consistency and joint asymptotic normality of joint estimates of certain center and scale sequences for a random variable in the domain of attraction of a stable law of index $0 < \alpha \leq 2$. Due to the specific restricted class of distributions which is being considered here, the techniques of Hahn, Kuelbs, and Weiner (1990a) simplify. Thus, a completely self-contained development and analysis of the proposed estimates will be provided here.

2. Notation and Preliminaries.

Auxiliary Functions and Properties: Let X be a nondegenerate random variable with distribution function F. Given $p > 0$, $t \geq 0$, define

(tail functions)

$$G(t) = P(|X| > t), \ G^{\pm}(t) = P(X > t) - P(X < -t), \qquad (2.1)$$

(censored moment functions)

$$M(p,t) = E\left[|X|^p \wedge t^p\right], \ M^{\pm}(p,t) = E[(|X|^p \wedge t^p)sgn(X)] \qquad (2.2)$$

(truncated moment functions)

$$\tilde{M}(p,t) = E\left[|X|^p I(|X| \leq t)\right], \ M^{\mp}(p,t) = E[|X|^p sgn(X)I(|X| \leq t)] \qquad (2.3)$$

Notice that
$$M(p,t) = \tilde{M}(p,t) + t^p G(t),$$
$$M^{\pm}(p,t) = \tilde{M}^{\pm}(p,t) + t^p G^{\pm}(t). \qquad (2.4)$$

We now summarize some important relations from Hahn, Kuelbs, and Weiner (1990a), Section 2: For $p \geq 1$ and $t \geq 0$,

$$M(p,t) = \int_0^t ps^{p-1} G(s)ds, \ M^{\pm}(p,t) = \int_0^t ps^{p-1} G^{\pm}(s)ds. \qquad (2.5)$$

For $p > 0$ and $t > 0$, define normalized functions

$$m(p,t) = t^{-p} M(p,t), \ \ m^{\pm}(t,p) = t^{-p} M^{\pm}(t,p) \qquad (2.6)$$

with similar definitions for $\tilde{m}$ and $\tilde{m}^{\pm}$. As $t \to \infty$, $m(p,t) \to 0$ with similar results for the other normalized functions. Note that for $p \le q$, $m(p,t) \ge m(q,t)$ since $\frac{|x|\wedge t}{t} \le 1$. Also, for $p \ge 1$ and $t \ge 0$,

$$m(p,t) = m(p,0) - \int_0^t p\frac{\tilde{m}(p,s)}{s}ds$$

$$m^{\pm}(p,t) = m^{\pm}(p,0) - \int_0^t p\frac{\tilde{m}^{\pm}(p,s)}{s}ds,$$

$$(2.7)$$

where

$$m(p,0) = m(p,0+) = G(0) = P(X \ne 0)$$

$$m^{\pm}(p,0) = m^{\pm}(p,0+) = G^{\pm}(0) = P(X > 0) - P(X < 0).$$

$$(2.8)$$

Define

$$\Delta = \inf\{t > 0: \quad G(t) < G(0) = P(X \ne 0)\}. \tag{2.9}$$

Then, given $p > 0, m(p,\cdot)$ is strictly decreasing to 0 on (Δ,∞), and therefore has continuous inverse $d(p,\cdot)$ satisfying

$$m(p,d(p,t)) = t \quad \text{for} \quad 0 < t < P(X \ne 0). \tag{2.10}$$

When $p = 2$, $d(2,\frac{1}{n})$ provides "classical" scalings for sums in the context of the central limit theorem.

We note, as is used several times in the sequel, that for $p > 0$,

$$z_n \sim y_n \Rightarrow m(p,z_n) \sim m(p,y_n), \tag{2.11}$$

with similar a statement holding for $m^{\pm}$.

3. The Estimates. Let X be a random variable with a regularly varying, balanced tail distribution. Specifically assume

$$\exists \, \alpha > 0: \quad \forall c > 0, \quad \lim_{t \to \infty} \frac{G(ct)}{G(t)} = c^{-\alpha} \tag{3.1}$$

and

$$\exists \, \beta \in (-1, 1): \quad \lim_{t \to \infty} \frac{G^{\pm}(t)}{G(t)} = \beta. \tag{3.2}$$

In practice our results only apply when a finite upper bound for α (e.g., $\alpha < T$) is known in advance. This would be the case, for example, when F is the domain of attraction of a stable law of index α, so that $0 < \alpha \leq 2$.

To motivate the estimates we propese, choose any $p > \alpha$. A classical criterion for (3.1) is Karamata's (cf. Feller (1971)):

$$\lim_{t \to \infty} \frac{G(t)}{m(p, t)} = \frac{p - \alpha}{p}. \tag{3.3}$$

(Equivalently, $\lim_{t \to \infty} \frac{\tilde{m}(p,t)}{m(p,t)} = \frac{\alpha}{q}$.) Thus, under (3.1) and (3.2), if $p > \alpha$, $q > \alpha$, then

$$\lim_{t \to \infty} \frac{m(p, t)}{m(q, t)} = \frac{q - \alpha}{q} \frac{p}{p - \alpha}$$

$$\lim_{t \to \infty} \frac{m^{\pm}(p, t)}{m(p, t)} = \beta, \tag{3.4}$$

where the latter equation holds due to $E|X|^p = \infty$ (since $p > \alpha$) and standard arguments. To identify scales $t \to \infty$ useful for insertion into (3.4), fix $q > \alpha$ and choose a real sequence $\{r_n\}$ satisfying

$$0 < r_n \to \infty, \quad \frac{r_n}{n} \to 0. \tag{3.5}$$

Since $P(X \neq 0) > 0$, we can uniquely define for all sufficiently large n scales $a_n = a_n(q) = d(q, r_n/n)$ satisfying

$$m(q, a_n) = \frac{r_n}{n}, \tag{3.6}$$

recalling (2.10). Since $r_n/n \to 0$, $a_n \to \infty$. Then, for any $p > \alpha$, (3.4) leads to

$$\theta_n \equiv \frac{n}{r_n} m(p, a_n) = \frac{m(p, a_n)}{m(q, a_n)} \to \frac{q - \alpha}{q} \frac{p}{p - \alpha} \equiv \theta. \tag{3.7}$$

Similarly, (3.4) leads to

$$\psi_n \equiv \frac{n}{r_n} m^{\pm}(p, a_n) = \frac{m^{\pm}(p, a_n)}{m(p, a_n)} \frac{m(p, a_n)}{m(q, a_n)} \to \beta \frac{q - \alpha}{q} \frac{p}{p - \alpha} \equiv \psi. \qquad (3.8)$$

Because of the smoothness of the normalized censored functions m and $m^{\pm}$ (cf. (2.7)), it is often possible to approximate the rates of convergence in (3.7) and (3.8); we show that these rates persist when the various quantities there are replaced by their empirical versions. Then reasonable estimates of (α, β) are easily determined and analyzed by means of those for (θ, ψ). Moreover, by varying p and q, one could consider the "best" estimates for this family.

Given the random sample $\mathcal{X}_n = \{X_1, ..., X_n\}$ from F, let $F_n = \frac{1}{n} \sum_{j=1}^{n} \delta_{X_j}$ denote the empirical distribution based on $\mathcal{X}_n$. Subscript by n the empirical versions based on F_n of the various quantities introduced in Section 2. For example, particularly important will be

$$G_n(t) = \frac{1}{n} \#\{j \le n : |X_j| > t\}$$

$$m_n(q, t) = \frac{1}{n} \sum_{j=1}^{n} \frac{|X_j|^p \wedge t^p}{t^p} \qquad (3.9)$$

$$m_n^{\pm}(p, t) = \frac{1}{n} \sum_{j=1}^{n} \frac{(|X_j|^p \wedge t^p)}{t^p} sgn(X_j).$$

An estimate $\hat{a}_n$ of a_n may be defined by

$$m_n(q, \hat{a}_n) = \frac{r_n}{n}. \qquad (3.10)$$

In Hahn, Kuelbs, and Weiner (1990a), the case $q = 2$ was studied exhaustively in connection with self-normaliza- tion of censored sums and sums-of-squares. Here, for more general q but under the more specific model (3.1), the elaborate argument in Hahn, Kuelbs, and Weiner (1990a) simplifies. Thus, using the same techniques, we can provide a full self-contained development here, for convenience and completeness.

We claim that almost surely, for all sufficiently large n, (3.10) does indeed uniquely define a scale estimate $\hat{a}_n > 0$ such that $\hat{a}_n \to \infty$. Let $C > 0$. Then, by the strong law of large numbers, $m_n(q, C) \to m(q, C) > 0$, a.s. Then, again almost surely, for all sufficiently large n, $m_n(q, C) \geq \frac{1}{2}m(2, C) > r_n/n$ so that, by monotonicity of $m_n(q, \cdot)$, we will have $0 < C < \hat{a}_n$. Since C was arbitrary, the claim follows.

Fix $q > \alpha$, $p > \alpha$ $(p \neq q)$, and define the transform $T : (0, \infty) \times (-1, 1) \to \mathbf{R}^2$ by

$$(\theta, \psi) = T(\alpha, \beta) = \left(\frac{q - \alpha}{q} \frac{p}{p - \alpha}, \beta \frac{q - \alpha}{q} \frac{p}{p - \alpha} \right). \tag{3.11}$$

To estimate (θ, ψ), it is natural (in view of (3.7)-(3.8)) to propose

$$\left(\hat{\theta}_n, \hat{\psi}_n \right) = \left(\frac{n}{r_n} m_n(p, \hat{a}_n), \frac{n}{r_n} m_n^{\pm}(p, \hat{a}_n) \right). \tag{3.12}$$

Our estimate of (α, β) is defined by inversion:

$$\left(\hat{\alpha}_n, \hat{\beta}_n \right) = T^{-1} \left(\hat{\theta}_n, \hat{\psi}_n \right) = \left(pq(\hat{\theta}_n - 1)/(q\hat{\theta}_n - p), \hat{\psi}_n/\hat{\theta}_n \right). \tag{3.13}$$

We will see, via the consistency of $\hat{\theta}_n$, that $(\hat{\alpha}_n, \hat{\beta}_n)$ is well-defined with probability tending to one.

Remark: For even greater flexibiliity, one could consider the alternative estimates $\hat{\psi}_n = \frac{n}{r_n} m_n^{\pm}(r, \hat{a}_n)$, where $q \neq r > \alpha$, but this idea will not be pursued here.

4. Consistency and Asymptotic Normality. Throughout this section, fix $q > \alpha$ and $p > \alpha$ $(p \neq q)$. Choose and fix the sequence $\{r_n\}$ such that $0 < r_n \to \infty$ and $r_n/n \to 0$. Define scales a_n by (3.6), and their estimates $\hat{a}_n$ by (3.10). Assume the model (3.1)-(3.2). Define the estimators $(\hat{\alpha}_n, \hat{\beta}_n)$ by (3.11)-(3.13).

The main objective of this section is to prove that our estimates are consistent and asymptotically normal (CAN) for (α, β). The key results are contained in Theorem 4.5 and Remark 4.40, with results specialized to just α appearing in Corollary 4.42. To state the first theorem, we need a little more notation.

For $t > \alpha$ put

$$v(t) = \frac{q - \alpha}{q} \frac{t}{t - \alpha}, \tag{4.1}$$

and define

$$\Sigma = \begin{pmatrix} v(2p) & \beta v(2p) & \beta v(p + q) \\ \beta v(2p) & v(2p) & v(p + q) \\ \beta v(p + q) & v(p + q) & v(2q) \end{pmatrix}. \tag{4.2}$$

Put

$$\Pi = \begin{bmatrix} \frac{p(q-p)}{q(p-\alpha)^2} & 0 \\ \frac{\beta p(q-p)}{(p-\alpha)^2} & \frac{q-\alpha}{q} \frac{p}{p-\alpha} \end{bmatrix}. \tag{4.3}$$

Finally, define (recalling (3.7)-(3.8))

$$(\alpha_n, \beta_n) = T(\theta_n, \psi_n). \tag{4.4}$$

Our main result is

Theorem 4.5. *The estimate* $(\hat{\alpha}_n, \hat{\beta}_n)$ *is CAN for* (α, β):

$$(\hat{\alpha}_n, \hat{\beta}_n) \xrightarrow{p} (\alpha, \beta) \tag{4.6}$$

$$\sqrt{r_n}(\hat{\alpha}_n - \alpha_n) \xrightarrow{D} N(0, \Pi \begin{pmatrix} 0 & 1 & -v(p) \\ 1 & 0 & -\beta v(p) \end{pmatrix} \Sigma \begin{pmatrix} 0 & 1 & -v(p) \\ 1 & 0 & -\beta v(p) \end{pmatrix}' \Pi'). \qquad (4.7)$$

To prove Theorem 4.5, we will first establish the easier corresponding result for $(\hat{\theta}_n, \hat{\psi}_n)$, and then apply a transform lemma. The discussion in Section 3 showed that the estimates $(\hat{\theta}_n, \hat{\psi}_n)$ can be analyzed directly by adapting techniques from Hahn, Kuelbs, and Weiner (1990a).

Proposition 4.8. *The estimate* $(\hat{\theta}_n, \hat{\psi}_n)$ *is CAN for* (θ, ψ):

$$(\hat{\theta}_n, \hat{\psi}_n) \xrightarrow{P} (\theta, \psi) \qquad (4.9)$$

$$\sqrt{r_n}(\hat{\theta}_n - \theta_n, \hat{\psi}_n - \psi_n) \to N(0, \begin{pmatrix} 0 & 1 & -v(p) \\ 1 & 0 & -\beta v(p) \end{pmatrix} \Sigma \begin{pmatrix} 0 & 1 & -v(p) \\ 1 & 0 & -\beta v(p) \end{pmatrix}'). \qquad (4.10)$$

Before proving the Proposition, we note how it leads directly to Theorem 4.5, by means of the following

Lemma 4.11. *(Transform Lemma) Let* $U : A \subset \mathbf{R}^2 \to \mathbf{R}^2$ *where* A *is open. Suppose that* $U \in C^1(A)$ *and* $y_0 \in A$. *If* $y_n \to y_0$ *and* $\sqrt{r_n}(Y_n - y_n) \xrightarrow{D} N(0, \Sigma_0)$, *then*

$$\sqrt{r_n}(U(Y_n) - U(y_n)) \xrightarrow{D} N(0, dU(y_0) \Sigma_0 (dU(y_0))'). \qquad (4.12)$$

Proof: A simple application of the Mean Value Theorem applied to U at y_n.

∎

Applying the Transform Lemma and Proposition 4.8 to $(\hat{\alpha}_n, \hat{\beta}_n) = T^{-1}(\hat{\theta}_n, \hat{\psi}_n)$, we need only note that $d(T^{-1})(\theta, \psi) = (dT(\alpha, \beta))^{-1} = \Pi$ to obtain Theorem 4.5.

Turning to the proof of Proposition 4.8, note, via (2.7),

$$\sqrt{r_n}(\hat{\theta}_n - \theta_n) = \frac{n}{\sqrt{r_n}}(m_n(p, \hat{a}_n) - m(p, a_n))$$

$$= \frac{n}{\sqrt{r_n}}(m_n(p, \hat{a}_n) - m_n(p, a_n)) + \frac{n}{\sqrt{r_n}}(m_n(p, a_n) - m(p, a_n))$$

$$\equiv C_n + A_n.$$

$$(4.13)$$

Thus, the analysis of $\sqrt{r_n}(\hat{\theta}_n - \theta_n)$ is reduced to the analysis of (C_n, A_n).

Similarly,

$$\sqrt{r_n}(\hat{\psi}_n - \psi_n) = \frac{n}{\sqrt{r_n}}(m_n^{\pm}(p, \hat{a}_n) - m_n^{\pm}(p, a_n)) + \frac{n}{\sqrt{r_n}}(m_n^{\pm}(p, a_n) - m^{\pm}(p, a_n))$$

$$\equiv D_n + A_n^{\pm},$$

$$(4.14)$$

Thus, the analysis of $\sqrt{r_n}(\hat{\psi}_n - \psi_n)$ is reduced to the analysis of $(D_{n,1}, A_n^{\pm})$.

The key to the (CAN) property of $(\hat{\theta}_n, \hat{\psi}_n)$ is the following lemma which yields immediately the asymptotic behaviors of A_n and $A_n^{\pm}$. The fact that it eventually yields everything is a consequence of the later lemmas which will establish that $C_n = -v(p)B_n + o_p(1)$ and $D_n = -\beta v(p)B_n + o_p(1)$.

Lemma 4.15. *Under the above assumptions,*

$$\mathcal{Z}_n \equiv (A_n^{\pm}, A_n, B_n) \equiv$$

$$\frac{n}{\sqrt{r_n}}(m_n^{\pm}(p, a_n) - m^{\pm}(p, a_n), m_n(p, a_n) - m(p, a_n), m_n(q, a_n) - m(q, a_n))$$

$$\xrightarrow{D} N(0, \Sigma),$$

where Σ is as in (4.2).

Proof. We utilize the Cramer-Wold device: Given $a, b, c \in \mathbf{R}$, consider the sum $(a, b, c)\mathcal{Z}'_n$. The appropriate triangular array to analyze is

$$\{y_{nj} : j \leq n\} \equiv \left\{ a\frac{|X_j|^p \wedge a_n^p}{\sqrt{r_n}a_n}sgn(X_j) + b\frac{|X_j|^p \wedge a_n^p}{\sqrt{r_n}a_n^p} + c\frac{|X_j|^q \wedge a_n^q}{\sqrt{r_n}a_n^q} : j \leq n \right\},$$

which is row-wise i.i.d., with $|y_{nj}| \leq \frac{1}{\sqrt{r_n}} \to 0$. The means and variances are easy to compute:

$$\sum_{j=1}^{n} Ey_{nj} = \frac{n}{\sqrt{r_n}}(a, b, c)(m^{\pm}(p, a_n), m(p, a_n), m(q, a_n))'$$

and

$$\sigma^2_{a,b,c,n} \equiv \sum_{j=1}^{n} \text{var}(y_{nj})$$

$$= a^2\frac{n}{a_n^{2p}r_n}\text{var}\left((|X|^p \wedge a_n^p)sgn(X)\right) + b^2\frac{n}{a_n^{2p}r_n}\text{var}(|X|^p \wedge a_n^p)$$

$$+ c^2\frac{n}{a_n^{2q}r_n}\text{var}(|X|^q \wedge a_n^q) + 2ab\frac{n}{a_n^{2p}r_n}\text{cov}((|X|^p \wedge a_n^p)sgn(X), |X|^p \wedge a_n^p) \quad (4.16)$$

$$+ 2ac\frac{n}{a_n^{p+q}r_n}\text{cov}((|X|^p \wedge a_n^p)sgn(X), |X|^q \wedge a_n^q)$$

$$+ 2bc\frac{n}{a_n^{p+q}r_n}\text{cov}(|X|^p \wedge a_n^p, |X|^q \wedge a_n^q).$$

Now for $EY^2 = \infty$, $(E(|Y| \wedge t))^2 = o(E(Y^2 \wedge t^2))$ as $t \to \infty$. Thus for $r > \alpha$, $M(r, a_n)^2 = o(M(2r, a_n))$. Also, note (via (3.6))

$$\frac{nM(p, a_n)M(q, a_n)}{r_n a_n^{p+q}} = \frac{n}{r_n}m(q, a_n)m(p, a_n) = m(p, a_n) \to 0.$$

Thus, utilizing the formulas $\text{var}Y = EY^2 - (EY)^2$ and $\text{cov}(Y_1, Y_2,) =$

$E(Y_1, Y_2) - EY_1 EY_2$, we have

$$\frac{n}{a_n^{2p} r_n} var((|X|^p \wedge a_n^p) sgn(X)) \sim \frac{n}{r_n} m(2p, a_n) = \frac{m(2p, a_n)}{m(q, a_n)} \to v(2p)$$

$$\frac{n}{a_n^{2p} r_n} var(|X|^p \wedge a_n^p) \sim \frac{n}{r_n} m(2p, a_n) = \frac{m(2p, a_n)}{m(q, a_n)} \to v(2p)$$

$$\frac{n}{a_n^{2q} r_n} var(|X|^q \wedge a_n^q) \sim \frac{n}{r_n} m(2q, a_n) = \frac{m(2q, a_n)}{m(q, a_n)} \to v(2q) \qquad (4.17)$$

$$\frac{n}{a_n^{2p} r_n} cov((|X|^p \wedge a_n^p) sgn(X), |X|^p \wedge a_n^p) \sim \frac{n}{r_n} m^{\pm}(2p, a_n) \to \beta v(2p)$$

$$\frac{n}{a_n^{p+q} r_n} cov((|X|^p \wedge a_n^p) sgn(X), |X|^q \wedge a_n^q) \sim \frac{n}{r_n} m^{\pm}(p+q, a_n) \to \beta v(p+q)$$

$$\frac{n}{a_n^{p+q} r_n} cov((|X|^p \wedge a_n^p), |X|^q \wedge a_n^q) \sim \frac{n}{r_n} m(p+q, a_n) \to v(p+q).$$

It follows that

$$\mathcal{L}((a, b, c)\mathcal{Z}_n' - E(a, b, c)\mathcal{Z}_n') = \mathcal{L}\left(\sum_{j=1}^n (y_{nj} - Ey_{nj})\right)$$

$$\to \mathcal{N}(0, a^2 v(2p) + b^2 v(2p) + c^2 v(2q) + 2ab\beta v(2p) + 2ac\beta v(p+q) + 2bcv(p+q))$$

$$= \mathcal{N}(0, (a, b, c)\Sigma(a, b, c)').$$

Thus, the lemma is proved. ∎

In particular,

$$B_n \equiv \frac{n}{\sqrt{r_n}}(m_n(q, a_n) - m(q, a_n)) \xrightarrow{D} \mathcal{N}(0, v(2q)). \qquad (4.18)$$

The analysis of C_n, D_n, and B_n is facilitated by converting to convenient integral representations.

$$B_n = \frac{n}{\sqrt{r_n}}(m_n(q, a_n) - m_n(q, \hat{a}_n)) \quad \text{(by (3.6) and (3.10))}$$

$$= \frac{n}{\sqrt{r_n}} \int_{a_n}^{\hat{a}_n} q\tilde{m}_n(q, s)\frac{ds}{s} \quad \text{(by (2.7))} \tag{4.19}$$

$$= \int_0^{(\frac{\hat{a}_n}{a_n}-1)\sqrt{r_n}} q\frac{n}{r_n}\tilde{m}_n\left(q, a_n\left(1 + \frac{u}{\sqrt{r_n}}\right)\right) \frac{du}{1 + \frac{u}{\sqrt{r_n}}},$$

$$C_n = \frac{n}{\sqrt{r_n}}(m_n(p, \hat{a}_n) - m_n(p, a_n))$$

$$= \frac{n}{\sqrt{r_n}} \int_{\hat{a}_n}^{a_n} p\tilde{m}_n(p, s)\frac{ds}{s} \tag{4.20}$$

$$= -\int_0^{(\frac{\hat{a}_n}{a_n}-1)\sqrt{r_n}} p\frac{n}{r_n}\tilde{m}_n(p, a_n(1 + \frac{u}{\sqrt{r_n}})) \frac{du}{1 + \frac{u}{\sqrt{r_n}}},$$

$$D_n = \frac{n}{\sqrt{r_n}}(m_n^{\pm}(p, \hat{a}_n) - m_n^{\pm}(p, a_n))$$

$$= -\frac{n}{\sqrt{r_n}} \int_{a_n}^{\hat{a}_n} p\tilde{m}_n^{\pm}(p, s)\frac{ds}{s} \tag{4.21}$$

$$= -\int_0^{(\frac{\hat{a}_n}{a_n}-1)\sqrt{r_n}} p\frac{n}{r_n}p\tilde{m}_n^{\pm}\left(p, a_n\left(1 + \frac{u}{\sqrt{r_n}}\right)\right) \frac{du}{1 + \frac{u}{\sqrt{r_n}}}.$$

If $\left\{\left(\frac{\hat{a}_n}{a_n} - 1\right)\sqrt{r_n}\right\}$ were tight, then the functions in the integrands which are constructed from the empirical distribution function could be replaced by their deterministic analogues as the following lemma indicates.

Lemma 4.22. Let $\rho_n \to \infty$ such that $\rho_n/\sqrt{r_n} \to 0$. Then for $s > \alpha$,

$$E \int_{-\rho_n}^{\rho_n} \left|\frac{n}{r_n}\left\{\tilde{m}_n\left(s, a_n\left(1 + \frac{u}{\sqrt{r_n}}\right)\right) - \tilde{m}\left(s, a_n\left(1 + \frac{u}{\sqrt{r_n}}\right)\right)\right\}\right| \frac{du}{1 + \frac{u}{\sqrt{r_n}}} \to 0. \tag{4.23}$$

Also, the same statement holds with $\tilde{m}^{\pm}$ replacing $\tilde{m}$.

Proof. First consider (4.23). By Fubini's Theorem, for all large n,

$$E \int_{-\rho_n}^{\rho_n} \left| \frac{n}{r_n} \left\{ \tilde{m}_n \left(s, a_n \left(1 + \frac{u}{\sqrt{r_n}} \right) \right) - \tilde{m} \left(s, a_n \left(1 + \frac{u}{\sqrt{r_n}} \right) \right) \right\} \frac{du}{1 + \frac{u}{\sqrt{r_n}}} \right|$$

$$= \int_{-\rho_n}^{\rho_n} \frac{n}{r_n} E \left| \tilde{m}_n \left(s, a_n \left(1 + \frac{u}{\sqrt{r_n}} \right) \right) - \tilde{m} \left(s, a_n \left(1 + \frac{u}{\sqrt{r_n}} \right) \right) \right| \frac{du}{1 + \frac{u}{\sqrt{r_n}}}$$

$$\leq \int_{-\rho_n}^{\rho_n} \frac{n}{r_n} \sqrt{var \, \tilde{m}_n \left(s, a_n \left(1 + \frac{u}{\sqrt{r_n}} \right) \right)} \frac{du}{1 + \frac{u}{\sqrt{r_n}}} \quad \text{by Cauchy} - \text{Schwarz}$$

$$\leq \int_{-\rho_n}^{\rho_n} \frac{2}{r_n} \sqrt{n\tilde{m} \left(2s, a_n \left(1 + \frac{u}{\sqrt{r_n}} \right) \right)} \, du$$

$$\leq \int_{-\rho_n}^{\rho_n} \frac{2}{r_n} \sqrt{nm \left(2s, a_n \left(1 + \frac{u}{\sqrt{r_n}} \right) \right)} \, du$$

$$\leq \int_{-\rho_n}^{\rho_n} \frac{2}{r_n} \sqrt{nm \left(s, a_n \left(1 + \frac{u}{\sqrt{r_n}} \right) \right)} \, du \qquad (4.24)$$

$$\leq \frac{2}{r_n} \int_{-\rho_n}^{\rho_n} \sqrt{nm \left(s, a_n \frac{(1 - \rho_n)}{\sqrt{r_n}} \right)} \, du \quad \text{since } m(s, \cdot) \text{ decreases}$$

$$= \frac{4\rho_n}{r_n} \sqrt{nm \left(s, a_n \frac{(1 - \rho_n)}{\sqrt{r_n}} \right)}$$

$$\sim \frac{4\rho_n}{r_n} \sqrt{nm(s, a_n)} \text{ (via (2.11))}$$

$$\sim \frac{4\rho_n}{\sqrt{r_n}} \sqrt{\frac{m(s, a_n)}{m(q, a_n)}} \text{ (by (3.6))}$$

$$\sim \frac{4\rho_n}{\sqrt{r_n}} v(s) \text{ (by (3.4) and (4.4))}$$

$$\longrightarrow 0,$$

since $\rho_n = o(\sqrt{r_n})$. A similar calculation establishes (4.23) with $\tilde{m}$ replaced by $\tilde{m}^{\pm}$. Thus the Lemma is proved. ∎

The above lemma is required several times, including in the proof of tightness of $\left\{\left(\frac{\hat{a}_n}{a_n} - 1\right)\sqrt{r_n}\right\}$.

Lemma 4.25. $\left\{\left(\frac{\hat{a}_n}{a_n} - 1\right)\sqrt{r_n}\right\}$ *is tight.*

Proof. Given $\rho_n \to \infty$ such that $\rho_n = o(\sqrt{r_n})$, note that since $\tilde{m}_n(q, \cdot) \geq 0$, if $\left(\frac{\hat{a}_n}{a_n} - 1\right)\sqrt{r_n} \geq \rho_n$, then (4.14) leads to

$$
\begin{aligned}
B_n &\geq \int_0^{\rho_n} q\frac{n}{r_n}\tilde{m}_n\left(q, a_n\left(1 + \frac{u}{\sqrt{r_n}}\right)\right)\frac{du}{1 + \frac{u}{\sqrt{r_n}}} \\
&= \int_0^{\rho_n} q\frac{n}{r_n}\tilde{m}\left(q, a_n\left(1 + \frac{u}{\sqrt{r_n}}\right)\right)\frac{du}{1 + \frac{u}{\sqrt{r_n}}} \\
&\quad + \int_0^{\rho_n} q\frac{n}{r_n}\left\{\tilde{m}_n\left(q, a_n\left(1 + \frac{u}{\sqrt{r_n}}\right)\right) - \tilde{m}\left(q, a_n\left(1 + \frac{u}{\sqrt{r_n}}\right)\right)\right\}\frac{du}{1 + \frac{u}{\sqrt{r_n}}} \\
&\equiv B_{n,1} + B_{n,2}.
\end{aligned}
$$

But

$$
B_{n,2} \xrightarrow{p} 0, \tag{4.26}
$$

due to Markov's inequality and (4.23). Moreover, from (3.3) we get $\tilde{m}(q, t)/m(q, t) \to \frac{\alpha}{q}$ as $t \to \infty$, so that

$$
\begin{aligned}
B_{n,1} &= \int_0^{\rho_n} q\frac{n}{r_n}\tilde{m}\left(q, a_n\left(1 + \frac{u}{\sqrt{r_n}}\right)\right)\frac{du}{1 + \frac{u}{\sqrt{r_n}}} \\
&\geq \int_0^{\rho_n} q\frac{n}{r_n}\frac{\tilde{M}(q, a_n)}{a_n^q\left(1 + \frac{u}{\sqrt{r_n}}\right)^q}\frac{du}{1 + \frac{u}{\sqrt{r_n}}} \quad \text{(since $\tilde{m}(q, \cdot)$ is nondecreasing)} \\
&\qquad\qquad\qquad\qquad\qquad\qquad\qquad\qquad\qquad\qquad\qquad\qquad\qquad\qquad (4.27) \\
&\geq \frac{1}{2}\int_0^{\rho_n} q\frac{n}{r_n}\tilde{m}(q, a_n)\,du \\
&\sim \frac{1}{2}\rho_n\alpha \to \infty.
\end{aligned}
$$

Thus,

$$
P\left(\left(\frac{\hat{a}_n}{a_n} - 1\right)\sqrt{r_n} \geq \rho_n\right) \leq P(B_n \geq B_{n,1} + B_{n,2}) \leq P\left(B_n \geq \frac{1}{3}\rho_n\alpha + o_p(1)\right) \to 0,
$$

since $\{B_n\}$ is tight by Lemma 4.25. A similar argument also shows that

$$P((\frac{\hat{a}_n}{a_n} - 1)\sqrt{r_n} \le -\rho_n) \to 0$$

for any $\rho_n \to \infty$ with $\rho_n = o(\sqrt{r_n})$. Thus, the lemma is proved. ∎

The next lemma shows that the deterministic versions of the integrands in (4.19)-(4.21) converge uniformly on $[-\rho_n, \rho_n]$.

Lemma 4.28. *Choose $\rho_n \to \infty$ such that $\rho_n/\sqrt{r_n} \to 0$. Let $U_n = [-\rho_n, \rho_n]$. Then*

$$\sup_{U_n} \left| \frac{n}{r_n} \tilde{m}(q, a_n \left(1 + \frac{u}{\sqrt{r_n}}\right)) - \frac{\alpha}{q} \right| \to 0 \tag{4.29}$$

$$\sup_{U_n} \left| \frac{n}{r_n} \tilde{m}(p, a_n \left(1 + \frac{u}{\sqrt{r_n}}\right)) - \frac{\alpha}{p} \frac{q-\alpha}{q} \frac{p}{p-\alpha} \right| \to 0 \tag{4.30}$$

$$\sup_{U_n} \left| \frac{n}{r_n} \tilde{m}^{\pm}(p, a_n \left(1 + \frac{u}{\sqrt{r_n}}\right)) - \frac{\alpha}{p} \beta \frac{q-\alpha}{q} \frac{p}{p-\alpha} \right| \to 0. \tag{4.31}$$

Proof. Again, we prove only (4.29) since the other two proofs are analogous. For $-\rho_n \le u \le \rho_n$, note that

$$\frac{n}{r_n} \tilde{m}(q, a_n \left(1 + \frac{u}{\sqrt{r_n}}\right)) \le \frac{n}{r_n} \frac{\tilde{M}(q, a_n \left(1 + \frac{\rho_n}{\sqrt{r_n}}\right))}{a_n^q \left(1 - \frac{\rho_n}{\sqrt{r_n}}\right)^q} \quad \text{(by monotonicity of } \tilde{M}(q, \cdot))$$

$$= \frac{n}{r_n} \tilde{m} \left(q, a_n \left(1 + \frac{\rho_n}{\sqrt{r_n}}\right)\right) \left(1 + \frac{\rho_n}{\sqrt{r_n}}\right)^q \left(1 - \frac{\rho_n}{\sqrt{r_n}}\right)^{-q}$$

$$\sim \frac{\alpha}{q} \frac{n}{r_n} m(q, a_n \left(1 + \frac{\rho_n}{\sqrt{r_n}}\right)) \quad \text{by (3.3) and } \rho_n/\sqrt{r_n} \to 0$$

$$\sim \frac{\alpha}{q} \frac{n}{r_n} m(q, a_n) \quad \text{by (2.11)}$$

$$= \frac{\alpha}{q} \quad \text{by (3.6).}$$

Similarly,

$$\frac{n}{r_n}\tilde{m}(q, a_n\left(1 + \frac{u}{\sqrt{r_n}}\right)) \geq \frac{n}{r_n}\tilde{m}(q, a_n\left(1 - \frac{\rho_n}{\sqrt{r_n}}\right))\left(1 - \frac{\rho_n}{\sqrt{r_n}}\right)^q\left(1 + \frac{\rho_n}{\sqrt{r_n}}\right)^{-q}$$
$$\sim \frac{\alpha}{q},$$

so that

$$\frac{n}{r_n}\tilde{m}(q, a_n\left(1 + \frac{u}{\sqrt{r_n}}\right)) \to \frac{\alpha}{q} \quad \text{uniformly on} \quad [-\rho_n, \rho_n]. \quad \blacksquare$$

Now we can relate C_n, D_n, and B_n.

Lemma 4.32.

$$B_n = \alpha\left(\frac{\hat{a}_n}{a_n} - 1\right)\sqrt{r_n} + o_p(1) \tag{4.33}$$

$$C_n = -\alpha v(p)\left(\frac{\hat{a}_n}{a_n} - 1\right)\sqrt{r_n} + o_p(1) = -v(p)B_n + o_p(1) \tag{4.34}$$

$$D_n = \beta\alpha v(p)\left(\frac{\hat{a}_n}{a_n} - 1\right)\sqrt{r_n} + o_p(1) = -\beta v(p)B_n + o_p(1). \tag{4.35}$$

Proof. Fixing $\rho_n \to \infty$ with $\rho_n/\sqrt{r_n} \to 0$, Lemma 4.25 allows (4.19) to be rewritten as

$$B_n = \int_0^{(|\frac{\hat{a}_n}{a_n} - 1|\sqrt{r_n})\wedge\rho_n)sgn(\frac{\hat{a}_n}{a_n} - 1)} q\frac{n}{r_n}\tilde{m}_n(q, a_n\left(1 + \frac{u}{\sqrt{r_n}}\right))\frac{du}{1 + \frac{u}{\sqrt{r_n}}} + o_p(1)$$

$$= \int_0^{(|\frac{\hat{a}_n}{a_n} - 1|\sqrt{r_n})\wedge\rho_n)sgn(\frac{\hat{a}_n}{a_n} - 1)} q\frac{n}{r_n}\tilde{m}(q, a_n\left(1 + \frac{u}{\sqrt{r_n}}\right))\frac{du}{1 + \frac{u}{\sqrt{r_n}}} + o_p(1)$$

by Lemma 4.22

$$= \alpha\left(\frac{\hat{a}_n}{a_n} - 1\right)\sqrt{r_n} + o_p(1) \quad \text{by (4.28).}$$

Thus, (4.33) holds.

By (4.20) and Lemma 4.22,

$$C_n = -1 \int_0^{(\frac{\hat{a}_n}{a_n}-1)\sqrt{r_n}} \frac{n}{r_n} p\tilde{m}(p, a_n(1 + \frac{u}{\sqrt{r_n}})) \; du + o_p(1)$$

$$= -\alpha v(p) \left(\frac{\hat{a}_n}{a_n} - 1 \right) \sqrt{r_n} + o_p(1) \quad \text{by (4.30)}$$

$$= -v(p)B_n + o_p(1).$$

Finally, (4.21) and Lemma 4.22 yield

$$D_n = -\int_0^{(\frac{\hat{a}_n}{a_n}-1)\sqrt{r_n}} \frac{n}{r_n} p\tilde{m}^{\pm}(p, a_n(1 + \frac{u}{\sqrt{r_n}})) \; du + o_p(1)$$

$$= -\beta v(p)\alpha \left(\frac{\hat{a}_n}{a_n} - 1 \right) \sqrt{r_n} + o_p(1)$$

$$= -\beta v(p)B_n + o_p(1),$$

and the lemma is proved. ∎

To complete the proof of Proposition 4.8, note that by (4.13) and Lemma 4.32

$$\sqrt{r_n}(\hat{\theta}_n - \theta_n) = C_n + A_n = -v(p)B_n + A_n + o_p(1)$$
$$= (0, 1, -v(p))Z'_n + o_p(1), \tag{4.36}$$

while (4.14) and the same lemma yield

$$\sqrt{r_n}(\hat{\psi}_n - \psi_n) = D_n + A_n^{\pm} = -\beta v(p)B_n + o_p(1)$$
$$= (1, 0, -\beta v(p))Z'_n + o_p(1). \tag{4.37}$$

Then

$$\sqrt{r_n}(\hat{\theta}_n - \theta_n, \hat{\psi}_n - \psi_n) = \begin{pmatrix} 0 & 1 & -v(p) \\ 1 & 0 & -\beta v(p) \end{pmatrix} Z'_n + o_p(1), \tag{4.38}$$

and so an application of Lemma 4.15 establishes the asymptotic normality assertion (4.10).

For consistency, simply note that by (3.7) and (3.8), $\theta_n \to \theta$ and $\psi_n \to \psi$. Thus

$$(\hat{\theta}_n - \theta_n), \hat{\psi}_n - \psi) = (\theta_n - \theta_n, \psi_n - \psi) + \frac{\sqrt{r_n}(\hat{\theta}_n - \theta_n, \hat{\psi}_n - \psi)}{\sqrt{r_n}} \qquad (4.39)$$
$$= o(1) + O_p(1/\sqrt{r_n}) \xrightarrow{p} 0,$$

thereby completing the proof of Proposition 4.8 (and hence also that of Theorem 4.5). ∎

Remark (4.40). When applying Theorem 4.5 in constructing confidence intervals for the parameter (α, β), it is desirable to have an empirically determinable asymptotic covariance. Since the estimate (α, β) is consistent, inspection of (4.1)–(4.3) and Theorem 4.5 makes it obvious that, because of the continuity of the covariances in (α, β) there, if (α, β) is replaced in these covariances by $(\hat{\alpha}_n, \hat{\beta}_n)$, the resulting empirical covariances will be consistent. By assumption, $|\beta| \neq 1$. Thus, it follows that these resulting empirical covariances eventually have a negative square root $\hat{\Gamma}_n$. In short, there are naturally defined empirical matrices $\hat{\Gamma}_n$ and a deterministic matrix Γ such that $\hat{\Gamma}_n \xrightarrow{p} \Gamma$ and

$$\sqrt{r_n}\, \hat{\Gamma}_n(\hat{\alpha}_n - \alpha_n, \hat{\beta}_n - \beta_n) \to \mathcal{N}(0, I), \qquad (4.41)$$

where I is the 2×2 identity. (This is the only place where $|\beta| \neq 1$ was actually used.)

Clearly the marginal behavior of $\hat{\alpha}_n$ in estimating α does not depend on the balance condition (3.2). Let $\hat{\Gamma}_n$, Γ be as above.

Corollary 4.42. *Assume only (3.1) and fix $p > \alpha$, $q > \alpha$. Then the estimate $\hat{\alpha}_n$ is CAN for α:*

$$\hat{\alpha}_n \xrightarrow{p} \alpha \qquad (4.43)$$

$$\sqrt{r_n}\hat{c}_n(\hat{\alpha}_n - \alpha) \xrightarrow{D} \mathcal{N}(0,1) \qquad (4.44)$$

where

$$\hat{c}_n \equiv (1,0)\hat{\Gamma}_n(1,0)' \xrightarrow{p} (1,0)\Gamma(1,0)' \equiv c. \qquad (4.45)$$

Asymptotic normality for the well-known Hill estimate of α has been established only under assumptions more stringent than (3.1). Consequently, Corollary 4.42 establishes that the estimate $\hat{\alpha}_n$ of α might, in some respects, have broader applicability than the Hill estimate.

Corollary 4.42 makes evaluating the Asymptotic Mean Square Error (AMSE) of $\hat{\alpha}_n$ easy: Let $g(x) = \frac{q-x}{q}\frac{p}{p-x}$. Then, separating variance and bias, and letting c_1, c_2 be the appropriate positive finite constants, we see that

$$AMSE(\hat{\alpha}_n) \sim c_1^2 r_n + (g^{-1}(\theta_n) - g^{-1}(\theta))^2$$
$$\sim c_1^2 r_n + c_2^2(\theta_n - \theta)^2 \qquad (4.46)$$
$$= c_1^2 r_n + c_2^2 \left\{ \frac{n}{r_n} m(p, a_n) - \frac{q-\alpha}{q}\frac{p}{p-\alpha} \right\}^2.$$

Therefore, the main effort in evaluating $AMSE(\hat{\alpha}_n)$ is the bias term, which is related to the rate of convergence in the Karamata relation (3.4). This rate is faster when $\{r_n\}$ grows more rapidly, but then the accompanying variance term which is proportional to r_n is inflated. The typical problem of choosing $\{r_n\}$ to balance variance/bias in $AMSE(\hat{\alpha}_n)$ is illustrated, for a particular submodel, in Section 5.

5. Optimal Rates in the Hall-Welsh Model. Restrict attention to the sub-model of (3.1) introduced by Hall and Welsh (1984). The estimate $\hat{\alpha}_n$ for α has already been shown to have CAN behavior universally throughout this large model. The purpose of this section is to show that $\hat{\alpha}_n$ can achieve an optimal rate of convergence for this model. These two properties recommend the estimate $\hat{\alpha}_n$ as an alternative alongside other well-known estimates.

Given positive constants α_0, C_0, ε, ρ and A, let $\mathcal{D} = \mathcal{D}(\alpha_0, C_0, \varepsilon, \rho, A)$ denote the set of all Lebesgue densities f on the positive half-line satisfying

$$f(x) = C\alpha x^{-\alpha-1}\{1 + r(x)\} \qquad (5.1)$$

where

$$|r(x)| \le Ax^\gamma, \ \gamma = \rho\alpha \qquad (5.2)$$

for some α and C such that $|\alpha - \alpha_0| < \varepsilon$, $|C - C_0| < \varepsilon$. Hall and Welsh proved that if for some estimate $\tilde{\alpha}_n$ of α and some $\alpha_0, C_0, \varepsilon, \rho, A$, we have

$$\liminf_{n \to \infty} \inf_{f \in \mathcal{D}} P_f(|\tilde{\alpha}_n - \alpha| \le \tau_n) = 1, \qquad (5.3)$$

then

$$\liminf_{n \to \infty} n^{\rho/(2\rho+1)} \tau_n = \infty. \qquad (5.4)$$

Thus, to demonstrate that $\hat{\alpha}_n$ can achieve an optimal rate of convergence to α in this sense, it is enought to find a class $\mathcal{D}$ and one version of $\hat{\alpha}_n$ such that (5.3) holds with $\tilde{\alpha}_n = \hat{\alpha}_n$ whenever $\{\tau_n\}$ satisfies (5.4).

Fix $0 < T < \infty$. Choose $\alpha_0 < T$ and then $\varepsilon > 0$ such that $\alpha_0 + \varepsilon < T$. Then choose $\rho > 0$ so that $\alpha(1 + \rho) < T$. Fix $C_0 > \varepsilon$, $A > 0$. Let $\mathcal{D} = \mathcal{D}(\alpha_0, C_0, \varepsilon, \rho, A)$

for these choices. Fix $p > \alpha$ and $q > \alpha$ $(p \neq q)$. We will show how to choose $\{r_n\}$ so that (5.3) holds with $\tilde{\alpha}_n = \hat{\alpha}_n$ whenever $\{\tau_n\}$ satisfies (5.4).

The choice of $\{r_n\}$ is motivated by (4.46). Specifically, we should arrange that $AMSE(\hat{\alpha}_n)$ be negligible compared to τ_n^2 for any τ_n satisfying (5.4). Since the variance contribution is on the order of $\frac{1}{r_n}$, clearly r_n should be of no smaller order than $n^{2\rho/(2\rho+1)}$. But, direct computations show that uniformly in $f \in \mathcal{D}$,

$$m(q,t) = C \frac{q}{q - \alpha} t^{-\alpha} \left\{1 + O(t^{-\gamma})\right\} \tag{5.5}$$

(here is where $q > \alpha + \gamma = \alpha(1 + \rho)$ is needed). It follows, again uniformly, that

$$a_n \sim \left(\frac{n}{r_n}\right)^{1/\alpha} \left(\frac{Cq}{q - \alpha}\right)^{1/\alpha} \tag{5.6}$$

and thus

$$\theta_n = \frac{n}{r_n} m(p, a_n) = \frac{q - \alpha}{q} \frac{p}{p - \alpha} \left\{1 + O\left(\frac{r_n}{n}\right)^\rho\right\}$$
$$= \theta + O\left(\left(\frac{r_n}{n}\right)^\rho\right) \tag{5.7}$$

(using $p > \alpha(1+\rho)$). Thus, uniformly in $f \in \mathcal{D}$, the squared bias term in $AMSE(\hat{\alpha}_n)$ in (4.46) has order no greater than $(r_n/n)^{2\rho}$. Setting the components of $AMSE(\hat{\alpha}_n)$ roughly equal, $(r_n/n)^{2\rho} \approx 1/r_n$, leads to the choice of $r_n \approx n^{2\rho/(1+2\rho)}$.

So choose $r_n = n^{2\rho/(1+2\rho)}$. Fix any $\{\tau_n\}$ satisfying (5.4); thus $\tau_n/\sqrt{r_n} \to \infty$. We claim that

$$\{\mathcal{L}_f(\sqrt{r_n}(\hat{\alpha}_n - \alpha_n)) : \quad f \in \mathcal{D}, \ n \geq 1\} \text{ is tight.} \tag{5.8}$$

This follows straightforwardly, albeit quite tediously, by verifying that in the proofs leading to Lemma 4.25, the important bounding quantities such as expectations and variances are bounded uniformly for $f \in \mathcal{D}$. The main ingredient is the easily verified uniformity of (3.1) for $f \in \mathcal{D}$, and hence also the uniformity of the first limit in (3.4). The uninstructive details have been omitted.

To deduce (5.3) from (5.5), it remains only to show that

$$\{\sqrt{r_n}(\alpha_n - \alpha): \quad f \in \mathcal{D}\} \quad \text{is uniformly bounded} \tag{5.9}$$

(recall α_n depends on f), that is, the bias terms in $AMSE(\hat{\alpha}_n)$ should remain uniformly bounded. But the uniformity in (5.7), combined with our choice of $r_n = n^{2\rho/(1+2\rho)}$, shows that (again, uniformly in $f \in \mathcal{D}$)

$$\sqrt{r_n}(\alpha_n - \alpha) \approx \sqrt{r_n}(\theta_n - \theta) \approx \sqrt{r_n}\left(\frac{r_n}{n}\right)^{\rho} = 1, \tag{5.10}$$

completing the verification of (5.3) for this choice of $\mathcal{D}$ and estimates $\hat{\alpha}_n$.

References

Csorgo, S., P. Deheuvels, and D. M. Mason (1985). Kernel estimates of the tail index of a distribution. *Ann. Statist.* **13**, 1050-1077.

Davis, R. and S. I. Resnick (1984). Tail estimates motivated by extreme value theory. *Ann. Statist.* **12**, 1467-1487.

Feller, W. (1971). *An Introduction to Probability Theory and its Applications*, Vol. 2. Wiley, New York.

Hahn, M. G., J. Kuelbs, and D. C. Weiner (1990a). The asymptotic joint distribution of self-normalized censored sums and sums-of-squares. *Ann. Probab* **18**, 1284-1341.

Hahn, M. G., J. Kuelbs, and D. C. Weiner (1990b). Asymptotic behavior of partial sums via trimming and self-normalization. In this volume.

Hall, P. and A. H. Welsh (1984). Best attainable rates of convergence for estimates of parameters of regular variation. *Ann. Statist.* **12**, 1079-1084.

Marjorie G. Hahn   Daniel C. Weiner
Department of Mathematics   Department of Mathematics
Tufts University   Boston University
Medford, MA 02155 USA   Boston, MA 02215 USA

CENTER, SCALE AND ASYMPTOTIC NORMALITY
FOR CENSORED SUMS OF
INDEPENDENT, NONIDENTICALLY DISTRIBUTED
RANDOM VARIABLES

Daniel Charles Weiner*

1. **Introduction.** This paper develops a new approach (called the Censored Centered Method, or CCM) to the problem of determining centers and scales for distributions, and features applications to asymptotic normality for censored sums of independent, generally nonidentically distributed random variables.

The development of this new approach to affine normalization was originally prompted by the need to refine the classical criteria for asymptotic normality of the centered and scaled sums (cf. Weiner (1990)). When these criteria fail, the possibility of asymptotic normality is prevented by the presence of summands of "excessive" size. Hence the question arises of how to obtain a more robust theory of asymptotic normality by somehow restricting or modifying the size of these extreme summands. Typically one restricts a large but relatively small number of these extremes ("intermediate" modification).

In the case of identically distributed summands, several approaches based on intermediate modification of the *magnitudes* of the extreme have been considered. Let us focus on two of these magnitude-based approaches in particular (Hahn and Kuelbs (1988), Hahn, Kuelbs and Weiner (1989)), where (among other results) deterministic levels tending to infinity were constructed whereat a prescribed number of

---

Supported in part by NSF grants DMS-86-03188 and DMS-88-96217.

leading summands were truncated $(x \to xI(|x| \le t))$ or censored $(x \to (|x| \wedge t)sgn(x))$ in magnitude. The resulting truncated or censored sums are always asymptotically normal, with centerings and scalings readily computed directly from the constructed truncation/censoring levels, regardless of the original distribution. (In the censored case empiricalized versions of their results are also obtained.)

In both of these approaches, success for their constructions depends in part on relations such as

$$(E(XI(|X| \le t)))^2 = o(E(X^2 I(|X| \le t)))$$

$$(E(|X| \wedge t))^2 = o(E(X^2 \wedge t^2)) \quad (t \to \infty), \tag{1.1}$$

which hold whenever $EX^2 = \infty$. These relations allow essential variance computations to be replaced by second moment computations based on the better-behaved and technically more convenient truncated/censored second moment functions. Moreover, these relations also show that in ranking summands by magnitude, i.e., distance from 0, it is reasonable to use 0 as the unchanging center, independent of the number of summands, a large departure from which justifies/requires modification of the offending summand.

Here, we seek to extend the deterministic censored results contained in Hahn, Kuelbs and Weiner (1989) to the case of nonidentically distributed summands.

Now in this general case, (1.1) may not hold uniformly in the given independent sequence. Thus, construction of suitable censoring levels is considerably complicated. Moreover, real (i.e., asymptotic-normality-preventing) extremity in the summands may not be appropriately determined by distance from the unchanging center 0, but rather from moving centers. (Consider an example $\{X_j - c_j\}$, where $\{X_j\}$ are symmetric, i.i.d., but $c_j \to \infty$ rapidly.) But use of the next "obvious" choice of a median of X_j as the center (e.g., Hahn and Klass (1981)), a large departure from which would constitute an "extremity," can be unsuitable in the general asymmetric case due to the difficulty in reconciling such centers to the necessary variance computations. In particular, after censoring summands about their medians, further recentering of the resulting sums would be required. (See the discussion in Hahn and Klass (1981), and also a related but distinct solution to this problem for uncensored sums in Weiner (1990).)

Here we develop a technique whereby suitable centerings, censoring levels, and scalings, are simultaneously constructed, based on a prescribed maximum number of leading terms to be censored. Under a minimal condition (without which the unscaled sums already converge), the resulting scaled sums of centered, censored summands are always asymptotically normal, and without further centering (Corollary 5.18). Thus, the centers constructed serve at once as centers of the intervals for the censoring, and as centering constants for the summands within the sum.

The author wishes to thank Professors J. Kuelbs and M. Hahn for many useful conversations, and for their steady encouragement. He wishes in addition to thank Renata Cioczek for corrections connected with (3.33) and (3.44) in an earlier version of this article.

Organization and Results

In Section 2 the necessary censored moment functions are introduced and analyzed.

In Section 3, the technical heart of the paper, the basic centering function γ is constructed and analyzed. The main idea is to arrange the center so that the resulting censored precentered variable has mean zero, viz.,

$$E[(|X - \gamma(t)| \wedge t)sgn(X - \gamma(t))] = 0, \tag{1.2}$$

for each censoring level $t > 0$. This is where the smoother nature of censoring (vis à vis truncation) is exploited to obtain the necessary (and computationally invaluable) smoothness properties of the corresponding precentered censored moment functions. The relation of γ to ordinary medians is also discussed. These results are summarized in Theorem 3.45.

In Section 4 the desired censoring levels, centerings, and scalings for the sums of censored precentered summands are constructed.

Our main results on asymptotic normality of censored sums are given in Section 5, specifically Theorem 5.6 and Corollary 5.18.

Finally, in Section 6, examples are presented to illustrate and illuminate the techniques and results.

2. Censoring and Associated Moment Functions. In this section we will introduce the censoring and truncating operations and investigate the properties of the associated moment functions.

Let X be a nondegenerate, real random variable with distribution function F.

We define the *censor* $c : [0, \infty) \times (-\infty, \infty) \to (-\infty, \infty)$ by

$$
\begin{aligned}
c(t, x) &= xI(|x| \le t) + t \text{ sgn } (x) \ I(|x| > t) \\
&= (|x| \wedge t) \text{ sgn } (x),
\end{aligned}
\tag{2.1}
$$

and the *truncator* $\tilde{c} : [0, \infty) \times (-\infty, \infty) \to (-\infty, \infty)$ by

$$
\tilde{c}(t, x) = xI(|x| \le t).
\tag{2.2}
$$

Define the censored and truncated expectation functions $e : [0, \infty)$

$\times(-\infty,\infty) \to (-\infty,\infty)$ and $\tilde{e} : [0,\infty) \times (-\infty,\infty) \to (-\infty,\infty)$ by

$$e(t,y) = Ec(t, X - y)$$

$$= E(X-y)I(|X-y| \le t) + t\{P(X-y > t) - P(X-y < -t)\} \tag{2.3}$$

and

$$\tilde{e}(t,y) = E\tilde{c}(t, X-y) = E(X-y)I(|X-y| \le t). \tag{2.4}$$

Clearly $|e(t,y)| \le t$, and furthermore

$$\lim_{y \to -\infty} e(t,y) = t, \quad \lim_{y \to \infty} e(t,y) = -t. \tag{2.5}$$

Similar facts hold for $\tilde{e}$, but the corresponding limits in (2.5) are both 0.

If $E|X| < \infty$, then $\lim_{t \to \infty} tP(|X| > t) = 0$, and we have
$\lim_{t \to \infty} e(t,y) = E(X-y)$
$= EX - y$, as well as $\lim_{t \to \infty} \tilde{e}(t,y) = EX - y$.

The next Proposition gives two important explicit representations for e. The proof is an elementary application of Fubini's Theorem. (Note also Pruitt (1981), lemma 2.1.)

Proposition 2.6.

For $t \ge 0$ and $-\infty < y < \infty$,

$$e(t,y) = t - \int_{-\infty}^{y} P(|X-s| \le t)ds \tag{2.7}$$

$$= \int_{0}^{t} \{P(X-y > s) - P(X-y < -s)\}ds. \tag{2.8}$$

The following Corollary is immediate:

Corollary 2.9.

For $t \geq 0$ and $-\infty < y < \infty$, the functions $e(t, \cdot)$ and $e(\cdot, y)$ are each absolutely continuous. Each obeys a Lipschitz condition with Lipschitz constant one; thus e is jointly uniformly continuous. Moreover, at every point (t_0, y_0) such that $t_0 > 0$ and $y_0 \pm t_0$ are points of continuity of F, we have

$$e_1(t_0, y_0) = \left.\frac{\partial e(t, y)}{\partial t}\right|_{(t_0, y_0)} = P(X - y_0 > t_0) - P(X - y_0 < -t_0) \qquad (2.10)$$

and

$$e_2(t_0, y_0) = \left.\frac{\partial e(t, y)}{\partial y}\right|_{(t_0, y_0)} = -P(|X - y_0| \leq t_0). \qquad (2.11)$$

Next, define the censored (truncated) second moment function $m(\tilde{m}) : [0, \infty) \times (-\infty, \infty) \to [0, \infty)$ by

$$\tilde{m}(t, y) = E\tilde{c}^2(t, X - y) = E(X - y)^2 I(|X - y| \leq t)$$

$$(2.12)$$

$$m(t, y) = Ec^2(t, X - y) = E\tilde{c}^2(t, X - y) + t^2 P(|X - y| > t) = E[(X - y)^2 \wedge t^2].$$

Clearly $m(t, y) \leq t^2$, $\tilde{m}(t, y) \leq t^2$, and

$$\lim_{|y| \to \infty} m(t, y) = t^2, \qquad (2.13)$$

while $\tilde{m}(t, y) \to 0$ as $|y| \to \infty$.

If $EX^2 < \infty$, then $\lim\limits_{t \to \infty} t^2 P(|X| > t) = 0$, and we have $\lim\limits_{t \to \infty} m(t, y)$
$= E(X - y)^2 = \lim\limits_{t \to \infty} \tilde{m}(t, y)$.

The next Proposition gives two important explicit representations for m, and relates m to $\tilde{e}$ (recall (2.4)). Its proof is another application of Fubini's Theorem. (Note also lemma 2.1 in Pruitt (1981).)

Proposition 2.14.

For $t \geq 0$ and $-\infty < y < \infty$,

$$m(t, y) = t^2 - 2 \int_{-\infty}^{y} \tilde{e}(t, s) ds \tag{2.15}$$

$$= 2 \int_{0}^{t} s P(|X - y| > s) ds. \tag{2.16}$$

The following Corollary is immediate:

Corollary 2.17.

For $t \geq 0$ and $-\infty < y < \infty$, the functions $m(t, \cdot)$ and $m(\cdot, t)$ are each absolutely continuous. Furthermore, m is jointly uniformly continuous on sets of form $[0, R] \times (-\infty, \infty)$, for $R < \infty$. If (t_0, y_0) is a point such that $t_0 > 0$ and $y_0 \pm t_0$ are points of continuity of F, then

$$m_1(t_0, y_0) = 2t_0 P(|X - y_0| > t_0) \tag{2.18}$$

and

$$m_2(t_0, y_0) = -2\tilde{e}(t_0, y_0). \tag{2.19}$$

Consider the problem of locating a suitable center for the distribution of X censored at the level t. To motivate the two distinct methods we will propose (one here, one elsewhere–cf. Weiner (1989)) suppose $EX^2 < \infty$ and take $t = \infty$. To define a center y for the distribution of X, one might require

$$\tilde{e}(\infty, y) = e(\infty, y) = E(X - y) = 0, \qquad (2.20)$$

in order that the centered variable $X - y$ have zero expectation, or one might require

$$m(\infty, y) = E(X - y)^2 = \inf_s E(X - s)^2 = \inf_s m(\infty, s), \qquad (2.21)$$

in order that the centered variable $X - y$ have minimized second moment among all shifted versions $X - s$ of X.

Of course, each of (2.20) and (2.21) leads to the same unique choice of $y = EX$, despite their a priori unrelated motivational origins. However, for finite t, the requirements analogous to (2.20) and (2.21), respectively, lead to distinct results.

Given a scale level $0 < t < \infty$, the attempt to locate the center of the censored distribution of X by solving

$$e(t, y) = Ec(t, X - y) = 0 \qquad (2.22)$$

will be called the *Censored Centered Method* (CCM). It will be developed and applied here. The simpler and less technical, but less powerful attempt to locate this center by minimizing, i.e., solving

$$m(t, y) = Ec^2(t, X - y) = \inf_s Ec^2(t, X - s) = \inf_s m(t, s) \qquad (2.23)$$

is called the *Minimum Moment Method* (MMM), and is developed and applied in Weiner (1990). Elsewhere we will show an important connection between these two methods in the study of the asymptotic distribution of sums using regular scaling sequences.

3. The Censored Centered Method Functions. In this section we will construct the functions needed for determining the precentered normalizing constants of the Censored Centered Method, and examine their regularity properties. The results are summarized in Theorem 3.45, below.

By (2.7), for each $t > 0$ the function $e(t, \cdot)$ is continuous. By (2.5),

$$t = \lim_{y \to -\infty} e(t, y) > \lim_{y \to \infty} e(t, y) = -t. \text{ Thus for each } t > 0, \text{ the equation}$$

$$E[(|X - y| \wedge t) sgn(X - y)] = Ec(t, X - y) = e(t, y) = 0 \qquad (3.1)$$

admits a solution y (depending on t).

We will show that for each t sufficiently large, the solution y of (3.1) is unique. Then we will establish the necessary regularity properties of the function of t defined implicitly by (3.1). These properties may be of considerable independent interest and in particular are useful in other problems where centering is a delicate matter.

Define $t_0 = t_0(F)$ by

$$t_0 = \sup\{t > 0 : \text{ There exists } y \text{ such that } e(t, y) = 0 \text{ and} \qquad (3.2)$$

$$P(|X - y| < t) = 0\},$$

with the convention $\sup \phi = 0$.

It is easy to see (recalling (2.3)) that $t_0 > 0$ if and only if the median of F is not unique. Hence we may always view t_0 as the radius of the "interval of medians," whose interior (empty when the median of F is unique) is massless under F. In particular, $t_0 < \infty$.

We claim that for $t \geq t_0$, the solution y of (3.1) is unique. Suppose to the contrary that for some $t \geq t_0, y_1 < y_2$ are each solutions of (3.1). From (2.7) we obtain

$$0 = e(t, y_1) - e(t, y_2)$$
$$= . \int_{y_1}^{y_2} P(|X - s| \leq t) ds. \tag{3.3}$$

Thus $P(|X - s| \leq t) = 0$ for $y_1 < s < y_2$. Because y_1 and y_2 each solve (3.1), we obtain $P(X \geq y_2 + t) = P(X \leq y_1 - t) = 1/2$. Choosing any $s \in (y_1, y_2)$, it follows we can find $t' > t \geq t_0$ such that $e(t', s) = 0$ yet $P(|X - s| < t') = 0$, contrary to (3.2). Thus for $t \geq t_0$, the solution y of (3.1) is indeed unique.

Let $T = (0, \infty) \cap [t_0, \infty)$. Then we can implicitly define a function $\gamma : T \rightarrow (-\infty, \infty)$ by the equation

$$e(t, \gamma(t)) = 0 \qquad\qquad (t \in T). \tag{3.4}$$

We wish to provide a canonical extension of γ to all of $[0, \infty)$. Let $\mu = \mu(F)$ denote the center of the interval of medians of F (taking μ to be the unique median of F when this exists). First assume that $t_0 > 0$. We see that $P(|X - \mu| < t_0) = 0$, since the interval of medians has massless interior under F, so that by the median property of μ we have $e(t, \mu) = 0$ for $0 \leq t \leq t_0$. The uniqueness of the solution (3.2) when $t = t_0$ now forces $\gamma(t_0) = \mu$. It is convenient, therefore, to define $\gamma(t) = \mu$ for $0 \leq t < t_0$, when $t_0 > 0$; in particular (3.4) now holds for all $t \geq 0$, and γ is continuous on $[0, t_0]$, when $t_0 > 0$. Now assume $t_0 = 0$. We will see, following (3.32) and following (3.42), that when $t_0 = 0$ we will be forced, by continuity considerations, to define $\gamma(0) = \mu$, the unique median of F. Therefore at this stage let us simply define $\gamma(0) = \mu$. Now we have defined γ on all of $[0, \infty)$, such that (3.1) holds for every $t \geq 0$ (the equation $e(0, \mu) = 0$ being obvious).

Now we will show γ is continuous on $(0, \infty)$. Since when $t_0 > 0$, γ is constant on $(0, t_0]$, we need only consider the case $t > 0, t \geq t_0$, and $0 < t_n \rightarrow t$ (regardless of whether $t_0 > 0$ or $t_0 = 0$). Suppose first that $\{\gamma(t_n)\}$ were unbounded. By passing to subsequences if necessary, we may suppose that $\gamma(t_n) \to -\infty$ (the case of $\gamma(t_n) \to \infty$ being perfectly analogous). We obtain the contradiction (using (2.5), (3.4) and the Lipschitz property (2.8) of e)

$$
\begin{aligned}
t &= |e(t_n, \gamma(t_n)) - t| \\
&\leq |e(t_n, \gamma(t_n)) - e(t, \gamma(t_n))| + |e(t, \gamma(t_n)) - t| \\
&\leq |t_n - t| + |e(t, \gamma(t_n)) - t| \\
&\to 0,
\end{aligned}
\tag{3.5}
$$

as $n \to \infty$. Hence $\{\gamma(t_n)\}$ is bounded. By passing to subsequences, we may suppose $\gamma(t_n) \to L \in (-\infty, \infty)$. But now joint continuity of e (Corollary 2.9) gives $0 = e(t_n, \gamma(t_n)) \to e(t, L)$. Uniqueness of solution of (3.4) now applies (since $t \geq t_0$) to force $L = \gamma(t)$, i.e., $\gamma(t_n) \to \gamma(t)$. Thus γ is indeed continuous on $(0, \infty)$. In fact if $t_0 > 0$, γ is continuous on $[0, \infty)$; when $t_0 = 0$, however, deeper properties of γ will be needed to show continuity at the origin (see following (3.27) and also following (3.42)).

The next claim is harder to establish, but is worth the effort due to the information which can be extracted from it, here and in other applications to centering/scaling problems. We claim that γ is absolutely continuous on $[0, \infty)$, and is differentiable except at most at countably many points there. We will show that whenever $\gamma'(t)$ exists, it satisfies $|\gamma'(t)| < 1$, and we will give a useful implicit formula for $\gamma'(t)$.

Here it is helpful to observe that $\gamma'(t)$ exists at each $t > t_0$ such that $\gamma(t) \pm t$ are points of continuity of F, and can be computed implicitly there. This fact follows from Corollary 2.9, together with the Implicit Function Theorem applied to (3.4) on T. (Note, for $t > t_0$, $P(|X - \gamma(t)| \leq t) > 0$, by (3.2).) Thus, if F is everywhere continuous, γ must be continuously differentiable. But the unnecessary restriction to continuous F, in order to avoid "technical difficulties," would prevent application of the CCM to rich classes of atomic distributions, where the associated functions are particularly easy to compute (see Section 6), and in particular would prevent consideration of empirical distributions based on a random sample, where statistical applications may be considered. Thus the technical, real analytic effort required is quite justified.

If $t_0 > 0$, clearly $\gamma'(t) = 0$ for $0 < t < t_0$, as γ is constant there. Thus we focus on $T = (0, \infty) \cap [t_0, \infty)$.

Recall the left-handed "Dini Derivates" of a function f at a point $t \in \mathbf{R}$:

$$D^- f(t) = \limsup_{h \uparrow 0} \frac{f(t+h) - f(t)}{h} \qquad (3.6)$$

and

$$D_- f(t) = \liminf_{h \uparrow 0} \frac{f(t+h) - f(t)}{h}. \qquad (3.7)$$

(There are also two right-handed derivates). The following lemma is not new, but is presented here for convenience and completeness.

Lemma 3.8.

Let $f : [a, b] \to \mathbf{R}$ be continuous, and suppose

$$-\infty < -M = \inf_{t \in (a,b]} D_- f(t) \leq \sup_{t \in (a,b]} D^- f(t) \leq M < \infty. \qquad (3.9)$$

Then f is Lipschitz on $[a, b]$: For $\alpha, \beta \in [a, b]$,

$$|f(\beta) - f(\alpha)| \leq M|\beta - \alpha|. \qquad (3.10)$$

In particular, f is absolutely continuous on $[a, b]$ with $|f'| \leq M$ a.e. on $[a, b]$.

Proof:

If $D_- f \geq \epsilon > 0$ on $[a,b]$, then f is nondecreasing on $[a,b]$, for if not, choose $a \leq \alpha < \beta \leq b$ with $f(\beta) < f(\alpha)$. Fix $f(\beta) < \rho < f(\alpha)$, and put $t^* = \inf\{\alpha < t < \beta : f(t) \leq \rho\}$. Then $\alpha < t^* < \beta$ by continuity, and $D_- f(t^*) \leq 0$, a contradiction.

If $D_- f \geq 0$, then for each $\epsilon > 0$, $D_-(f(x) + \epsilon x) \geq \epsilon > 0$, so $f(x) + \epsilon x$ is nondecreasing. Then for $a \leq \alpha \leq \beta \leq b$, $0 \leq f(\beta) + \epsilon \beta - (f(\alpha) + \epsilon \alpha) = f(\beta) - f(\alpha) + \epsilon(\beta - \alpha)$. Let $\epsilon \downarrow 0$.

Finally, under (3.9), $D_-(Mx + f(x)) \geq 0$ and $D_-(Mx - f(x)) = M - D^- f(x) \geq 0$, so that each of $Mx \pm f(x)$ is nondecreasing. This leads directly to (3.10). ∎

To apply the lemma to γ, we will verify that for every $t > t_0$,

$$-1 < D_- \gamma(t) \leq D^- \gamma(t) < 1. \tag{3.11}$$

We study the difference quotients in (3.6)-(3.7) via the equation (cf. Proposition 2.6)

$$\begin{aligned}
0 &= e(t + h, \gamma(t + h)) - e(t, \gamma(t)) \\
&= e(t + h, \gamma(t + h)) - e(t, \gamma(t + h)) \\
&\quad + e(t, \gamma(t + h)) - e(t, \gamma(t)) \\
&= \int_t^{t+h} \{P(X > \gamma(t + h) + s) - P(X < \gamma(t + h) - s)\} ds \\
&\quad - \int_{\gamma(t)}^{\gamma(t+h)} P(|X - s| \leq t) ds.
\end{aligned} \tag{3.12}$$

Fix $t > t_0$ and let

$$q(h) = \frac{\gamma(t + h) - \gamma(t)}{h}, \tag{3.13}$$

for $h \neq 0$.

If $0 \neq h_n \to 0$ satisfies $\gamma(t + h_n) = \gamma(t)$, then $\lim_{n \to \infty} q(h_n) = 0$. Thus, to verify (3.11) it suffices to consider sequences $0 \neq h_n \to 0$ such that $\gamma(t + h_n) \neq \gamma(t)$, and we restrict ourselves to this case in all the following analysis.

Now since $t > t_0$, we have $P(|X - \gamma(t)| < t) > 0$. As γ is continuous, we have $\gamma(t + h_n) \to \gamma(t)$ if $h_n \to 0$. Thus if $h_n \to 0$,

$$1 \geq \liminf_{n \to \infty} (\gamma(t + h_n) - \gamma(t))^{-1} \int_{\gamma(t)}^{\gamma(t+h_n)} P(|X - s| \leq t) ds \qquad (3.14)$$

$$\geq \liminf_{s \to \gamma(t)} P(|X - s| \leq t)$$

$$\geq P(|X - \gamma(t)| < t)$$

$$> 0.$$

It follows that if L is any limit point of $q(h)$ as $h \to 0$, say $L = \lim_{n \to \infty} q(h_n)$ (a priori, possibly infinite), then L can be evaluated via

$$L = \lim_{n \to \infty} \frac{h_n^{-1} \int_t^{t+h_n} \{P(X > \gamma(t + h_n) + s) - P(X < \gamma(t + h_n) - s)\} ds}{(\gamma(t + h_n) - \gamma(t))^{-1} \int_{\gamma(t)}^{\gamma(t+h_n)} \{P(X \leq s + t) - P(X < s - t)\} ds} \qquad (3.15)$$

due to (3.12). In particular, (3.14) shows that L is always finite.

To verify (3.11), there are two cases of limit points L of $q(h)$ as $h \uparrow 0$ to consider, namely $L > 0$ and $L < 0$. We will carefully analyze the former and briefly sketch the latter, as these two cases are handled similarly.

Case 1. $L < 0$, so there exist $0 > h_n \uparrow 0$ with $q(h_n) \to L$ and $\gamma(t + h_n) > \gamma(t)$. By continuity, $\gamma(t) < \gamma(t + h_n) \to \gamma(t)$, so that

$$\Lambda = \lim_{n \to \infty} (\gamma(t + h_n) - \gamma(t))^{-1} \int_{\gamma(t)}^{\gamma(t+h_n)} \{P(X \leq s + t) - P(X < s - t)\} ds \qquad (3.16)$$

$$= \lim_{u \downarrow \gamma(t)} \{P(X \le u + t) - P(X < u - t)\}$$

$$= P(X \le \gamma(t) + t) - P(X \le \gamma(t) - t)$$

$$= P(-t < X - \gamma(t) \le t).$$

Also,

$$\lambda_2 = \lim_{n \to \infty} h_n^{-1} \int_t^{t+h_n} P(X < \gamma(t + h_n) - s) ds$$

(3.17)

$$= \lim_{u \downarrow \gamma(t) - t} P(X < u) = P(X \le \gamma(t) - t),$$

since for $t + h_n < s < t$, we have

$$P(X \le \gamma(t) - t) \le P(X < \gamma(t + h_n) - t)$$

$$\le P(X < \gamma(t + h_n) - s) \le P(X < \gamma(t + h_n) - t - h_n) \qquad (3.18)$$

and

$$\limsup_{n \to \infty} P(X < \gamma(t + h_n) - t - h_n) \le P(X \le \gamma(t) - t). \qquad (3.19)$$

Since

$$L = \lim_{n \to \infty} q(h_n) \qquad (3.20)$$

exists, (3.15)-(3.17) now force

$$\lambda_1 = \lim_{n \to \infty} h_n^{-1} \int_t^{t+h_n} P(X > \gamma(t + h_n) + s) ds \qquad (3.21)$$

to exist. But, arguing as above, λ_1 necessarily obeys $P(X > \gamma(t) + t) \le \lambda_1 \le P(X \ge \gamma(t) + t)$.

But, recalling (2.3) and (3.4), observe the following important identity, useful later and valid for every $t > 0$:

$$P(X > \gamma(t) + t) - P(X < \gamma(t) - t) = -t^{-1} E(X - \gamma(t)) I(|X - \gamma(t)| \le t) \qquad (3.22)$$

Writing $\lambda_1 = P(X > \gamma(t) + t) + uP(X = \gamma(t) + t)$, where $0 \le u \le 1$, we obtain from (3.15)-(3.19)

$$
\begin{aligned}
L &= \frac{\lambda_1 - \lambda_2}{\Lambda} = \frac{P(X > \gamma(t) + t) - P(X \le \gamma(t) - t) + uP(X = \gamma(t) + t)}{P(-t < X - \gamma(t) \le t)} \\
&= \frac{P(X > \gamma(t) + t) - P(X < \gamma(t) - t) + uP(X = \gamma(t) + t) - P(X = \gamma(t) - t)}{P(-t < X - \gamma(t) \le t)} \\
&= -\left\{ \frac{E(X - \gamma(t))I(-t \le X - \gamma(t) \le t) - tuP(X = \gamma(t) + t) + tP(X = \gamma(t) - t)}{tP(-t < X - \gamma(t) \le t)} \right\} \\
&= -\left\{ \frac{E(X - \gamma(t))I(-t < X - \gamma(t) \le t)}{tP(-t < X - \gamma(t) \le t)} \right\} + \frac{uP(X = \gamma(t) + t)}{P(-t < X - \gamma(t) \le t)} \\
&\ge -\left\{ \frac{E(X - \gamma(t))I(-t < X - \gamma(t) \le t)}{tP(-t < X - \gamma(t) \le t)} \right\} \\
&> -1,
\end{aligned}
\tag{3.23}
$$

because $P(|X - \gamma(t)| < t) > 0$, due to (3.2). This establishes the leftmost inequality in (3.11).

Case 2. $L > 0$, so there exist $0 > h_n \uparrow 0$ with $q(h_n) \to L$ and $\gamma(t + h_n) < \gamma(t)$. Here, the analogue of (3.16) holds:

$$
\Lambda = P(-t \le X - \gamma(t) < t)
\tag{3.24}
$$

Calculations similar to those above lead (for some $0 \le u \le 1$) to

$$
\begin{aligned}
L &= -\left\{ \frac{E(X - \gamma(t))I(-t \le X - \gamma(t) < t)}{tP(-t \le X - \gamma(t) < t)} \right\} - \frac{uP(X = \gamma(t) - t)}{P(-t \le X - \gamma(t) < t)} \\
&\le -\left\{ \frac{E(X - \gamma(t))I(-t \le X - \gamma(t) < t)}{tP(-t \le X - \gamma(t) < t)} \right\} \\
&< 1,
\end{aligned}
\tag{3.25}
$$

for similar reasons. This establishes the rightmost inequality in (3.11).

Thus (3.11) holds for every $t > t_0$. Lemma 3.8 now implies γ is absolutely continuous on every compact subset of (t_0, ∞), and satisfies $|\gamma'| < 1$ almost everywhere on (t_0, ∞). Of course when $t_0 > 0$, $\gamma' = 0$ on $(0, t_0)$, and $D^-\gamma(t_0) = 0$ (γ being constant on $(0, t_0]$). Thus $|\gamma'(t)| < 1$ for every $t > 0$ such that $\gamma'(t)$ exists.

It remains to consider absolute continuity near t_0. If $t_0 > 0$, fix $a > t_0$. We have, by absolute continuity,

$$\gamma(t) = \gamma(a) - \int_t^a \gamma'(s)ds, \tag{3.26}$$

for every $t > t_0$. Since $|\gamma'(s)| \leq 1$ for almost every $s > t_0$, the Bounded Convergence Theorem and continuity of γ at t_0 give

$$\begin{aligned}\gamma(t_0) &= \lim_{t \downarrow t_0} \gamma(t) \\ &= \lim_{t \downarrow t_0}\{\gamma(a) - \int_t^a \gamma'(s)ds\} \\ &= \gamma(a) - \int_{t_0}^a \gamma'(s)ds.\end{aligned} \tag{3.27}$$

Thus γ is absolutely continuous on $[t_0, \infty)$, and constantly $\gamma(t_0)$ on $[0, t_0]$. Hence γ is absolutely continuous on all of $[0, \infty)$, when $t_0 > 0$.

When $t_0 = 0$, the argument in (3.27) allows us to define $\gamma(0) = \gamma(a) - \int_0^a \gamma'(s)ds$ and have γ continuous at 0 and hence absolutely continuous on $[0, \infty)$. (Later we will see that this definition is consistent with our earlier definition of $\gamma(0) = \mu$, the unique median of F – see following (3.42)).

Preceeding (3.6) we noted that γ is differentiable at $t > t_0$ provided $\gamma(t) \pm t$ are points of continuity of F. We claim that since $|\gamma'(s)| < 1$ for almost every $s > 0$, only at most for countably many $t > 0$ can $\gamma(t) + t$ or $\gamma(t) - t$ be discontinuity points of F. For, using absolute continuity of γ, we see that the functions $a(t) = \gamma(t) + t$ and $b(t) = \gamma(t) - t$ are each *strictly* monotone on $(0, \infty)$. If we let D be the set of discontinuity points of F, we see that since D is countable, so must be $A = a^{-1}(D) \cup b^{-1}(D)$, by strict monotonicity. Thus the claim is established, and γ' exists (and satisfies $|\gamma'| < 1$) off an at most countable subset A of $(0, \infty)$.

Thus, for $t > t_0$ with $t \notin A$, reference to (3.15) and (3.22) shows

$$\gamma'(t) = \frac{P(X > \gamma(t) + t) - P(X < \gamma(t) - t)}{P(|X - \gamma(t)| \leq t)}$$
$$= -\frac{E(X - \gamma(t))I(|X - \gamma(t)| \leq t)}{tP(|X - \gamma(t)| \leq t)}, \tag{3.28}$$

while of course $\gamma'(t) = 0$ for $0 < t < t_0$, if $t_0 > 0$.

We will have need of the functions $v : [0, \infty) \to [0, \infty)$ and $w : [0, \infty) \to (0, 1]$ defined by

$$v(t) = m(t, \gamma(t)) = Ec^2(t, X - \gamma(t)) = E(X - \gamma(t))^2 \wedge t^2 \tag{3.29}$$

and

$$w(t) = v(t)/t^2, \ t > 0 \tag{3.30}$$

for use in determining our precentered scaling constants. (We will define $w(0)$ later.)

Clearly $v(t) > 0$ for $t > 0$, lest X be degenerate.

Now Corollary 2.7, (3.28), the Chain Rule and the definition of A show that for $t_0 < t \notin A$, we have

$$v'(t) = m_1(t, \gamma(t)) + m_2(t, \gamma(t))\gamma'(t)$$
$$= 2t P(|X - \gamma(t)| > t) + \frac{2(E(X - \gamma(t))I(|X - \gamma(t)| \leq t))^2}{tP(|X - \gamma(t)| \leq t)} \tag{3.31}$$
$$\geq 0.$$

If $t_0 > 0$, we note that for $0 \leq t \leq t_0$, we have $v(t) = t^2$, since for $0 < t < t_0$, $P(|X - \gamma(t)| \leq t) = P(|X - \gamma(t_0)| \leq t) \leq P(|X - \gamma(t_0)| < t_0) = 0$, as we saw following

(3.4). Now, γ and m being continuous, we see that v is continuous on $[0, \infty)$, and differentiable off the countable subset A of $(0, \infty)$. Theorem 264 in Kestelman (1937) (or direct verification of the Lipschitz property) guarantees v is absolutely continuous on $[0, \infty)$. Moreover, (3.31) now shows that v is nondecreasing on $[0, \infty)$, and strictly increasing if X is essentially unbounded.

It follows w is absolutely continuous on $(0, \infty)$.

If $t_0 < t \notin A$, the quotient rule (3.31), and the fact $P(|X - \gamma(t)| \leq t) > 0$ via (3.2), yield, by the Cauchy-Schwarz inequality applied to $(X - \gamma(t)) I(|X - \gamma(t)| \leq t) \cdot I(|X - \gamma(t)| \leq t)$,

$$
\begin{aligned}
t^4 w'(t) &= t^2 v'(t) - 2tv(t) \\
&= 2t^3 P(|X - \gamma(t)| > t) + \frac{2t(E(X - \gamma(t)) I(|X - \gamma(t)| \leq t))^2}{P(|X - \gamma(t)| \leq t)} \\
&\quad - 2t^3 P(|X - \gamma(t)| > t) - 2t \, E(X - \gamma(t))^2 I(|X - \gamma(t)| \leq t) \\
&\leq 2t E(X - \gamma(t))^2 I(|X - \gamma(t)| \leq t) \\
&\quad - 2t E(X - \gamma(t))^2 I(|X - \gamma(t)| \leq t) \\
&= 0,
\end{aligned}
\tag{3.32}
$$

so that w is nonincreasing on (t_0, ∞).

If $t_0 > 0$, then for $0 < t \leq t_0$, $w(t) = v(t)/t^2 = 1$, so we may define $w(0) = 1$ and have w absolutely continuous on $[0, \infty)$.

If $t_0 = 0$, we may define $w(0) = \lim_{t \to 0} w(t) \leq 1$, since w is nonincreasing, and have w absolutely continuous on $[0, \infty)$, using the Monotone Convergence Theorem in a calculation similar to that in (3.27). (See also following (3.42).)

We claim w is eventually *strictly* decreasing. Now for $t_0 < t \notin A$, the Cauchy-Schwarz inequality in (3.32) will fail to be strict exactly when, on $[|X - \gamma(t)| \leq t]$,

$X - \gamma(t)$ is constant, i.e., $[\gamma(t) - t, \gamma(t) + t]$ contains exactly one atom of F and no other mass under F, since $P(|X - \gamma(t)| \leq t) > 0$. Since, as we have seen, each of $\gamma(t) \pm t$ is strictly monotone on $[0, \infty)$, with $t - |\gamma(t)| \to \infty$ as $t \to \infty$, it follows that

$$t_1 = t_1(F) = \sup\{t \geq t_0 : [\gamma(t) - t, \gamma(t) + t] \text{ contains exactly one atom}$$

(3.33)

$$\text{and no other mass under } F\}$$

is finite. (Here $\sup \emptyset \equiv t_0$, and recall F is nondegenerate.) Thus, w will be strictly decreasing on $(t_1(F), \infty)$. Putting (temporarily)

$$t_1' = \inf\{t \geq t_0 : [\gamma(t) - t, \gamma(t) + t] \text{ contains exactly one atom}$$

(3.34)

$$\text{and no other mass under } F\},$$

with $\inf \emptyset = 0$, note that necessarily $P(|X - \gamma(t_1')| < t') = 0$, so that $t_1' \leq t_0$, yet if $t_1' > 0$, $P(|X - \gamma(t_1')| \leq t_1') > 0$, so that $t_1' \geq t_0$, hence $t_1' = t_0$. Thus on $[t_0, t_1]$, w is constant (since $w' = 0$ there), while on $[0, t_0]$, w is constant, so that by continuity, w is constant on all of $[0, t_1(F)]$. Finally, if $t_1' = 0$, clearly $t_0 = 0$, and then $\gamma(0)$ is necessarily (see following (3.42)) the (unique) median of F.

We will need the fact that $w(t) \to 0$, as $t \to \infty$. First, we claim that $\gamma(t)/t \to 0$, as $t \to \infty$. If so, then $|X - \gamma(t)|/t \to 0$ almost surely, as $t \to \infty$, and then the Dominated Convergence Theorem applies to give

$$w(t) = E((\frac{X - \gamma(t)}{t})^2 \wedge 1) \to 0 \quad (t \to \infty).$$

(3.35)

To check $\gamma(t)/t \to 0$, we need only show $t - |\gamma(t)| \to \infty$, for then the first part of (3.28) shows

$$
\begin{aligned}
|\gamma'(t)| &\leq \frac{P(|X - \gamma(t)| > t)}{P(|X - \gamma(t)| \leq t)} \\
&\leq \frac{P(|X| > t - |\gamma(t)|)}{P(|X|X \leq t - |\gamma(t)|)} \\
&\to 0,
\end{aligned}
\tag{3.36}
$$

as $t \to \infty$.

If $t - |\gamma(t)| \to \infty$ fails, then (by absolute continuity) there must exist $t_0 < t_n \to \infty$ such that $t_n \not\in A$ and $|\gamma'(t_n)| \to 1$, lest there exist $B < 1$ with $\limsup_{t\to\infty} |\gamma'(t)| < B$, implying for all large t that $t - |\gamma'(t)| \geq t - Bt = (1 - B)t \to \infty$.

Since w is eventually strictly decreasing, we know $0 \leq \lim_{t\to\infty} w(t) = C < 1$ exists. But then the Cauchy-Schwarz inequality gives

$$
\begin{aligned}
w(t_n) &= P(|X - \gamma(t_n)| > t_n) + t_n^{-2} E(X - \gamma(t_n))^2 I(|X - \gamma(t_n)| \leq t_n) \tag{3.37} \\
&\geq P(|X - \gamma(t_n)| > t_n) + \frac{(E(X - \gamma(t_n))I(|X - \gamma(t_n)| \leq t_n))^2}{t_n^2 P(|X - \gamma(t_n)| \leq t_n)} \\
&= P(|X - \gamma(t_n)| > t_n) + \left(\frac{E(X - \gamma(t_n))I(|X - \gamma(t_n)| \leq t_n)}{t_n P(|X - \gamma(t_n)| \leq t_n)} \right)^2 \\
&\qquad P(|X - \gamma(t_n)| \leq t_n) \\
&= P(|X - \gamma(t_n)| > t_n) + (\gamma'(t_n))^2 P(|X - \gamma(t_n)| \leq t_n) \\
&= 1 + P(|X - \gamma(t_n)| \leq t_n)\{\gamma'(t_n)^2 - 1\} \\
&\to 1 > C,
\end{aligned}
$$

as $n \to \infty$, a contradiction.

Thus (3.35) holds. As a byproduct we learn that $t - |\gamma(t)| \uparrow \infty$ strictly, as $t \to \infty$, and from (3.36) we also obtain

$$
\lim_{t\to\infty} \gamma(t)/t = 0.
\tag{3.38}
$$

As we will see, v and w will be useful in determining the precentered normalizing constants for the Censored Centered Method.

Let us examine the case $E|X| < \infty$. If $\limsup_{t \to \infty} |\gamma(t)| = \infty$, choose $t_n \to \infty$ with $\gamma(t_n) \to \infty$ (the case $\gamma(t_n) \to -\infty$ being similar). Fix $L > 0$. Now we saw following (2.5) that $e(t_n, L) \to EX - L$. By (2.7) the function $e(t, \cdot)$ is nonincreasing for every t, so we have, for all n large enough that $\gamma(t_n) \geq L$,

$$0 = e(t_n, \gamma(t_n)) \leq e(t_n, L) \to EX - L. \qquad (3.39)$$

Letting $L \to \infty$ gives $EX = \infty$, a contradiction.

Thus $\limsup_{t \to \infty} |\gamma(t)| < \infty$. Let $t_n \to \infty$ such that $\gamma(t_n) \to a$, say. Then, since $e(t_n, \cdot)$ is Lipschitz with constant one for every n (Corollary 2.9),

$$|e(t_n, a)| = |e(t_n, \gamma(t_n)) - e(t_n, a)| \qquad (3.40)$$

$$\leq |\gamma(t_n) - a| \to 0,$$

as $n \to \infty$. But, following (2.5) we saw $e(t_n, a) \to EX - a$. Thus $a = EX$, so that when $E|X| < \infty$,

$$\lim_{t \to \infty} \gamma(t) = EX. \qquad (3.41)$$

If $EX^2 < \infty$, then (3.41) and Dominated Convergence give us

$$\lim_{t \to \infty} v(t) = \lim_{t \to \infty} E(X - \gamma(t))^2 \wedge t^2 = var\, X. \qquad (3.42)$$

Two useful properties of the CCM centering function γ support its use as a "natural" centering. First, if X is symmetric about c, then we may take $\gamma \equiv c$, and indeed we *must* take $\gamma(t) = c$, for $t \geq t_0$ as defined in (3.2).

The second property concerns the connection between γ and the median(s) of F. We have conveniently defined $\gamma(0) = \mu$, the center of the interval of medians of F, when $t_0 > 0$. When $t_0 = 0$, we defined $\gamma(0)$ by continuity following (3.27). We now show that when $t_0 = 0$, this definition is consistent with the original one $\gamma(0) = \mu$, the unique median of F.

We have seen that the functions $a(t) = \gamma(t) + t$ and $b(t) = \gamma(t) - t$ are each strictly monotone on $(0, \infty)$. As $t \downarrow 0$, by continuity of γ at 0 we have $a(t) \downarrow \gamma(0)$, $b(t) \uparrow \gamma(0)$. It follows $P(|X - \gamma(t)| \leq t) \to P(X = \gamma(0))$, as $t \downarrow 0$, while $P(X > \gamma(t) + t) \to P(X > \gamma(0))$ and $P(X < \gamma(t) - t) \to P(X < \gamma(0))$. Suppose $\gamma(0)$ is a point of continuity of F. Since in (3.28), $|\gamma'(t)| < 1$ remains bounded as $t \downarrow 0$, from $P(|X - \gamma(t)| \leq t) \to 0$ we are forced to conclude that $P(X > \gamma(0)) - P(X < \gamma(0)) = 0$. Together with $P(X > \gamma(0)) + P(X < \gamma(0)) + P(X = \gamma(0)) = 1$, this leads to $P(X > \gamma(0)) = P(X < \gamma(0)) = 1/2$, or $\gamma(0) = \mu$, the unique median of F. But if $P(X = \gamma(0)) > 0$ we have

$$1 \geq \left| \frac{P(X > \gamma(0)) - P(X < \gamma(0))}{1 - P(X > \gamma(0)) - P(X < \gamma(0))} \right|. \tag{3.43}$$

Suppose, without loss of generality, that $P(X > \gamma(0)) \geq P(X < \gamma(0))$. Then $P(X \geq \gamma(0)) \geq P(X \leq \gamma(0))$, so that $P(X \geq \gamma(0)) \geq 1/2$. However, rearranging inequality (3.43) shows that $P(X > \gamma(0)) \leq 1/2$. Thus $P(X \leq \gamma(0)) \geq 1/2$. It follows that $\gamma(0)$ is the unique median μ of F, as claimed.

Inspecting (3.28), letting $t \downarrow 0$ in (3.35), using Dominated Convergence and recalling that $w(0) = w(0+)$ exists a priori due to monotonicity, we obtain the existence of $\gamma'(0+)$ and also

$$w(0) = \gamma'(0+)^2 P(X = \gamma(0)) + P(X \neq \gamma(0))$$
$$= \gamma'(0+)^2 P(X = \mu) + P(X \neq \mu). \tag{3.44}$$

Clearly, $w(0) < 1$ is possible only when F has an atomic, hence unique, median μ.

We summarize the CCM method in

Theorem 3.45.

Let $X \sim F$ nondegenerate.

There exists a function $\gamma : [0, \infty) \to (-\infty, \infty)$, such that for every $t \geq 0$,

$$E[(|X - \gamma(t)| \wedge t)sgn(X - \gamma(t))] = Ec(t, X - \gamma(t)) = 0. \qquad (3.46)$$

There exists $t_0 = t_0(F) \in [0, \infty)$ (given by (3.2)) such that (3.46) defines γ uniquely on $[t_0, \infty)$, and such that we may take $\gamma \equiv \mu$ on $[0, t_0]$, where μ is the (central) median of F. In particular, $\gamma(0) = \mu$, regardless of $t_0 \geq 0$.

The resulting function γ is absolutely continuous on $[0, \infty)$, and differentiable off a countable subset of $(0, \infty)$, with derivative γ' given by (3.28) and satisfying

$$|\gamma'(t)| < 1 \quad (t > 0) \qquad (3.47)$$

whenever $\gamma'(t)$ exists.

This function γ satisfies

$$\lim_{t \to \infty} \frac{\gamma(t)}{t} = 0. \qquad (3.48)$$

The function $v : [0, \infty) \to [0, \infty)$ defined by

$$v(t) = var[(|X - \gamma(t)| \wedge t)sgn(X - \gamma(t))] = E[(X - \gamma(t))^2 \wedge t^2] = t^2 P(|X \quad \gamma(t)| > t)$$

$$\qquad (3.49)$$

$$+ E(X - \gamma(t))^2 I(|X - \gamma(t)| \leq t)$$

is absolutely continuous, differentiable off a countable set, and nondecreasing, with $v(t) > 0$ for $t > 0$, and with derivative given by (3.31).

The function $w : [0, \infty) \to (0, \infty)$ given by

$$w(t) = \frac{v(t)}{t^2} = E[\frac{(X - \gamma(t))^2 \wedge t^2}{t^2}], \quad t > 0, \tag{3.50}$$

and defined by continuity at $t = 0$, is absolutely continuous and nonincreasing on $[0, \infty)$, and differentiable off a countable set, with $w(t) > 0$ for $t > 0$, and with derivative given in (3.32).

   There exists $t_1 = t_1(F)$ (given by (3.33)) such that $w = w(0)$ (given by (3.44)) on $[0, t_1]$, and such that w is strictly decreasing on $[t_1, \infty)$. We have

$$\lim_{t \to \infty} w(t) = 0. \tag{3.51}$$

If $E|X| < \infty$, then γ satisfies

$$\lim_{t \to \infty} \gamma(t) = EX. \tag{3.52}$$

If $EX^2 < \infty$, then v satisfies

$$\lim_{t \to \infty} v(t) = var X. \tag{3.53}$$

If X is symmetric about c, then we may take

$$\gamma \equiv c \tag{3.54}$$

on $[0, \infty)$.

4. The Precentered Scale Sequences. Let $\{X_j : j \geq 1\}$ be an independent sequence of nondegenerate real random variables. Subscript by j the various functions, introduced in the previous sections, corresponding to X_j.

In this section, given a function h satisfying certain regularity conditions, we will construct sequences of constants increasing strictly to infinity, which will serve to measure scale relative to h. The result is Proposition 4.9, below. Various choices of h will lead to scale sequences appropriate to (i) normalize suitably centered partial sums derived from $\{X_j\}$, (ii) serve as censoring levels for precentered summands derived from $\{X_j\}$, and (iii) determine normalizing constants for partial sums of precentered, censored summands derived from $\{X_j\}$.

Our only assumption on the sequence $\{X_j\}$, aside from nondegeneracy of each variable, is that

$$\sum_{j=1}^{\infty} \inf_y E(X_j - y)^2 \wedge t^2 = \infty, \tag{4.1}$$

for some $t > 0$. This assumption of "nondegeneracy" for the sequence is truly minimal: It was seen in Weiner (1990) that the failure of (4.1) for any $t > 0$ is equivalent to the almost sure convergence of the suitably centered partial sums,

$$\sum_{j=1}^{n} X_j - y_n \tag{4.2}$$

to a nondegenerate random variable, in which case there is neither need nor interest in scaling the sums.

Now (4.1) for some (hence every) $t > 0$ implies

$$\forall t > 0 : \sum_{j=1}^{\infty} w_j(t) = \infty. \tag{4.3}$$

Thus in the sequel we will assume (4.1) and hence also (4.3) hold for every $t > 0$, and in particular we can assume that no sequence of form (4.2) converges, so that scaling is actually required. (It will be easily seen, moreover, that (4.3) alone is enough to insure the validity of the theorems in this and the following sections.)

Let $h : (0, \infty) \to (0, \infty)$ be continuous and satisfy

$$h(t)/t^2 \text{ is nondecreasing on } (0, \infty). \tag{4.4}$$

Define $C_n : (0, \infty) \to (0, \infty)$ by

$$C_n(t) = \sum_{j=1}^{n} v_j(t)/h(t)$$
$$= (t^2/h(t)) \sum_{j=1}^{n} w_j(t), \tag{4.5}$$

We will show that, for all sufficiently large n, we can uniquely solve the equation $C_n(t) = 1$; then we will examine the solution sequence.

Now for each $t > 0$,

$$\lim_{n \to \infty} C_n(t) = \sum_{j=1}^{\infty} w_j(t) = \infty, \tag{4.6}$$

by (4.3). For each $n \geq 1$,

$$\lim_{t \to \infty} C_n(t) = 0, \tag{4.7}$$

by (3.51).

Let $T_1 = t_1(F_1)$, where F_1 is the distribution function of X_1 (recall (3.33)). Then for each $n \geq 1$, $C_n(\cdot)$ is strictly decreasing on $[T_1, \infty)$ by (4.4),(4.5) and the fact that w_1 is strictly decreasing there.

Finally, since v_j is strictly positive for each $j \geq 1$, we see that for each $t > 0$ and $n > m$,

$$C_n(t) > C_m(t) \tag{4.8}$$

Choose any $T > T_1$. By (4.6), there exists N so that $C_N(T) > 1$. By (4.7) and continuity and strict decrease of and $C_N(\cdot)$ on $[T, \infty)$, there exist unique $c_N > T$ such that $C_N(c_N) = 1$.

Given $n > N$, (4.8) shows that $C_n(c_N) > 1$ Thus an argument similar to that above shows there exist unique numbers $c_n > c_N$ $C_n(c_n) = 1$. It is clear, by the same argument, that the sequence $\{c_n : n \geq N\}$ is strictly increasing. Moreover, we can show that this sequence is unbounded: Given $M > 0$, one finds $N_1 \geq N$ so that $C_{N_1}(M) > 1$, by (4.6); then $c_{N_1} > M$.

We have $\lim_{n \to \infty} c_n = \infty$.

We summarize these results in

Proposition 4.9.

Assume (4.1) holds for some (hence every) $t > 0$.

Given $h : (0, \infty) \to (0, \infty)$ continuous and satisfying (4.4), there exists N such that for each $n \geq N$, the equations

$$\sum_{j=1}^{n} E(X_j - \gamma_j(c_n))^2 \wedge c_n^2 = h(c_n) \tag{4.10}$$

uniquely determine positive numbers c_n. The sequence $\{c_n : n \geq N\}$ is strictly increasing, with

$$\lim_{n \to \infty} c_n = \infty. \tag{4.11}$$

We remark that equations (4.10) in Proposition 4.9 clarify and justify the apellation "precentered" for the scale sequence $\{c_n\}$.

Remark 4.12.

Taking $h(t) = \tau t^2$, for various $\tau > 0$, in Proposition 4.9, we obtain constants $\{c_n = c_n(\tau)\}$ and centerings $\{\gamma_j(c_n(\tau))\}$ suitable for considering criteria for convergence in distribution of affine-normalized (uncensored) partial sums of independent (not necessarily identically distributed) random variables in the context of the classical Central Limit Theorem. In Weiner (1990) a parallel approach to this problem is utilized using the Minimized Moment Method (rather than the CCM method being developed here) with $\tau = 1$ in the case of a standard normal limiting distribution. As will be discussed elsewhere, either method may be used for this problem, but the CCM approach appears to be the technically stronger and computationally more convenient and informative one, especially in the case of a general infinitely divisible (nonnormal) limiting distribution.

Since occasionally here we shall have need of the sequences corresponding to the "standard" choice $h(t) = t^2$, it is worth distinguishing them here.

Corollary 4.13.

Assume (4.1) holds for some (hence every) $t > 0$.

For sufficiently large n, the equation

$$\sum_{j=1}^{n} E(X_j - \gamma_j(a_n))^2 \wedge a_n^2 = a_n^2 \qquad (4.14)$$

uniquely determines $\{a_n\}$ such $\{a_n\}$ is strictly increasing to infinity. Given any h as in Proposition 4.9, with $\{c_n\}$ determined by (4.10), we have

$$a_n > c_n, \qquad (4.15)$$

for all sufficiently large n.

5. Asymptotic Normality for Censored Sums. In this section we will give the main theorem on asymptotic normality for normalized sums of precentered, censored summands derived from an independent sequence $\{X_j\}$ of nondegenerate random variables satisfying (4.1) for some (hence every) $t > 0$. Then we will consider relatively how much of the sequence $\{X_j : j \leq n\}$ actually requires censoring in order to obtain the asymptotic normality, and at the same time somewhat quantify the failure of asymptotic normality for the uncensored sums.

Fix a function h as in (4.4), and (assuming (4.1)) construct $\{c_n\}$ as in (4.10).

A key point in the results here is the dual and simultaneous function of the unbounded scale constants $\{c_n\}$ (as constructed in Section 4) both as censoring levels (to control "excessively" large magnitudes) and as normalizers of the censored sums (in the form $\sqrt{h(c_n)}$). Thus, the function h serves to relate the total variance $h(c_n)$ of the "censored sample"

$$\chi_n = \{(|X_j - \gamma_j(c_n)| \wedge c_n)sgn(X_j - \gamma_j(c_n)) : j \leq n\}$$

$$(5.1)$$

$$= \{c(X_j - \gamma_j(c_n), c_n) : j \leq n\}$$

to the censoring level c_n. By choosing h in advance, in such a way that the total censored standard deviation of the sample (i.e. , the standard deviation of the sum of χ_n) of (5.1) grows faster than the censoring levels c_n, we may obtain asymptotic normality for the sums of χ_n derived from *any* sequence $\{X_j\}$ "nondegenerate" in the truly minimal sense of (4.1). Indeed, we will show (Corollary 5.18) that asymptotic normality may be obtained by altering as few terms in the sum (but tending to infinity) as may be desired. Because of the convenient (with respect to expectation and variance) properties of the constants constructed through h, the proofs here will be easy and short, and the relationships among the various controlling quantities (and our flexibility in controlling them) will be made transparent.

Given $t > 0$ and $n \geq 1$, define

$$S_n(t) = \sum_{j=1}^{n} c(t, X_j - \gamma_j(t)) = \sum_{j=1}^{n}(|X_j - \gamma_j(t)| \wedge t)sgn(X_j - \gamma_j(t)). \qquad (5.2)$$

Given h as in (4.4), define (with $\{c_n\}$ as in (4.10) and $n \geq N$) the h-censored sum by

$$S_n^h = S_n(c_n) = \sum_{j=1}^{n}(|X_j - \gamma_j(c_n)| \wedge c_n)sgn(X_j - \gamma_j(c_n)). \qquad (5.3)$$

To count "excessively large" magnitudes relative to h, given $t > 0$, put

$$T_n(t) = \#\{j \leq n : |X_j - \gamma_j(t)| > t\}, \quad T_n^h = T_n(c_n). \qquad (5.4)$$

Then T_n^h counts the number of precentered summands actually altered by the censoring in (5.3).

Consider the special case $h(t) = t^2$. (Recall Corollary 4.13.) It is easy to see that S_n^h is asymptotically normal if and only if

$$\forall \epsilon > 0 : T_n(\epsilon\sqrt{h(a_n)}) = T_n(c a_n) \to 0 \qquad (5.5)$$

(Elsewhere, in connection with Weiner (1990), it is being shown that (5.5) is both necessary and sufficient for the ordinary, uncensored sums $S_n = \sum_{j=1}^{n} X_j$ to be asymptotically normal.) Under (5.5), one easily has $S_n^h/\sqrt{h(a_n)} = S_n^h/a_n \xrightarrow{D} N(0,1)$. But when (5.5) fails, there are "excessively" large summands (i.e. , terms with magnitudes comparable to the normalization of the sum) in the sum S_n^h, even *after* the censoring. Here, the censoring levels are too large compared to the normalizations of the sum.

By choosing larger h, we obtain smaller censoring levels c_n which are negligible compared to the normalizations $\sqrt{h(c_n)}$, thus actually controlling the asymptotic-normality-damaging excessive magnitudes by means of the censoring. In Corollary 5.18 we show that in order to induce the desired asymptotic normality, it always suffices to actually censor only an asymptotically vanishing proportion of the summands.

Here is our main result:

Theorem 5.6.

Let $\{X_j\}$ be independent and satisfy (4.1) for some $t > 0$.

Fix $h : (0,\infty) \to (0,\infty)$ such that $h(t)/t^2$ is nondecreasing and such that

$$\lim_{t \to \infty} h(t)/t^2 = \infty. \tag{5.7}$$

Then, the h-centered sum S_n^h is asymptotically normal:

$$S_n^h / \sqrt{h(c_n)} \xrightarrow{D} N(0,1). \tag{5.8}$$

Moreover, for each $\lambda > 1$,

$$P(T_n^h > \lambda h(c_n)/c_n^2) \to 0. \tag{5.9}$$

Proof:

The proof is straightforward by now, due to the construction of the various constants.

We verify the Criteria for Normal Convergence (see, e.g. Loéve (1977), p.328). For $n \geq N$ and $j \leq n$, put $X_{nj} = c(c_n, X_j - \gamma_j(c_n))/\beta_n$, where $\beta_n^2 = h(c_n)$. Then

$$|X_{nj}| \leq c_n/\beta_n = (c_n^2/h(c_n))^{1/2} \to 0, \tag{5.10}$$

by (5.7) and (4.11). Now (5.10) is the key to the whole proof.

Thus, given $\epsilon > 0$, we can choose $N_0 > N$, such that for $n \geq N_0$ and $j \leq n, |X_{nj}| \leq \epsilon$. We have, for $n \geq N_0$,

$$\sum_{j=1}^{n} P(|X_{nj}| > \epsilon) = 0 \tag{5.11}$$

$$\sum_{j=1}^{n} EX_{nj}I(|X_{nj}| \leq \epsilon) = \sum_{j=1}^{n} Ec(c_n, X_j - \gamma_j(c_n))/\beta_n = 0 \tag{5.12}$$

and

$$\sum_{j=1}^{n} var X_{nj}I(|X_{nj}| \leq \epsilon) = \beta_n^{-2} \sum_{j=1}^{n} Ec^2(c_n, X_j - \gamma_j(c_n))$$

$$\tag{5.13}$$

$$= h(c_n)^{-2} \sum_{j=1}^{n} E(X_j - \gamma_j(c_n))^2 \wedge c_n^2 = 1,$$

by (3.4) and (4.10). Thus the Criteria for Normal Convergence are satisfied, and thus

$$\sum_{j=1}^{n} X_{nj} = \frac{1}{\beta_n} \sum_{j=1}^{n} c(c_n, X_j - \gamma_j(c_n)) \xrightarrow{D} N(0,1), \tag{5.14}$$

which is to say (5.8) holds.

For (5.9), fix $\lambda > 1$. Now by (4.10) and independence,

$$ET_n^h = E\#\{j \leq n : |X_j - \gamma_j(c_n)| > c_n\} = \sum_{j=1}^{n} P(|X_j - \gamma_j(c_n)| > c_n)$$

$$\leq \frac{1}{c_n^2} \sum_{j=1}^{n} E(X_j - \gamma_j(c_n))^2 \wedge c_n^2 = \frac{h(c_n)}{c_n^2}. \tag{5.15}$$

Similarly, independence gives

$$var T_n^h = \sum_{j=1}^{n} P(|X_j - \gamma_j(c_n)| > c_n)(P(|X_j - \gamma_j(c_n)| \leq c_n))$$

$$\leq \sum_{j=1}^{n} P(|X_j - \gamma_j(c_n)| > c_n) \leq h(c_n)/c_n^2, \tag{5.16}$$

Hence, by Chebyshev's inequality and (4.11),(5.7),

$$P(T_n^h > \frac{\lambda h(c_n)}{c_n^2}) = P(T_n^h - ET_n^h > (\frac{\lambda h(c_n)}{c_n^2} - ET_n^h))$$

$$\leq \frac{var T_n^h}{(\frac{\lambda h(c_n)}{c_n^2} - ET_n^h)^2} \leq \frac{h(c_n)/c_n^2}{(\frac{\lambda h(c_n)}{c_n^2} - ET_n^h)^2}$$

$$\tag{5.17}$$

$$\leq (c_n^2/h(c_n))(\lambda - 1)^{-2} \to 0.$$

Thus (5.9), and hence also the Theorem itself, is proved. ∎

To exhibit the flexibility and the role of h, we note the following:

Corollary 5.18.

Suppose $\{X_j\}$ is an independent nondegenerate sequence satisfying (4.1). Then, given nondecreasing $\{r_n\}$ sucht that $r_n \to \infty$, there exists $h : (0,\infty) \to (0,\infty)$ as in Theorem 5.6 such that

$$S_n^h / \sqrt{h(c_n)} \xrightarrow{D} N(0,1), \tag{5.19}$$

and for which

$$P(T_n^h \geq r_n) \to 0, \tag{5.20}$$

where $\{c_n\}$ is determined by (4.10).

Proof:

Recalling Corollary 4.13 and having determine $\{a_n\}$, fix $\lambda > 1$, define $g(a_n) = r_n/\lambda$, and interpolate linearly on $[a_n, a_{n+1}]$. Put $h(t) = t^2 g(t)$, so that h satisfies the requirements of Theorem 5.6 (note that g is nondecreasing). Then (5.19) is immediate, while (5.20) follows from (5.10) and (4.15) by noting $\lambda h(c_n)/c_n^2 = \lambda g(c_n) \leq \lambda g(a_n) = \lambda(r_n/\lambda) = r_n$. ∎

Note that we may choose $r_n \to \infty$ as slowly as we like; in particular, we may arrange $r_n/n \to 0$.

Remark 5.21.

Specialize to the case of identically distributed summands.

The proof of Corollary 5.18 is not the most direct construction of h in this case. In the i.i.d. setting, of course, Corollary 5.18 is practically equivalent to the corresponding result in Proposition 4.3 in Hahn, Kuelbs and Weiner (1989), since for one fixed nondegenerate variable X we have $\gamma(c_n)/c_n \to 0$ due to (3.48) (so that relative to the censoring levels, the centers move only negligibly), and condition (4.1) is trivially satisfied. To obtain this connection (and compare results) directly, given $r_n \to \infty$ with $r_n/n \to 0$, simply solve

$$nw(c_n) = r_n, \tag{5.22}$$

and note that the proof of Theorem 5.6 applies to give

$$\frac{\sum_{j=1}^{n}(|X_j - \gamma(c_n)| \wedge c_n)sgn(X_j - \gamma(c_n))}{\sqrt{r_n}c_n} \xrightarrow{D} N(0,1), \tag{5.23}$$

and given $\lambda > 1$,

$$P(\#\{j \leq n : |X_j - \gamma(c_n)| > c_n\} > \lambda r_n) \to 0. \tag{5.24}$$

6. Examples.

Example 6.1.

In order to generate interesting examples of CCM centering functions γ, we will concretely identify the class of all such functions.

Let $X \sim F$ nondegenerate.

For $t \geq 0$, define events $A_t = [X - \gamma(t) > t]$ and $B_t = [X - \gamma(t) < -t]$. Recalling the strict monotonicity of $a(t) = \gamma(t) + t$ and $b(t) = \gamma(t) - t$, we see that the event families $\{A_t : t \geq 0\}$ and $\{B_t : t \geq 0\}$ are each nonincreasing. Moreover, the functions $G^+ : [0,\infty) \to [0,1]$ and $G^- : [0,\infty) \to [0,1]$ defined by $G^+(t) = P(A_t)$ and $G^-(t) = P(B_t)$ are each nonincreasing, right continuous, and vanishing at infinity. Thus the function $\Gamma : (-\infty,\infty) \to [0,1]$ given by

$$\Gamma(t) = \begin{cases} G^-(t+) & t < 0 \\ G^-(0) + P(X = \mu), & t = 0 \\ 1 - G^+(t), & t > 0 \end{cases} \tag{6.2}$$

is a distribution function. The one-sided tail functions of Γ are G^+ and G^-; we have already seen the usefulness of the tail-sum corresponding to Γ, namely $G(t) = G^+(t)$ ⊢

$G^-(t) = P(|X - \gamma(t)| > t)$. It is easy to check that $t_0(\Gamma) = t_0(F)$ and $\mu(\Gamma) = 0$. Moreover, due to $|\gamma'(t)| < 1$ and (3.28), we have

$$\left| \frac{G^+(t) - G^-(t)}{1 - G(t)} \right| < 1 \qquad (t > t_0). \tag{6.3}$$

We note that many of the monotonicity properties given here are useful in other problems where centering is a delicate matter.

Thus each nondegenerate distribution F determines what we will term a "corrected" distribution Γ: Specifically, Γ has central median at 0, and has tails $G^+(t) = 1 - \Gamma(t)$, $G^-(t) = \Gamma(-t-)$, and $G(t) = G^+(t) + G^-(t)$ which satisfy (6.3) for $t > t_0 = t_0(\Gamma)$.

We will show, conversely, that every "corrected" distribution Γ (in the sense above) arises from some nondegenerate distribution F (i.e., distributions F exist with $t_0(F) = t_0(\Gamma)$ such that the construction leading to (6.2) holds). Moreover, $\gamma = \gamma(F)$ can be explicitly produced in terms of Γ.

So, given a distribution Γ with central median zero and with tails G^+, G^-, G satisfying (6.3) for $t > t_0 = t_0(\Gamma)$, let μ be any number we desire as the central median of the distribution F we wish to construct.

For $t \geq t_0$, define

$$\gamma(t) = \mu + \int_{t_0}^t \frac{G^+(s) - G^-(s)}{1 - G(s)} ds, \tag{6.4}$$

and define $\gamma(t) = \mu$ for $0 \leq t \leq t_0$ if $t_0 > 0$. By (6.3), γ is absolutely continuous on $[0, \infty)$ with $|\gamma'(t)| < 1$ (for all but countably many values of t), and of course $\gamma(0) = \mu$. Thus the functions $a(t) = \gamma(t) + t$ and $b(t) = \gamma(t) - t$ are each absolutely continuous and strictly monotone on $[0, \infty)$. We have $a(0) = b(0) = \gamma(0) = \mu$. We now define

$F : (-\infty, \infty) \to [0, 1]$ by

$$F(t) = \begin{cases} \Gamma(-b^{-1}(t)), & t < \mu \\ \Gamma(a^{-1}(t)), & t \geq \mu. \end{cases} \tag{6.5}$$

Evidently F is actually a distribution; moreover, it is easy to check that $t_0(F) = t_0(\Gamma)$ and that the central median of F is μ, using (6.4)-(6.5).

It remains to check that $\gamma(F)$ is actually the function γ defined by (6.4). It suffices to check that $e(t, \gamma(t)) = 0$ for $t > 0$, recalling (3.4). Now $e(t, \gamma(t))$ is continuous in t; certainly $\lim_{t \to 0} e(t, \gamma(t)) = e(0, \gamma(0)) = 0$, and so it will suffice to demonstrate that $\frac{d}{dt} e(t, \gamma(t)) = 0$ for all but countably many values of t. Using the properties consequent on (6.4)-(6.5) along with Corollary 2.9 and the chain rule, we have

$$
\begin{aligned}
\frac{d}{dt} e(t, \gamma(t)) &= e_1(t, \gamma(t)) + e_2(t, \gamma(t)) \gamma'(t) \\
&= 1 - F(\gamma(t) + t) - F(\gamma(t) - t-) \\
&\quad - \{F(\gamma(t) + t) - F(\gamma(t) - t-)\} \left(\frac{G^+(t) - G^-(t)}{1 - G(t)} \right) \\
&= 1 - F(a(t)) - F(b(t)-) \\
&\quad - \{F(a(t)) - F(b(t)-)\} \left(\frac{G^+(t) - G^-(t)}{1 - G(t)} \right) \\
&= 1 - \Gamma(t) - \Gamma(-t-) \\
&\quad - \{\Gamma(t) - \Gamma(-t-)\} \left(\frac{1 - \Gamma(t) - \Gamma(-t-)}{\Gamma(t) - \Gamma(-t-)} \right) \\
&= 0,
\end{aligned}
\tag{6.6}
$$

for all but countably many $t > 0$. Hence $\gamma = \gamma(F)$, as desired.

The most important feature of this derivation is that every function γ of form (6.4), for a corrected Γ, is actually a CCM centering function, and of course only γ of this form are CCM functions, as we saw at the beginning. For a nontrivial concrete example of one such centering function, start with a distribution Γ having tail functions

G^+ and G^- given by

$$G^+(t) = \begin{cases} 1/2t, t \geq 1 \\ 1/2, 0 \leq t < 1 \end{cases} , \quad G^-(t) = \begin{cases} 1/2t^2, \ t \geq 1 \\ 1/2, \quad 0 \leq t < 1 \end{cases} \tag{6.7}$$

Here $t_0(\Gamma) = 1$ and 0 is the central median of γ. Moreover, Γ is corrected because the quotient in (6.3) is easily computed to be $1/(2t + 1) < 1$, for $t > t_0 = 1$. Given any number μ, we find from (6.4) that

$$\gamma(t) = \begin{cases} \mu, \qquad\qquad 0 \leq t \leq 1 \\ \mu + \frac{1}{2}\log(\frac{2t+1}{3}), \quad t > 1. \end{cases} \tag{6.8}$$

This particular example yields an eventually strictly increasing γ tending to infinity.

Example 6.9.

Here we illustrate the implemention of Theorem 5.6 in a situation where ordinary (uncensored) asymptotic normality is impossible to obtain.

Define $\alpha_j = 2^{j/2}, p_j = 1/(j+1)^2$, and $b_j = (2j)!$, for $j \geq 1$. Construct, on the same probability space, independent sequences $\{\tau_j\}$ and $\{\sigma_j\}$ which are independent of each other and satisfy $P(\tau_j = 1) = p_j, P(\tau_j = 0) = 1 - p_j$, and $P(\sigma_j = 1) = P(\sigma_j = -1) = 1/2$. Let $Y_j = \alpha_j \sigma_j$, and take $X_j = b_j$ on $[\tau_j = 1]$ and $X_j = Y_j$ on $[\tau_j = 0]$. Since

$$\sum_{j=1}^{\infty} P(X_j \neq Y_j) \leq \sum_{j=1}^{\infty} p_j < \infty, \tag{6.10}$$

we can analyze the partial sums $S_n = \sum_{j=1}^{n} X_j$ via the sums $T_n = \sum_{j=1}^{n} Y_j$.

Let $s_n^2 = varT_n + 2 = \alpha_{n+1}^2$. Then

$$T_n/s_n = \sum_{j=1}^{n} \alpha_j \sigma_j / \alpha_{n+1} = \sum_{j=1}^{n} \sigma_j / \alpha_{n+1-j} \tag{6.11}$$

where the distribution F of S is nondegenerate since var $S = \sum_{j=1}^{\infty} \alpha_j^{-2} = 1$. F is nonnormal: Indeed, F is clearly *uniform* on $[-1,1]$. Now (6.10) and $s_n \to \infty$ shows $S_n/s_n \xrightarrow{D} F$; in particular $\{S_n\}$ and $\{T_n\}$ are not asymptotically normal.

Now, by symmetry and $s_n^2 \to \infty$, the preceding remarks imply that no sequence of form $\{T_n - e_n\}$ can converge almost surely. But then (6.10) guarantees that neither can any sequence of form $\{S_n - e_n\}$ converge almost surely. The results in Weiner (1990, Section 4) now imply that (4.1) holds for $\{X_j\}$, for every $t > 0$.

Let $h(t) = (t^2 \log(t^2)) \wedge t^2$, where the logarithm is to the base two. In the sequel we will let $\{\gamma_j\}, \{v_j\}$, etc., be the auxiliary functions associated with the sequence $\{X_j\}$.

Now $t_0(X_j) = 0$ and $\mu_j = \mu(X_j) = \alpha_j$. Viewing t as time, consideration of (3.4) and use of the elementary properties of γ_j allow us to compute the time of introduction of the atoms at $-\alpha_j$ and b_j (in that order) into the central interval $[\gamma_j(t) - t, \gamma_j(t) + t]$; we note that $\gamma_j(0) = \mu_j = \alpha_j$ is always present, due to the Lipschitz property of γ_j. We can thus straight-forwardly compute

$$\gamma_j(t) = \begin{cases} \alpha_j - t(1 - 3p_j), 0 \le t < \alpha_j(1 - p_j)/(1 - 2p_j) \\ tp_j/(1 - p_j), \alpha_j(1 - p_j)/(1 - 2p_j) \le t \le b_j(1 - p_j) \\ b_j p_j \quad , t \ge b_j(1 - p_j). \end{cases} \tag{6.12}$$

It follows

$$v_j(t) = \begin{cases} t^2(1 - 3p_j + 4p_j^2)/(1 - p_j), 0 \le t \le \alpha_j(1 - p_j)/(1 - 2p_j) \\ (t^2 p_j + \alpha_j^2(1 - p_j)^2)/(1 - p_j), \alpha_j(1 - p_j)/(1 - 2p_j) \le t \le b_j \\ (1 - p_j)(\alpha_j^2 + b_j^2 p_j), t \ge b_j(1 - p_j). \end{cases} \tag{6.13}$$

(Incidentally, using (6.13) and the facts $\alpha_j \ge 1$ and $p_n \to 0$, it is now trivial to verify (4.3) directly for $t = 1$, which would be enough for Theorem 5.6 to apply.)

Now, for our choice of h, there exist $c_n \uparrow \infty$ satisfying (4.10). To find them, put

$$\delta_n = \sup\{j \ge 1 : c_n \ge \alpha_j(1 - p_j)/(1 - 2p_j)\} \wedge n \tag{6.14}$$

$$\Delta_n = \sup\{j \geq 1 : c_n \geq b_j(1 - p_j)\} \wedge n. \tag{6.15}$$

Note $n \geq \delta_n \geq \Delta_n \to \infty$ and $n - \delta_n \to \infty$; observe $b_{\Delta_n} \leq c_n/(1 - p_{\Delta_n}) \sim c_n$. We may write $c_n^2 = \theta_n \alpha_{\Delta_n}^2$, where $\{\theta_n\}$ is bounded with limits in $[1,2]$.

Let $\{\lambda_n\}$, $\{\lambda_n'\}$, etc., denote generic sequences tending to one. From (4.10) we get, as $n \to \infty$,

$$c_n^2 \log(c_n^2) = \sum_{j \leq \Delta_n} (1 - p_j)(\alpha_j^2 + b_j^2 p_j) + \sum_{\Delta_n < j \leq \delta_n} (c_n^2 p_j + \alpha_j^2(1 - p_j)^2)/(1 - p_j)$$

$$+ \sum_{\delta_n < j \leq n} c_n^2(1 - 3p_j + 4p_j^2)/(1 - p_j)$$

$$= \sum_{j \leq \delta_n} \alpha_j^2(1 - p_j) + \sum_{j \leq \Delta_n} (1 - p_j)p_j b_j^2$$

$$+ c_n^2 \left\{ \sum_{\delta_n < j \leq n} (1 - 3p_j + 4p_j^2)/(1 - p_j) + \sum_{\Delta_n < j \leq \delta_n} p_j/(1 - p_j) \right\}$$

$$= 2\lambda_n \alpha_{\delta_n}^2 + \lambda_n' b_{\Delta_n}^2 p_{\Delta_n} + c_n^2(n - \delta_n + o(1))$$

$$= 2\lambda_n c_n^2/\theta_n + (n - \delta_n)c_n^2 + o(c_n^2),$$

which leads to

$$\delta_n + \theta_n = \log c_n^2 = 2\lambda_n/\theta_n + (n - \delta_n) + o(1), \tag{6.17}$$

forcing

$$\delta_n + \theta_n = (n + 2\lambda_n/\theta_n + \log \theta_n)/2 + o(1). \tag{6.18}$$

Analysis of $f(x) = \log x + 2/x$ over $[1,2]$ shows that there is a bounded sequence $\{t_n\}$, all of whose limits belong to $[1, \sqrt{2}]$, such that

$$c_n^2 = t_n 2^{(n+1)/2}. \tag{6.19}$$

Now, recalling the development of (6.12), in particular the time of introduction of each new atom into the central interval, we can see that

$$P(|X_j - \gamma_j(c_n)| > c_n) = \begin{cases} 0, & j \leq \Delta_n \\ b_j, & \Delta_n < j \leq \delta_n \\ (1 + p_j)/2, & \delta_n < j \leq n. \end{cases} \qquad (6.20)$$

Now $h(c_n)/c_n^2 = \log c_n^2 \sim \delta_n \sim n/2$ by (6.18), so (6.20) leads to

$$\lim_{n \to \infty} \sum_{j=1}^{n} c_n^2 P(|X_j - \gamma_j(c_n)| > c_n)/h(c_n) = 1/2. \qquad (6.21)$$

Then Theorem 5.6 gives

$$\frac{S_n}{\sqrt{h(c_n)}} \approx \frac{S_n(\sqrt{t_n}2^{(n+1)/4})}{\sqrt{t_n}2^{(n+1)/4}\sqrt{n/2}} \xrightarrow{D} N(0,1), \qquad (6.22)$$

Moreover, the proof of that Theorem easily leads here to

$$\frac{T_n^h}{h(c_n)/c_n^2} \sim \frac{T_n^h}{n/2} \xrightarrow{P} 1, \qquad (6.23)$$

so that asymptotically half of the sequence $\{X_j - \gamma_j(c_n) : j \leq n\}$ actually requires censoring in this example.

BIBLIOGRAPHY

[1] Hahn, M.G. and M. Klass (1981). The multidimensional central limit theorem for arrays normed by affine transformations. *Ann. Probab.* **9**, 611-623.

[2] Hahn, M.G. and J. Kuelbs (1988). Universal asymptotic normality for conditionally trimmed sums. *Stat. & Probab. Letters* **7**, 9-15.

[3] Hahn, M.G., J. Kuelbs and D.C. Weiner (1989). The asymptotic joint distribution of self-normalized censored sums and sums-of-squares. To appear in *Ann. Probab.*

[4] Loève, M. (1977). *Probability Theory*. Springer-Verlag, New York.

[5] Pruitt, W. (1981). General one-sided laws of the iterated logarithm. *Ann. Probab.* **9**, 1-49.

[6] Weiner, D.C. (1990). Center, scale and asymptotic normality for sums of independent random variables. *Proc.* VII *Conf. Probab. in Banach Spaces* (Oberwohlfach, 1988), Birkhäuser, 287-307.

Department of Mathematics
Boston University
Boston, MA 02215
USA

A REVIEW OF SOME ASYMPTOTIC PROPERTIES OF TRIMMED SUMS OF MULTIVARIATE DATA.

R.A. Maller

1. Introduction

'Trimming' in this article will be used to describe the idea of removing points of a sample (usually 'extreme points') in order to improve the properties of estimators based on the sample. We will also consider procedures in which sample points are weighted so that the influence of the extremes is reduced. The sample will always consist of n independent and identically distributed (iid) random vectors in $\mathbb{R}^d$, and our emphasis will be on the (asymptotic) behaviour of that portion of it which remains after deleting or downweighting the extremes, rather than in the behaviour of the extremes themselves. The object of interest will be the estimation of location or scale of the distribution of the sample, or more precisely, in the 'robust' estimation of location and scale due to removal or downweighting of extremes. Thus we are led to the investigation of the asymptotic properties of 'trimmed sums' (or 'trimmed means' or 'robust variance matrices') in $\mathbb{R}^d$.

In 1-dimension, the study of trimmed sums is well advanced, as is amply demonstrated by the other articles in this volume and their references, but there seems to have been little systematic study of the higher dimensional case. The difficulties are fairly obvious: there are many ways of defining extremes of a multivariate sample, some of which no doubt are more meaningful for given data sets and distributions than for others. Conversely, with the freedom available in higher dimensions, one can probably always display data sets for which any given procedure will produce misleading results. These practical disadvantages can however be viewed positively, at least by theorists; the degrees of freedom available in multivariate samples allow for the development of radically new approaches, and provide a fertile ground for experimentation, particularly for the application of higher dimensional geometric techniques.

Although we are far from formulating all of the problems of robust location and scale estimation of multivariate data in this review, we can attempt to survey what is known in a class of procedures, and compare it with what might be useful. In the process, connections between statistics, asymptotic probability theory, and geometrical probability suggest themselves naturally; these, we hope, will attract practitioners in these fields to further a fascinating and useful research area.

2.   Overview

The sample will always be denoted by $X_1, \ldots, X_n$, which are points in $\mathbb{R}^d$ representing observations on a random variable X. Points in $\mathbb{R}^d$ will also be denoted as column vectors with d components. An obvious way to proceed is to trim the point furthest from the origin, in other words that X_i for which $|X_i|$ is largest. Call this point $X_n^{(1)}$, the first extreme, then define $X_n^{(2)}$, $X_n^{(3)}$, etc, as the next most distant points from 0. This version of multivariate extremes is a generalisation of 'absolute value trimming' in 1-dimension which we could call 'spherical extremes';

it leads to a relatively straightforward theory of spherical trimming which forms one member of a class of trimming procedures which is discussed in Section 3. (We usually assume continuity properties on the distribution of X to ensure that $X_n^{(i)}$ are uniquely defined although this is often not necessary.)

A general approach to the asymptotics of trimmed sums may be formulated as follows. Supposing a fixed number (r, say) of extremes have been defined in some way, let $^{(r)}S_n$ denote the summation of the remaining $n - r$ sample points. In investigating the asymptotic properties of $^{(r)}S_n$, the behaviour of r is crucial. If $r(1 \leq r \leq n)$ is kept fixed as $n \to \infty$, we speak of 'light trimming'. If $r = [n\alpha]$ (or $r/n \to \alpha$), we are in the case of 'heavy trimming', where a proportion of the sample is trimmed, and if $r = r_n$ but $r_n/n \to 0$, we call the trimming 'intermediate'. These terms are not in common usage in statistics, since the concepts seem not to have been considered there; but it is now well established that the cases lead to radically different behaviours of $^{(r)}S_n$ as $n \to \infty$.

In fact, typical behaviour seems to be as follows, at least regarding convergence in distribution. Light trimming makes 'no difference' asymptotically in that the trimmed or untrimmed sums converge to normality together (after affine normalisation, ie appropriate centering and norming). Heavy trimming 'always' gives good behaviour, eg a limiting distribution which may be the normal or a mixture of normals. Intermediate trimming produces a range of possibilities depending on the tail behaviour of the underlying distribution. We hasten to add that exceptions may be found to these 'rules', but they provide a useful starting point for analysis.

Spherical trimming and straightforward generalisations of it (trimming by cuboids, ellipsoids, etc, defined without reference to the sample) have an obvious disadvantage; they depend strongly on the origin from which the data is measured. Trying to avoid this leads in some interesting directions. One idea is to allow a sphere to 'float' about the sample and determine extremes as those points on the surface of the sphere of minimum volume which encloses the data. Having removed this 'layer' of extremes, we can define second, third,... layers; and rather than spheres, any family of convex sets may be used. This gives extremes and the corresponding trimmed sums defined in terms of 'minimum covering sets'. The price to be paid for freedom from a fixed origin is that the number of extremes in a layer will be random. Nevertheless some of the theory goes through, as we show in Section 4.

In the search for other affine equivariant (location and scale invariant) estimators we consider in Section 5 'projection pursuit' estimators, whereby a multivariate sample is projected onto a 1-dimensional subspace, trimming or other robust procedures are performed in this space, and the results are optimised in some way over all possible subspaces. There are statistical reasons (related to concepts of 'breakdown') for expecting estimates such as these to have very good properties, and certainly they are attractive intuitively. The asymptotic theory however is not well developed.

Section 6 considers 'convex hull' trimming. As its name suggests, it consists simply of trimming (or 'peeling') the convex hull of the sample points layer by layer. The little that is known of this procedure is summarised in Section 6, together with reference to related procedures such as the 'depth peeling' of Tukey.

Section 7 contains a brief discussion of another class of estimators related to the 'M-estimators' of Huber and to the 'random means' studied by Bickel and Shorack. Here the idea is quite general and consists in essence of 'testing' a point and excluding it from the trimmed sum if it is judged to be too discrepant from the rest of the sample. A great variety of procedures arise in this way.

3.   Spherical etc.

We use this relatively simple setup to illustrate a traditional method of attack on the asymptotics of a trimmed sum, which involves the use of some classical methods of the limit theory of sums and extremes. The essential ideas have been known in the theory of order statistics for a long time; for 1-dimensional applications in our context, early use of them was made by Arov and Bobrov(1960), Bickel(1965).

Let $\{S(y)\}, y > 0$, be a fixed family of subsets of $\mathbb{R}^d$, indexed by the non-negative real y, for which $S(y)$ and $\partial S(y)$, the boundary of $S(y)$, are bounded and measurable, satisfying

$$\{0\} = S(0) \subseteq S(y_1) \subseteq S(y_2) \subseteq S(+\infty) = \mathbb{R}^d$$

for $0 \leq y_1 \leq y_2 \leq +\infty$. Assume $S(y)$ increase continuously in the sense that for $x \neq 0$ in $\mathbb{R}^d$ there is a unique $y > 0$ such that $x \in \partial S(y)$; this y can be taken to be the 'discrepancy' of x from 0. Assuming also that F, the distribution of X_1, is continuous with respect to $S(y)$ in the sense that $F(\partial S(y)) = 0$, the family $\{S(y)\}$ can be used to order (a.s. uniquely) the sample $X_1, \ldots, X_n$, according to decreasing discrepancy from 0, resulting in $X_n^{(1)}, \ldots, X_n^{(n)}$.

Simple examples of $\{S(y)\}$ are spheres $S(y) = \{x : |\ x\ | \leq y\}$, cuboids, ellipsoids, etc, or $S(y)$ may be generated by $S(y) = yS$, where S is a fixed convex compact set of nonzero measure containing 0 ('Minkowski distance'). The ordering here is essentially 1-dimensional and the distribution of the extremes is governed by the 1-dimensional continuous distribution function

$$h(y) = P\{X \in S(y)\}, y > 0.$$

As with 1-dimensional order statistics, $X_n^{(j)}$ satisfy a Markov property of the form: if $y_1 \geq y_2 \ldots \geq y_r$, then

$$\{X_n^{(j)}, r+1 \leq j \leq n \mid X_n^{(k)} \in \partial S(y_k), 1 \leq k \leq r\}$$
$$\stackrel{D}{=} \{X_n^{(j)}, r+1 \leq j \leq n \mid X_n^{(r)} \in \partial S(y_r)\}$$
$$\stackrel{D}{=} \{X_{n-r}^{(j)}(y_r), 1 \leq j \leq n-r\}$$

where the notation denotes equality of multivariate distributions, conditioned as indicated. Here $X_i(y)$ are random variables having the distribution of $(X \mid X \in S(y))$, and $X_n^{(j)}(y_r)$ is the ordering induced on $X_1(y_r), \ldots X_n(y_r)$ by $\{S(y)\}$. The required integral property is easily obtained from this as:

$$P\{^{(r)}S_n \in A\} = \int_0^\infty P\{S_{n-r}(y) \in A\}dP\{X_n^{(r)} \in S(y)\} \qquad (3.1)$$

where $^{(r)}S_n = X_n^{(r+1)} + \ldots + X_n^{(n)}$ is the trimmed sum, $S_{n-r}(y)$ is the sum of $n - r$ iid copies of $X(y)$, and A is a Borel set in $\mathbb{R}^d$.

Formula (3.1) is an attractive starting point for analysis of $^{(r)}S_n$, whose distribution is represented as the mixture of the distributions of $S(y)$ and of the r-th order statistic $X_n^{(r)}$. Consider for example the heavy trimming case when $r = [n\alpha], 0 < \alpha < 1$. Then $X_n^{(r)}$, like its 1-dimensional counterpart, behaves well; in fact, if Φ is the Gaussian distribution function, as $n \to \infty$,

$$P\{X_n^{[n\alpha]} \in S(a + yn^{-1/2})\} \to \Phi\{yh'(a)(\alpha(1-\alpha))^{-1/2}\}$$

when $h'(a)$ exists and is nonzero. Since $S_{n-[n\alpha]}(y)$ is asymptotically normal after appropriate norming and centering, it is plausible from (3.1) that the same is true of $^{[n\alpha]}S_n$, and this and various related results are proved in Maller(1988a). Furthermore (3.1) is clearly well adapted to studying the rate of convergence of $^{[n\alpha]}S_n$ to its limit, and this is also looked at in Maller(1988a); the higher dimensional rate of convergence ideas are taken from Bhattacharya(1977) and Sweeting(1977). What results from all this is a generalisation of 1-dimension heavy trimming results due to Bickel(1965), Stigler(1973), Griffin and Pruitt(1987,1989), Csörgő, Haeusler, and Mason(1988); see also Hall(1984) for rates of convergence. After publication of Maller(1988a), I received a copy of Egorov and Nevzorov(1981), in which a similar method is also used to obtain rates of convergence of multivariate trimmed sums to normality, when the extremes are defined in terms of convex functionals of the data.

Again, (3.1) is a good starting point for investigation of light or intermediate trimming, but we defer a discussion of this to the next section where it occurs in a more complex form. Similarly (3.1) or similar versions are used by Griffin and Pruitt(1987), for 1-dimensional intermediate trimming results. On the other hand an entirely different approach to the 1-dimensional problem is taken by Csörgő, Csörgő, Haeusler and Mason(1986), Csörgő, Haeusler, and Mason(1986, 1988, 1990), using empirical distribution functions; see Section 6 for some discussion of this.

A major drawback of this type of trimming is its dependence on the choice of origin. In fact it only seems a sensible procedure (and the asymptotic results are certainly simpler to formulate) if the distribution function F is symmetric (possibly after translation by a fixed constant) with respect to $S(y)$ at least in the sense that $E(X \mid X \in S(y)) = 0$ for $y > 0$. This is not to preclude its use , possibly after initial centering and/or symmetrising, just as in 1-dimension, trimming by absolute values (according to the extremes of $\mid X_i \mid$) is more than just a probabilistic convenience (see eg Shorack(1974 example 3)). However the search for location invariant methods seems called for by (statistical) applications, and we turn to such a method next.

4. Minimum Covering Sets

Consider n random points in 2 dimensions with continuous distribution function, and let R_1 be the smallest circle which encloses them. As in Daniels(1952), the circle will pass through (a.s.) 2 or 3 of the sample points, which we may call the first layer of extremes. Remove them from the sample and repeat to obtain a second layer specified by a smallest circle R_2 , then iterate the procedure to obtain R_3, etc. But there is no need to restrict ourselves to 2 dimensions or to circles; in general if $\{R\}$ is a class of convex sets, layers containing (random) numbers of d-dimensional points can be defined. After removing a specified number of such layers, and a random number $\tilde{v}$ of points, ('vertices'), let $^{(v)}S_n$ be the sum of the remaining points, upon which we can base a robust estimate of the location of F.

Such procedures have been used (and recommended) in statistics. Titterington(1975,1978) and Silverman and Titterington(1980) investigate the case when $\{R\}$ is the class of d-dimensional ellipsoids, producing the 'minimum covering ellipsoid'. In one of those elegant dualities which connect two seemingly disparate fields, Sibson(1972) and Titterington(1978) proved a conjecture of Silvey to show that the minimum covering ellipse is the solution to an optimal design problem.

Let us assume two properties: that the covering sets and the distribution function F of the sample are such that the points in the layers are defined uniquely (a kind of continuity property); and that the number of points removed at each stage is bounded by K, say, with probability approaching 1, i.e. $P(\tilde{v} > K) \to 0$ as $n \to \infty$. Titterington(1978) shows that both properties hold for the minimum covering ellipse if F is continuous; the number of trimmed points in each layer is between $d + 1$ and $(d + 3)/2$ almost surely.

Another version of this type of trimming, called polyhedral or polygonal trimming trimming is used in Maller(1990b); there we take R to be the (not necessarily bounded) convex polyhedrum with sides perpendicular to a fixed set of directions $u_1, u_2, \ldots, u_K$, in $\mathbb{R}^d$. If F is continuous the required properties are clearly satisfied.

Following the prescription outlined in Section 3, we try to find analogues of the Markov property and the integral representation (3.1). In fact this can be done sufficiently explicitly to prove the light trimming result, that $^{(v)}S_n$ is asymptotically normal (after norming and centering) if and only if the same is true of S_n, the ordinary sample sum. Let us sketch the main steps in the proof of this, since they suggest that the light trimming result will hold very generally.

Denote by $X_{i_1}, \ldots, X_{i_v}$, the points removed form the sample, and by R_v the minimum covering member of $\{R\}$ determined (a.s.) by the remaining points, when

$\tilde{v} = v$. The trimmed sum is

$$^{(v)}S_n = \sum_{i=1}^{n} X_i - \sum_{j=1}^{v} X_{i_j}.$$

The distribution of

$$\{X_i, i \notin (i_1, \ldots, i_v)\} \quad \text{conditional on} \quad X_{i_1}, \ldots, X_{i_v}, \text{ and } \tilde{v} = v \qquad (4.1)$$

is the same as the distribution of $n - v$ iid copies of $X(R_v)$, where $X(R) \overset{D}{=} (X \mid X \in R)$. From this follows the integral representation (c.f.(3.1))

$$P\{^{(v)}S_n \in A\} = \sum_{v=1}^{n} \int_{A_v} P\{S_{n-v}(R_v) \in A\} dP_v\{x_1, \ldots, x_v\} \qquad (4.2)$$

where A_v represents the constraints on the trimmed points $X_{i_1}, \ldots, X_{i_v}$ and P_v is the joint distribution of $X_{i_1}, \ldots, X_{i_v}$ and $\tilde{v}$.

From (4.2) it can be proved that $T_n(^{(v)}S_n - A_n)$ converges in distribution to normality iff $T_n(S_n - A_n)$ does so, for nonstochastic sequences of vectors A_n, and nonsingular $d \times d$ matrices T_n (and the constants may be chosen to be the same in each case) which is what we refer to as the 'light trimming' result. In fact we reduce this to a 1-dimensional problem by a 'Cramer-Wold' device of projecting onto 1 dimension: if u is an arbitrary unit vector in $\mathbb{R}^d$, (4.2) gives for $x > 0$

$$8P\{| u^T T_n(^{(v)}S_n - A_n)| > x\} \geq P\{\sup_{(v)} | u^T T_n X |> 5x\}$$

by ordinary (1-dimensional) probability arguments such as in Maller(1982). Here $\sup_{(v)} |u^T T_n X|$ means the maximum value of $|u^T T_n X_i|$ for those X_i remaining after the $\tilde{v}$ extremes have been removed and 'T' denotes the transpose of a vector. Since at most K terms can be removed in this way, (with probability approaching 1) we obtain

$$8P\{|u^T T_n(^{(v)}S_n - A_n)| > x\} \geq P\{|u^T T_n X|_n^{(n-K)} > 5x\} + o(1)$$

where $|u^T T_n X|_n^{(i)}$ are the order statistics of $|u^T T_n X_i|$. Thus convergence of $T_n(^{(v)}S_n - A_n)$ to normality implies a bound of $0(exp(-cx^2))$ on the tail of the distribution of $| u^T T_n X|_n^{(n-K)}$, and a lemma in Maller(1990b) shows that this implies a bound of the same order on $|u^T T_n X|_n^{(n)}$, the largest of $|u^T T_n X_i|$. This in turn gives a bound of the same order on $u^T T_n(S_n - A_n)$ via

$$|u^T T_n^{(v)} S_n - u^T T_n S_n| \leq K|u^T T_n X|_n^{(n)} \tag{4.3}$$

(with probability approaching 1, since $\tilde{v} \leq K$ with probability approaching 1.)

A result of Sato(1973) states that the tail of an infinitely divisible distribution with nonzero Lévy measure ν can decrease no faster than $0(exp(-cxlog(x))$ as $x \to \infty$ for some $c > 0$. Taking subsequential limits of $u^T T_n(S_n - A_n)$, which are infinitely divisible random variables with Lévy tails ν, and deducing that nonzero ν is impossible in view of the bound $0(exp(-cx^2))$ on the tail of $u^T T_n(S_n - A_n)$, it is then easy to conclude that $T_n(S_n - A_n)$ converges to normality, ie, to a nondegenerate infinitely divisible random variable with zero Lévy tail.

Conversely, if $T_n(S_n - A_n)$ is asymptotically normal, then $\max_{1 \leq i \leq n} |T_n X_i| \xrightarrow{P} 0$ by standard arguments, and so $T_n(^{(v)}S_n - A_n)$ is asymptotically normal by (4.3).

When $d = 1$, the choice $R = \{$closed intervals$\}$ trims equal numbers of large and small values from the sample; trimming from one end or another is obtained by taking intervals of the form $(-\infty, x]$ or $[x, +\infty)$, while defining the layers in terms of

$|X_i|$ with $R = \{$intervals of the form $[0, x]\}$ gives absolute value trimming. The light trimming result for the last case was given in Maller(1982) when the distribution function of X_i is continuous and symmetric, and in general by Mori(1984). For the other cases in 1-dimension see Csörgő, Hauesler and Mason(1988,1990).

The mode of convergence used in the above result, ie after normalisation by a matrix T_n of the centered vector $S_n - A_n$, was introduced by Hahn and Klass(1980,1981), as a general formulation for asymptotic normality of sums of iid random vectors. It was also used by Griffin(1986) to study stochastic compactness in $\mathbb{R}^d$ of such sums. Further discussion is given in Maller(1990b).

Proving a heavy trimming result in the context of this section presents a nice challenge. We ask for the limiting distribution of $^{(v)}S_n$, after stripping by layers a proportion as close as possible to $[n\alpha]$ of points from the sample. The representation (4.2) would likely be of use here but needs to be supplemented with a study of the number of points removed. We conjecture that order $n^{1/2}$ convergence would result in most cases. Note that $^{(v)}S_n$, if trimming were done by minimum covering ellipsoids, would then differ from the following estimator suggested in essence by Rousseauw(1986): find the smallest ellipse which contains $[n\alpha]$ of the observations , then sum these points. When $\alpha = 1/2$ and $d = 1$ this kind of estimator has order $n^{1/3}$ limiting behaviour rather than $n^{1/2}$; see Davies(1987) for a discussion and modification of some of these ideas, also Grubel(1988) for the (order $n^{1/3}$) behaviour of a scale estimator based on similar methods.

5. Projection Pursuit

In 1-dimension extremes of a sample are easy to define, and a very fruitful idea in statistics is to project higher dimensional data onto the one dimensional subspace defined by a unit direction vector u, define a statistic in terms of the order statistics (say) of the resulting 1-dimensional points $u^T X_i, 1 \le i \le n$, then optimise in some way over u. By these means a number of new multivariate statistical methods, not just related to robust estimation of location and scale, have been suggested in recent years. Important papers promoting this 'projection pursuit' idea are Friedman and Tukey(1974), Huber(1985), Jones and Sibson (1987), Diaconis and Friedman(1984).

We consider here a definition of location and scale estimators which is a slight generalisation of one due to Donoho, Huber and Stahel (see Maller(1990a) for references). We begin with a robust measure of the discrepancy of a projected point $u^T x$ defined by:

$$\hat{d}^2(x,u) = \{u^T x - \hat{\mu}_\alpha(u)\}^2 / \hat{\sigma}_\alpha^2(u)$$

where $\hat{\mu}_\alpha(u)$ and $\hat{\sigma}_\alpha(u)$ are preliminary (robust) estimates of location and scale along direction u, and x is any point in $\mathbb{R}^d$. Then taking

$$\hat{d}(x) = \sup_u \hat{d}(x,u) \tag{5.1}$$

where u ranges over unit vectors in $\mathbb{R}^d$, gives a direction-wise maximum discrepancy of point x. As a robust estimate of multivariate location in the sample $X_1, \ldots, X_n$, we can now define (when I denotes the indicator function and the denominator is nonzero)

$$\hat{\mu}_n = \sum_{i=1}^n X_i I(\hat{d}_i < a) / \sum_{i=1}^n I(\hat{d}_i < a) \tag{5.2}$$

where $\hat{d}_i = \hat{d}(X_i)$ and a is a fixed constant. For the preliminary estimators $\hat{\mu}_\alpha(u)$ and $\hat{\sigma}_\alpha^2(u)$ we can use any 1-dimensional estimators; previous experience as in Section 3 tells us that taking them to be the heavily trimmed mean and variance in direction u will at least guarantee an asymptotic limit for them , so we let $(u^T X)_n^{(i)}$ be the order statistics of $u^T X_i$ and define

$$\hat{\mu}_\alpha = \sum_{i=[n\alpha]+1}^{n-[n\alpha]} (u^T X)_n^{(i)}/(n - 2[n\alpha]), \qquad (5.3)$$

$$\hat{\sigma}_\alpha^2 = \sum_{i=[n\alpha]+1}^{n-[n\alpha]} \{(u^T X)_n^{(i)} - \hat{\mu}_\alpha(u)\}^2/(n - 2[n\alpha]). \qquad (5.4)$$

Now a new problem arises; we intend to take the supremum over u as in (5.1), so we will require uniform convergence over directions u in an asymptotic analysis of (5.3) and (5.4). This is available in fact from the Vapnik-Cervonenkis theory (see Shorack and Wellner(1985)), and it seems plausible then that $\hat{d}_i$ is asymptotically equivalent to

$$d_i^2 = \sup_u (u^T X_i - u^T \mu)^2/\sigma_\alpha^2(u)$$

$$= (X - \mu)^T \Sigma_\alpha^{-1}(X - \mu)$$

where $u^T \mu$ and $\sigma_\alpha^2(u) = u^T \Sigma_\alpha u$ are the limits of $\hat{\mu}_\alpha(u)$ and $\hat{\sigma}_\alpha^2(u)$, if indeed it turns out that they are expressible in this form for some vector μ and $d \times d$ nonsingular matrix Σ_α. It would follow that the limiting behaviour of $\hat{\mu}_n$ in (5.2) might be similar to that of

$$\mu_n = \sum_{i=1}^{n} X_i I(d_i < a)/\sum_{i=1}^{n} I(d_i < a) \qquad (5.5)$$

which is easily worked out since (5.5) involves only independent and identically distributed random vectors.

The above optimistic sketch can in fact be followed through at least in respect to the consistency (a.s. convergence of $\hat{\mu}_n$), but not necessarily in respect of possible asymptotic distributions of $\hat{\mu}_n$. The first question is: what do $\hat{\mu}_n$ and a scale estimate $\hat{\Sigma}_n$ (defined in a similar way) actually estimate? A rather general and highly useful model for 'what we are trying to estimate' in robust statistical analysis was formulated by Collins(1976,1982) as a class of distributions 'elliptically symmetric in a central region'. The distribution function F of the sample is assumed to depend on $x \in \mathbb{R}^d$ only via $(x - \mu)\Sigma^{-1}(x - \mu)$ for values of this expression less than R for some given $R > 0$, and to be arbitrary otherwise, for some μ and Σ. This provides us with location and scale parameters (a nonsingular $d \times d$ matrix)to be estimated, and allows for the possibility of wild values or outliers occurring outside a central symmetric region. We need to assume somewhat more than this, in fact that the distribution is 'projection symmetric' in a central region in the sense that $u^T \Sigma^{-1/2}(X - \mu)$ is symmetrically distributed about 0 in a central region. This is more general than requiring spherical symmetry of the distribution or 'centro symmetry' (Huber(1985)), ie that $X - \mu$ and $-(X - \mu)$ have the same distribution, but it does imply symmetry of a kind outside the central region, as pointed out by a referee.

Under this assumption, $\hat{\mu}_n$ is consistent for μ, ie $\hat{\mu}_n \to \mu$ a.s., provided $P(d(X) = a) = 0$ and $P(d(X) < a) > 0$; and $\hat{\Sigma}_n$ is consistent for Σ apart from a multiplicative factor. These results are proved (in a slightly different setup involving a continuous weighting function rather than the indicator function in (5.2)) in Maller(1990a). The problem of finding the asymptotic distribution of $\hat{\mu}_n$ has not been solved for $d > 1$; preliminary calculations suggest that the convergence is $0(n^{1/2})$ but that the limiting distribution is not normal although it may be expressed as a mixture of normals. In 1-dimension, the limiting distribution is however normal, as can be proved by methods related to those of Shorack(1974).

6.   Convex Hull Trimming

The study of the extreme points of the convex hull (CH) of points in $\mathbb{R}^d$ forms an elegant and difficult topic in probability theory whose usefulness is reflected in applications suggested for statistics and in many other areas of Mathematics, especially to linear programming. Clearly most reasonable methods of defining extremes in a multivariate sample will designate a subset, at least, of the extremes of the CH, and the idea of 'peeling' the CH, ie stripping off outer layers, seems to have first been advanced by Tukey (Green(1981)).

An obvious proposal then is to define $^{(r)}S_n$ as the sum of the points remaining after peeling r layers of the CH, and ask for the asymptotic behaviour of $^{(r)}S_n$. More specifically, following the program outlined in Section 4, we can ask for light, medium and heavy trimming results.

One approach is to approximate the CH extremes by the polygonal extremes as defined in Section 4, from which it is possible to prove the light trimming result, that $^{(1)}S_n$ is asymptotically normal iff S_n is, (after affine normalisation), for data from an elliptically symmetric distribution with a relatively 'heavy' tail, more precisely, such that $P\{|X| > x\}$ is regularly varying with negative index as $x \to \infty$. This is useful since we expect outliers to be generated by some such model, but after the results of Section 4, we can conjecture that this result should hold without any symmetry or tail restrictions on F. Referees advanced a simpler proof of the above under the same restrictions, essentially by reducing the problem to spherical trimming and using the 1-dimensional techniques of Maller(1982) and Mori(1984). Interested readers may be able to construct this proof for themselves. There are not, so far, analogues of medium or heavy trimming results for the CH trimmed sum, and such would certainly be of interest. Small simulations are relatively easy to do, and readers with a personal computer are invited to try their own.

The study of the convex hull extremes themselves has attracted some powerful probabilists, and some nice results have been obtained, usually for $d = 2$ and assuming F to be elliptically symmetric. Of great interest is the behaviour of the expected number of extremes, $E(V_n)$, for which see Carnal(1970), Raynaud(1970) and their references. These authors show among other things that $E(V_n)$ is smaller for heavier tailed distributions. It is not hard to give a fairly simple characterisation (in terms of functions of the tail of F) of when $E(V_n)$ remains bounded as $n \to \infty$, which is to appear elsewhere. For other results see Davis, Mulrow and Resnick(1987),Groeneboom(1988), Aldous et al(1991).

Peeling the CH has been criticised as a statistical procedure by some (eg Silverman and Titterington(1980)) on the grounds that in higher dimensions 'too many' points are removed (Johnstone(1987)). What is required is a 'finer trim' , and the methods of Section 4, such as minimum covering ellipsoid or polyhedral trimming, provide this. Another suggestion called 'depth trimming', also due to Tukey(1947), (see also Wald(1943)) is outlined in Donoho and Huber(1983); it again designates a subset of the CH extremes (at least for the first layer), but looks even more complicated to handle theoretically.

7.   'Random Means'

Another way of removing the dependence on the origin of measurement is as follows. Let $\hat{\mu}_n(1)$ and $\hat{\Sigma}_n(1)$ be preliminary (robust) estimates of location and scale, $\hat{\Sigma}_n(1)$ being a symmetric nonsingular $d \times d$ matrix. A class of location estimators is given by

$$\hat{\mu}_n(2) = \sum_{i=1}^{n} X_i I(X_i \in \hat{R}_n) / \sum_{i=1}^{n} I(X_i \in \hat{R}_n) \tag{7.1}$$

where $\hat{R}_n$ is a subset of $\mathbb{R}^d$ possibly depending on the sample, $X_1, \ldots, X_n$, as the hat is meant to indicate. For example, take

$$\hat{R}_n = \{X_i : [X_i - \hat{\mu}_n(1)]^T \hat{\Sigma}_n^{-1}(1)[X_i - \hat{\mu}_n(1)] < a\} \tag{7.2}$$

for some constant a. Similarly, letting

$$\hat{\Sigma}_n(2) = \sum_{i=1}^{n} \{X_i - \hat{\mu}_n(1)\}\{X_i - \hat{\mu}_n(1)\}^T I(X_i \in \hat{R}_n) / \sum_{i=1}^{n} I(X_i \in \hat{R}_n) \tag{7.3}$$

we can ask for the asymptotic behaviour of $\hat{\mu}_n(2)$ and $\hat{\Sigma}_n(2)$. In 1-dimension these types of estimators have been explored by Shorack(1974), (see also Mason and Shorack (1990)), using empirical distribution function methods, but not, to my knowledge, in higher dimensions, where the empirical distribution function methods seem ill-adapted.

The behaviour of these quantities will of course depend on that of $\hat{\mu}_n(1)$ and $\hat{\Sigma}_n(1)$, for which we could use any of the other types of estimators in previous sections of this paper. Estimators like (7.1) are only 'trimmed means' for certain choices of $\hat{R}_n$; taking for example

$$\hat{R}_n = \{X_i : |X_i| < |X|_n^{n-r}\}$$

where $|X|_n^{(i)}$ are the order statistics of $|X_i|$, $1 \leq i \leq n$, recovers the spherical trimming defined in Section 2. Also related is a class of trimmed/truncated (or 'censored') estimators considered by Hahn, Kuelbs and Zamur(1987).

Suppose now that (7.1)-(7.3) are iterated by replacing $\hat{\mu}_n(1)$ and $\hat{\mu}_n(2)$ by $\hat{\mu}_n(2)$ and $\hat{\mu}_n(3)$ and similarly for $\hat{\Sigma}_n(1)$ and $\hat{\Sigma}_n(2)$. Continuing this process suggests the study of location and scale estimators satisfying, for example,

$$\hat{\mu} = \sum_{i=1}^{n} X_i I(d(X_i) < a) / \sum_{i=1}^{n} I(d(X_i) < a) \qquad (7.4)$$

$$\hat{\Sigma} = \sum_{i=1}^{n} \{X - \hat{\mu}\}\{X - \hat{\mu}\}^T I(d(X_i) < a) / \sum_{i=1}^{n} I(d(X_i) < a) \qquad (7.5)$$

which define $\hat{\mu}$ and $\hat{\Sigma}$ implicitly via

$$d^2(X_i) = \{X_i - \hat{\mu}\}^T \hat{\Sigma}^{-1} \{X_i - \hat{\mu}\}.$$

We have thus arrived at some 'M-Estimators' similar to those studied by Maronna (1976) (he used a continuous weight function rather than the indicator $I(.)$).

Maronna proves the existence and uniqueness of solutions to (7.4)-(7.5) under fairly heavy restrictions on F, and that the resulting estimators converge a.s. to quantities which satisfy analogues of (7.4) and (7.5) 'in the limit', ie after replacing summation by expectation. He also proves asymptotic normality under some conditions. Maronna and others have however expressed doubts about the statistical usefulness of these estimators and alternatives have been suggested by Tyler(1981).

Somewhat related to this class are trimmed estimators that can be constructed by a procedure similar to one discussed by Barnett and Lewis(1984, page 210). Let $\hat{\mu}_{(i)}$ and $\hat{\Sigma}_{(i)}$ be (robust) estimators of location and scale defined on the sample omitting X_i, and assume that $\hat{\Sigma}_{(i)}$ is nonsingular. Use

$$d^2(X_i) = [X_i - \hat{\mu}_{(i)}]^T \hat{\Sigma}_{(i)}^{-1} [X_i - \hat{\mu}_{(i)}]$$

as a discrepancy measure to order the sample, ie, let $X_n^{(1)}$ be the point for which $d(X_i)$ is largest, remove this point from the sample, then repeat to obtain $X_n^{(2)}$, etc. Let $^{(r)}S_n$ be the sum of the points remaining after r points have been removed. To my knowledge, properties of this estimator have not been investigated, but they might be good even for choices such as that of the ordinary mean and covariance matrix (of the sample omitting X_i) for $\hat{\mu}_{(i)}$ and $\hat{\Sigma}_{(i)}$, because the 'self normalising' effect of dividing out the sample (co)variance will confer some stability at least against long-tailedness, eg Maller(1981). Alternatively of course one could use any other robust estimators for them. There are connections here with the ideas of jacknifing and perhaps bootstrapping.

8.   Discussion

Research into trimmed sums lies at the intersection of at least three areas which are under vigorous current development. Firstly, the computational aspects of most of the methods discussed herein are nontrivial; by this I mean simply the numerical calculation of the estimator (if unique!) from the n sample points. For example, a good numerical procedure is not at present available for calculating the projection pursuit estimator defined in Section 5; the difficulty is in performing the maximisation in (5.1) in high dimensions. Again, computation of a convex hull in high dimensions is currently computationally highly expensive, as any Operations Researcher will testify; let alone some of the more complex procedures such as depth trimming or minimum covering sets. (For a discussion of minimum covering ellipsoids see Silverman and Titterington(1980)). Research into improved methods of calculation would be of great value.

Note that thoughout our discussion the emphasis has been on the 'one- sample' location and scale problem. In higher dimensions (perhaps not so much in 1-dimension) this is not a trivial statistical consideration since multidimension exploratory techniques often work by looking for directions of maximum variation. But the many sample (eg discrimination and allocation procedures) or (multivariate) regression techniques also need 'robustification'; see Maller(1990a) and references for a projection pursuit discriminant analysis, and Welsh(1987), and Ruppert and Carroll(1980) for example, for trimming in linear regression. Again,to use perhaps an appropriate expression, the surfaces have hardly been scratched.

A second, related, aspect is that computer simulations of these techniques are likely to be of great value in understanding their properties with or without accompanying theoretical results. Of course we must be able to calculate the estimators before we can simulate them. This connects us with another currently lively area in statistics; the viewing and analysis of multivariate data by projecting it onto a 1, 2, or 3-dimensional subspace via computer graphics, and searching within that subspace for 'interesting' properties or aspects of the sample. Suppose, for example, we could view a 3-dimensional projection via computer graphics, strip an outer layer as defined by one of our techniques, and view the effect on the sample. Repeat this a number of times for exploratory data analysis and/or let $n \to \infty$ at some rate for 'theoretical' simulation. We have moved beyond trimmed sums here to looking at the entire 'trimmed sample' ; still, we will always want to estimate location and scale.

Returning to our metaphor of interesting areas, a third of course is the probability theory itself, which has demonstrably found important application in this
subject. Most of our work in this area will be asymptotic; so what are the practical
implications of asymptotic results? Having thus formulated an old question, not yet
(and perhaps never to be) answered, we can at least see that the approach we have
suggested, of considering behaviour in terms of light, medium or heavy trimming,
brings us to the very essence of this matter. Because whether trimming is 'light'
or otherwise is completely an asymptotic property which is meaningless for a fixed
sample- yet produces vastly differing results (at least in known cases) as $n \to \infty$.

To expand on this, I have sampled opinions of statisticians on the light trimming
result, that removing a fixed number of extremes from a sample (in 1-dimension)
cannot produce asymptotic normality of the normed and centered sum unless the
untrimmed sample has this property. They usually reply that this is 'obvious' but
on reflection agree that it requires proof. And although it seems counter-intuitive,
there may well be more complex forms of trimming (in higher dimensions) for which
the light trimming result is not true! If so, it would certainly be interesting to find
an example of this kind of behaviour.

Indeed, from another point of view, especially in higher dimensions, one gets
the impression that that it is very much light trimming (or peeling) that statisticians
have had in mind. Often they talk of peeling just one or two layers. Conversely, in
1-dimension, 'trimming' in statistics has almost always been taken to mean what
we are calling 'heavy' trimming, without the acknowledgement of the possibility of

other kinds. But although trimming 10 percent from a sample of 100 points may seem reasonable, routinely discarding 100,000 points from a sample of 1,000,000 (easily obtainable these days with computers) does not. There seems to be a clear role here for 'intermediate' trimming and the adaptive methods discussed earlier in this Chapter and elsewhere in this book.

To conclude, we mention a few areas with which this report has not been concerned. We have omitted discussion of properties of the multivariate extremes themselves (however defined), whereas of course this is an important and interesting research area in itself. See de Haan and Resnick(1977), Resnick(1988), Barnett and Lewis(1984), among many others.

We have not endeavoured to survey the (enormous) robustness literature as any practitioner in that area will be aware, and merely refer to books like Huber(1985), Hampel et al (1986),etc, for introduction and usage of the terms 'breakdown', 'influence functions', etc.

A small but growing area, ultimately to be of great importance I believe, is that concerned with 'self-normalisation' , which was mentioned briefly in Section 7. The key idea is that normalising the sample mean (say) by the sample standard deviation confers some robustness in itself against long tails; extending this idea to trimming leads to estimates perhaps related to those of Section 7, where a point is first 'tested' for inclusion in the sample and omitted if it fails the test. (See Jaeckel (1971) for another kind of modification). These kinds of techniques perhaps come closest to emulating in an objective way what the statistician does in practice, and higher dimensional methods would be well worth developing. For various self normalising results in 1 dimension see Csörgö and Horvath(1988), Griffin(1989), Hahn, Kuelbs, and Weiner(1989a,b), Maller(1981), Huber(1970).

Another area we have not touched on is modes of convergence of trimmed sums other than in distribution, to normality. The almost sure (strong law and iterated log-type laws) behaviour has been well studied in 1-dimension: Griffin and Kuelbs(1989a,b), Hahn and Kuelbs(1989), Maller(1984), Maller(1988b), Mason(1982a,b), Hauesler(1990), Hauesler and Mason(1987,1990). Early results in this area are those of Feller(1968) and Mori(1976,1977). Multivariate versions of these make sense and would keep a collection of probabilists in work for some time and may even have applications, perhaps to do with computer imaging of random processes and sets (see Matheron(1975) for random sets).

There are of course other possible weak convergence behaviours such as to non-normal laws, perhaps via subsequences; in fact the whole of the classical limit theory of normed sums is in the process of being generalised, as some of the other chapters in this volume testify (See also Mori(1981) and Pruitt(1988)). And is there an application for 'trimmed' renewal theory, 'trimmed' branching processes, etc? If this sounds fanciful, for an example of the justification of some pretty theoretical results in martingales and counting processes we need look no further than the (giant) statistical subject of survival analysis- a study, from one point of view, of the order statistics of a sample after trimming or censoring.

REFERENCES

ALDOUS, D.J., FRISTEDT, B., GRIFFIN, P.S., and PRUITT, W.E.(1991) The number of extreme points in the convex hull of a random sample (preprint).

AROV, D.Z. AND BOBROV, A.A.(1960) The extreme terms of a sample and their role in the sum of independent variables. *Theor. Prob. Appl.5, 377-396.*

BARNETT, V.D. AND LEWIS, T.(1984) *Outliers in Statistical Data.* 2nd Ed, Wiley, New York.

BHATTACHARYA, R.N.(1977) Refinements of the multidimensional central limit theorem and applications. *Ann. Probab.,* 5 1-27.

BICKEL, P.J.(1965) On some robust estimates of location. *Ann. Math. Statist.,* 36, 847-848.

CARNAL, H.(1970) Die konvexe Hulle von n rotationssymmetrisch verteilten Punkten. *Z. Wahrscheinlichkeitstheorie verw. Geb,* 15, 168-179.

COLLINS, J.R.(1976) Robust estimation of a location parameter in the presence of asymmetry. *Ann. Statist.,* 4, 68-85.

COLLINS, J.R.(1982) Robust M-estimators of location vectors. *J. Mult. Anal.,* 12, 480-492.

CSÖRGŐ, M., CSÖRGŐ, S., HÓRVATH, L. AND MASON, D.M.(1986) Normal and stable convergence of integral functions of the empirical distribution function.*Ann. Prob.,* 14, 86-118.

CSÖRGŐ, S., HÓRVATH, L. AND MASON, D.M.(1986) What portion of a sample makes a partial sum asymptotically stable or normal? *Prob. Theor. Rel. Fields.* 72, 1-16.

CSÖRGŐ, S., HAEUSLER, E. AND MASON, D.M.(1988) The asymptotic distribution of trimmed sums. *Ann. Prob.*, 16, 672-699.

CSÖRGŐ, S., AND HÓRVATH, L.(1988) Asymptotic representations of self normalised sums. *Prob. and Math. Statistics*, 9.1, 15-24.

CSÖRGŐ, S., HAEUSLER, E. AND MASON, D.M.(1990) A probabilistic approach to the asymptotic distribution of sums of independent, identically distributed random variables (Preprint).

DANIELS, H.E.(1952) The covering circle of a sample from a circular normal distribution. *Biometrika* 39, 137-143.

DAVIES, P.L.(1987) Asymptotic behaviour of S-estimates of multivariate location parameters and dispersion matrices. *Ann. Statist.* 15, 1269-1292.

DAVIS, R., MULROW, E., AND RESNICK, S.(1987) The convex hull of a random sample in R^2. *Stochastic Models* 3, 1-29.

DE HAAN, L. AND RESNICK, S.I.(1977) Limit theory for multivariate sample extremes. *Z. Wahrscheinlichkeitstheorie verw. Geb.*, 40, 317-337.

DIACONIS, P. AND FREEDMAN, D.(1984) Asymptotics of graphical projection pursuit. *Ann. Statist.* 12, 793-815.

DONOHO, D. AND HUBER, P.(1983) The notion of breakdown point. In: *A Festschrift for Erich Lehmann, Bickel, Doksum Eds, (Wadsworth)*.

EGOROV, V.A., AND NEVZOROV, V.B.(1981) On a rate of convergence to a normal law of sums of induced order statistics. Notes of the Science Seminars of LOMI, 108, 45-46.

FELLER, W. (1968) An extension of the law of the iterated logarithm to variables without variance. *J. Math. Mech.* 18, 343-355.

FRIEDMAN, J. AND TUKEY, J.W.(1974) A projection pursuit algorithm for exploratory data analysis. *IEEE Transactions on Computers*, C-23, 881-889.

GREEN, P.J.(1981) Peeling bivariate data. In: *Interpreting Multivariate Data* (ed. V Barnett) Wiley, 3-18.

GRIFFIN, P.S.(1986) Matrix normalised sums of independent identically distributed random vectors. *Ann. Prob.*, 14, 224-246.

GRIFFIN, P.S.(1989) Asymptotic normality of self-normalised sums. (preprint).

GRIFFIN, P.S. AND PRUITT, W.E.(1987) The central limit problem for trimmed sums. *Math. Proc. Camb. Phil. Soc*, 102, 329-349.

GRIFFIN, P.S. AND PRUITT, W.E.(1989) Asymptotic normality and subsequential limits of trimmed sums. *Ann. Prob.* 17, 1186-1210.

GRIFFIN, P.S. AND KUELBS, J.(1989a) Self normalised laws of the iterated logarithm. (Preprint).

GRIFFIN, P.S. AND KUELBS, J.(1989b) Some extensions of the LIL via self normalisations *Ann. Prob.* 17, 1571-1601.

GROENEBOOM, P.(1988) Limit theorems for convex hulls. *Prob. Theor. Related Fields*, 79, 327-368.

GRUBEL, R.(1988) The length of the shorth. *Ann. Statist.*, 16, 619-628.

HAEUSLER, E. AND MASON, D.M.(1987) Laws of the iterated logarithm for sums of the middle portion of the sample. *Math. Proc. Camb. Phil. Soc.*, 101, 301-312.

HAEUSLER, E. AND MASON, D.M.(1990) A law of the iterated logarithm for modulus trimming. (Preprint).

HAEUSLER, E.(1990) Laws of the iterated logarithm for sums of order statistics from a distribution with a regularly varying upper tail. (Preprint).

HAHN, M.G. AND KLASS, M.J.(1980) Matrix normalisation of sums of random vectors in the domain of attraction of the multivariate normal. *Ann. Prob.*, 8, 262-280.

HAHN, M.G. AND KLASS, M.J.(1981) The multidimensional central limit theorem for arrays normed by affine transformations. *Ann. Prob.*, 9, 611-623.

HAHN, M.G., KUELBS, J. AND SAMUR, J.D.(1987) Asymptotic normality of trimmed sums of ϕ-mixing random variables. *Ann. Prob.*, 15, 1395-1418.

HAHN, M.G., AND KUELBS, J.(1989) Asymptotic normality and the LIL for trimmed sums: the general case. *J. Theoret. Prob.* 3, 137-168.

HAHN, M.G., KUELBS, J. AND WEINER, D.C.(1989a) The asymptotic joint distribution of self normalised censored sums and sums-of-squares. (Preprint).

HAHN, M.G., KUELBS, J. AND WEINER, D.C.(1989b) The asymptotic distribution of magnitude Winsorised sums via self-normalisation. *J. Theor. Prob.*, 3, 137-168.

HALL, P.(1984). On the influence of extremes on the rate of convergence in the central limit theorem. *Ann. Prob.*, 12, 154-172.

HAMPEL, F.R., ROUSSEEUW, P.J., RONCHETTI, E.M. AND STAHEL, W.A. (1986). *Robust Statistics- the Approach based on Influence Functions.* Wiley, New York.

HUBER, P.J.(1985) Projection Pursuit. *Ann. Statist,*, 13, 435-522.

HUBER, P.J.(1970) Studentizing robust estimates. *In: Nonparametric Techniques in Statistical Inference*, M.L. Puri., Cambridge Univ. Press.

JAECKEL, L.A.(1971) Some flexible estimates of location. *Ann. Math. Statist.*, 42, 1540-1552.

JOHNSTONE, I.(1987) Discussion to: Jones, M.C., and Sibson, R.: What is projection pursuit? *J. R. Statist. Soc.*, 150, 1-36.

JONES, M.C. AND SIBSON, R.(1987) What is projection pursuit? *J. R. Statist. Soc. A,* 150 1-36.

MALLER, R.A.(1981) A theorem on products of random variables, with application to regression. *Aust. J. Statist.*, 23, 25-37.

MALLER, R.A.(1982) Asymptotic normality of lightly trimmed means - a converse. *Math. Proc. Camb. Phil. Soc.*, 92, 535-545.

MALLER, R.A.(1984) Relative stability of trimmed sums.*Z. Wahrscheinlichkeitstheorie verw. Geb.*, 66, 61-80.

MALLER, R.A.(1988a) Asymptotic normality of trimmed means in higher dimensions. *Ann. Prob.*, 16, 1608-1622.

MALLER, R.A.(1988b) A functional law of the iterated logarithm for distributions in the domain of partial attraction of the normal distribution. *Stoch. Proc. Appl.*, 27, 179-194.

MALLER, R.A.(1990a) Some consistency results on projection pursuit estimators of location and scale. *Canad. J. Statist.*, 17, 81-90.

MALLER, R.A.(1990b) Defining extremes and trimming by minimum covering sets. *Stoch. Proc. Appl.* 35,

MARONNA, R.A. (1976) Robust M-estimators of multivariate location and scatter. *Ann. Statist.*1, 51-67.

MASON, D.M. (1982a) Laws of large numbers for sums of extreme values. *Ann. Prob.* 10, 754-764.

MASON, D.M. (1982b) Some characterisations of strong laws for linear functions of order statistics. *Ann. Prob.* 10, 1051-1057.

MASON, D.M. AND SHORACK, G.R.(1990) Necessary and sufficient conditions for asymptotic normality of L-statistics. (preprint).

MATHERON, G.(1975) *Random Sets and Integral Geometry*, Wiley, New York.

MORI, T.(1976) The strong law of large numbers when extreme terms are excluded from sums. *Z. Wahrscheinlichkeitstheorie verw. Geb.*, 36, 189-194.

MORI, T.(1977) Stability for sums of iid random variables when extreme terms are excluded. *Z. Wahrscheinlichkeitstheorie verw. Geb.*, 40, 159-167

MORI. T.(1981) The relation of sums and extremes of random variables. Session summary booklet: invited papers, Buenos-Aires Session, Nov 30-Dec 11, 1981 (International Statistical Institute).

MORI, T.(1984) On the limit distributions of lightly trimmed sums. *Math. Proc. Camb. Phil. Soc.*, 96, 507-516.

PRUITT, W.E.(1988) Sums of independent random variables with the extreme terms excluded. In: *Probability and Statistics*, Essays in honour of Franklin A. Graybill, J. N. Srivastava, Ed, North Holland.

RAYNAUD, H.(1970) Sur l'envelope convexe des nuages de points aleatoires dans R^n. *J. Appl. Prob.*, 7, 35-48.

RESNICK, S.(1988) Association and multivariate extreme value distributions. In: *Studies in Modelling and Statistical Science, C.C. Heyde, Ed. Aust. J. Statist. 30A, 261-271.*

ROUSSEEUW, (1986) Multivariate estimation with high breakdown point. In: *Mathematical Statistics and Its Applications*, Grossman, Vincze and Wertz, Eds, Reidal, Dordrecht, 283-297.

RUPPERT, D. AND CARROLL, R.J.(1980) Trimmed least squares estimation in the linear model. *J. Amer. Statist. Assoc.*, 75, 828-297.

SATO, K.(1973) A note on infinitely divisible distributions and their Lévy measures. *Sci. Rep. Tokyo Kyoiku Daigaku Sect A*, 12, 101-109.

SHORACK, G.(1974) Random means. *Ann. Statist.*, 2, 661-675.

SHORACK, G.R. AND WELLNER, J.A.(1985) *Empirical Processes with Application to Statistics.* Wiley, N.Y.

SIBSON, R.(1972) Discussion of a paper by H.P. Wynn. *J. Roy Statist. Soc.*, B 34, 181-183.

SILVERMAN, B.W. AND TITTERINGTON, D.M.(1980) Minimum covering ellipses. *Siam J. Sci. Statist. Comput.*, 1, 401-409.

STIGLER, S.M.(1973) The asymptotic distribution of the trimmed mean. *Ann. Statist.*, 1, 472-477.

SWEETING, T.J.(1977) Speeds of convergence for the multidimensional central limit theorem. *Ann. Prob.*, 5, 28-41.

TITTERINGTON, D.M.(1975) Optimal design: some geometrical aspects of D-optimality. *Biometrika*, 62, 313-320.

TITTERINGTON, D.M.(1978) Estimation of correlation coefficients by ellipsoidal trimming. *Appl. Statist.*, 27, 227-234.

TUKEY, J.W.(1947) Nonparametric estimation II. Statistically equivalent blocks and tolerance regions in the continuous case. *Ann Math. Statist.*, 18, 529-539.

TYLER, D.E.(1981) Asymptotic inference for eigenvectors. *Ann Statist.*, 9, 725-736.

WALD, A.(1943) An extension of Wilks' method for setting tolerance limits. *Ann.*

Math. Statist., 14, 45-55.

WELSH, A.H.(1987) The trimmed mean in the linear model. *Ann. Statist.*, 15, 20-36.

Department of Mathematics

The University of Western Australia

Nedlands Western Australia 6009

THE QUANTILE-TRANSFORM APPROACH TO THE ASYMPTOTIC DISTRIBUTION OF MODULUS TRIMMED SUMS

Sándor Csörgő,* Erich Haeusler and David M. Mason †

1. Introduction and statement of results. Let $X, X_1, X_2, \ldots$, be independent random variables with a common non-degenerate distribution function F which is *symmetric about zero*. For $1 \leq j \leq n$, set

$$m_n(j) = \#\{1 \leq i \leq n : |X_i| > |X_j| \text{ or } (|X_i| = |X_j| \text{ and } i \leq j)\},$$

and let $^{(k)}X_n = X_j$ if $m_n(j) = k$, $1 \leq k \leq n$. Thus $^{(k)}X_n$ is the kth largest random variable in absolute value among $X_1, \ldots, X_n$, with ties broken according to the order in which the sample occurs. Consider the 'modulus' trimmed sums defined for $0 \leq k \leq n-1$ as

$$^{(k)}S_n = {}^{(k+1)}X_n + \cdots + {}^{(n)}X_n.$$

Following a preliminary investigation by Pruitt [7], Griffin and Pruitt [4] undertook a detailed study of the asymptotic distribution of $^{(k_n)}S_n$, when $\{k_n\}$ is a sequence of integers such that $0 \leq k_n \leq n-1$ for $n \geq 1$ and both

(1.1) $$k_n \to \infty \text{ and } k_n/n \to 0 \text{ as } n \to \infty.$$

The aim of the present paper is to illuminate their main results by means of our quantile-transform–weak-approximation approach.

---

* Partially supported by the Hungarian National Foundation for Scientific Research, Grants 1808/86 and 457/88

† Partially supported by a Fulbright Grant and NSF Grant #DMS-8803209

Part II

The Quantile-Transform-Empirical-Process
Approach to Trimming

based on the sample $X_1, \ldots, X_n$ and $U_{1,n} \leq \cdots \leq U_{n,n}$ denote the order statistics pertaining to $U_1, \ldots, U_n$, then

$$(1.2) \qquad \{X_{j,n} : 1 \leq j \leq n, \ n \geq 1\} \stackrel{\mathcal{D}}{=} \{Q(U_{j,n}) : 1 \leq j \leq n, \ n \geq 1\}$$

and hence, introducing the truncated means

$$(1.3) \qquad \mu_n(m+1, \ n-(k+1)) = n \int_{(m+1)/n}^{1-(k+1)/n} Q(u+) \, du,$$

as a natural centering sequence, we have

$$\sum_{j=m+1}^{n-k} X_{j,n} - \mu_n(m+1, \ n-(k+1)) \stackrel{\mathcal{D}}{=} \sum_{j=m+1}^{n-k} Q(U_{j,n}) - \mu_n(m+1, \ n-(k+1))$$

$$(1.4) \qquad = \left\{ Q(U_{m+1,n}) + n \int_{(m+1)/n}^{U_{m+1,n}} \left(G_n(s) - \frac{m+1}{n} \right) dQ(s) \right\}$$

$$+ n \int_{(m+1)/n}^{1-(k+1)/n} (s - G_n(s)) \, dQ(s)$$

$$+ \left\{ Q(U_{n-k,n}) + n \int_{U_{n-k-1,n}}^{1-(k+1)/n} \left(G_n(s) - \frac{n-k-1}{n} \right) dQ(s) \right\}$$

for any integers $m, k \geq 0$ such that $m+1 < n-k$.

Relations (1.1) and (1.4) suggest the feasibility of a probabilistic approach to the problem of the asymptotic distribution of sums of independent, identically distributed random variables, or more generally, of the corresponding trimmed sums and sums of extreme values to be based on properties of Q and the asymptotic behavior of G_n, rather than on characteristic functions or other transforms of F. It is the aim of the present survey to describe the main features of such an approach. Naturally enough, the analytic conditions for convergence in distribution that this method yields are all expressed in terms of the quantile function Q.

On the technical side, it turns out that a very convenient way to exploit the representation (1.4) is by means of in probability approximations of the various parts of the (trimmed) sums appearing on the right–hand side (1.4). Such approximation techniques can be used because we are completely free to choose the underlying space $(\Omega, \mathcal{F}, P)$. This probability space will be the one constructed in [4,5]. It carries two independent sequences $\{Y_n^{(j)}, n \geq 1\}$, $j = 1, 2$, of independent, exponentially distributed random variables with mean 1 and a sequence $\{B_n(s), 0 \leq s \leq 1; \ n \geq 1\}$ of Brownian bridges with the properties that we describe now. For each $n \geq 2$, let

$$Y_j(n) = \begin{cases} Y_j^{(1)}, & j = 1, \ldots, [n/2], \\ Y_{n+2-j}^{(2)}, & j = [n/2] + 1, \ldots, n+1, \end{cases}$$

and for $k = 1, \ldots, n+1$, write

$$S_k(n) = \sum_{j=1}^{k} Y_j(n).$$

Then the ratios $U_{k,n} = S_k(n)/S_{n+1}(n)$, $k = 1, \ldots, n$, have the same joint distribution as the order statistics of n independent uniform $(0,1)$ random variables, and for the corresponding (left–continuous version of the) empirical distribution function

$$G_n^{(1)}(s) = n^{-1} \sum_{j=1}^{n} I(U_{j,n} < s), \quad 0 \leq s \leq 1,$$

where $I(\cdot)$ is the indicator function, and the empirical quantile function

$$U_n(s) = \begin{cases} U_{k,n}, & (k-1)/n < s \leq k/n; \ k = 1, \ldots, n, \\ U_{1,n}, & s = 0, \end{cases}$$

we have the weighted weak (i.e., in probability) approximations

(1.5) $$\sup_{1/n \leq s \leq 1-1/n} \frac{|n^{1/2}(G_n^{(1)}(s) - s) - B_n(s)|}{(s(1-s))^{1/2-\nu}} = O_p(n^{-\nu})$$

for any fixed $\nu \in [0, 1/4)$ and

$$(1.6) \qquad \sup_{1/n \le s \le 1-1/n} \frac{|n^{1/2}(s - U_n(s)) - B_n(s)|}{(s(1-s))^{1/2-\nu}} = O_p(n^{-\nu})$$

for any fixed $\nu \in [0, 1/2)$ as $n \to \infty$. As a consequence of approximation (1.5) we will see later that, after appropriate normalization, "middle terms" like

$$n \int_{(m+1)/n}^{1-(k+1)/n} (s - G_n(s)) \, dQ(s)$$

in (1.4) usually contribute a normal component to the limiting distribution of the whole (trimmed) sums. The two "extreme terms"

$$Q(U_{m+1,n}) + n \int_{(m+1)/n}^{U_{m+1,n}} \left(G_n(s) - \frac{m+1}{n} \right) dQ(s)$$

and

$$Q(U_{n-k,n}) + n \int_{U_{n-k-1,n}}^{1-(k+1)/n} \left(G_n(s) - \frac{n-k-1}{n} \right) dQ(s)$$

on the right–hand side of (1.4) can be rewritten in terms of the random functions $n\, G_n^{(j)}(t/n)$, $j = 1, 2$, where $G_n^{(1)}$ is already defined above and where

$$G_n^{(2)}(s) = n^{-1} \sum_{j=1}^{n} I(1 - U_{n+1-j,n} < s), \quad 0 \le s \le 1,$$

is the empirical distribution function obtained by "counting down" from 1. Then the two independent Poisson processes

$$N_j(t) = \sum_{k=1}^{\infty} I(S_k^{(j)} < t), \quad 0 \le t < \infty, \quad j = 1, 2,$$

associated with the two independent jump–point sequences $S_k^{(j)} = Y_1^{(j)} + \cdots + Y_k^{(j)}$, $j = 1, 2$, are close enough approximations to the random functions $n\, G_n^{(j)}(t/n)$, $j = 1, 2$, to determine the contributions of the above "extreme terms" to the asymptotic distributions of the whole (trimmed) sums.

The quantile–function representation (1.4) and the examination of the middle and extreme terms as indicated above constitute the basis of our approach to the asymptotic distribution of sums of order statistics. The quantile–transform method itself has long been in use in statistical theory and scattered applications of it can be found also in probability. Here we don't aim at giving any bibliography of this method; a good source for its earlier use is the book [82]. It is the approximation results in (1.5) and (1.6) from empirical processes theory in combination with Poisson approximation techniques for extremes that has made this old method especially feasible for our handling of problems of the asymptotic distribution of various sums of order statistics. The combination of all these techniques is what is referred to in the title of the present survey.

In [4], the original proofs of the weighted approximation results in (1.5) and (1.6) were quite involved since they used a refinement of the Komlós–Major–Tusnády type inequality for the uniform quantile process. (For a parallel approach see Mason and van Zwet [75].) Simpler direct proofs have been found by Mason [69] and M. Csörgő and Horváth [7]. The unpublished report of Mason [69], dated January, 1986, is reproduced in this volume for the benefit of the reader.

The approach touched upon above was first used in [4] and [5] to obtain probabilistic proofs of the sufficiency parts of the normal and stable convergence criteria, respectively, for the whole sums $\sum_{j=1}^{n} X_j$. The effect on the asymptotic distribution of trimming off a fixed number m of the smallest and a fixed number k of the largest summands, i.e. the investigation of the lightly trimmed sums $T_n(m,k) = \sum_{j=m+1}^{n-k} X_{j,n}$, was already considered in [5] under the (quantile equivalent of the) classical stable convergence criterion. This line of research goes back

to Darling [30] and Arov and Bobrov [1], with later contributions by Hall [57],
Teugels [86], Maller [65], Mori [76], Egorov [35] and Vinogradov and Godovan'chuk
[88]. (Again, we don't intend to compile full bibliographies of the problems considered.). The earlier literature is concentrated almost exclusively on trimmed sums
where summands with largest absolute values are discarded. We shall refer to this
kind of trimming as modulus or magnitude trimming in the sequel, as opposed to
our natural–order, or simply natural, trimming described above.

The paper [20] has initiated the study of two problems. One was the problem of
the asymptotic distribution of moderately trimmed sums $T_n(k_n, k_n)$, where $k_n \to \infty$
as $n \to \infty$ such that $k_n/n \to 0$, the other one was the same problem for the
corresponding extreme sums $T_n(0, n - k_n)$ and $T_n(n - k_n, 0)$. In [20], the first
problem was looked at under the restrictive initial assumption that F belonged to
the domain of attraction of a normal or a non–normal stable distribution, while
the second one only in the non–normal stable domain. Later, the second problem
concerning extreme sums was solved in [22] for all F with regularly varying tails and,
extending a result in [21], Lo [62] determined the asymptotic distribution of extreme
sums for all F which are in the domain of attraction of a Gumbel distribution in
the sense of extreme value theory. All these papers use the probabilistic method.

The method itself has been perfected in the three papers [16, 17, 18], where
a general pattern of necessity proofs has also been worked out, which together
constitute a general unified theory of the asymptotic distribution of sums of order
statistics. The next three sections are devoted to a sketch of this theory according
to [16], [17] and [18], respectively. In this sketch we cannot go into the technical
details of our approach. Instead we refer the reader to the paper [19] in the present

volume, where the asymptotic distribution of modulus trimmed sums is considered. It is shown in detail how our method works in this particular situation, which provides the reader with a good introduction to the technical side of the matter. The main results of the three papers [16, 17, 18] are illustrated in [13], also in the present volume, on a single, sufficiently complex example. This example is the famous Petersburg game, and the obtained results are of interest in themselves. Note that Sections 2, 3 and 4 in [13] match Sections 2, 3 and 4 in the present survey.

When treating the classical problem of the asymptotic distribution of full sums in [16], we were led to a probabilistic representation of a general infinitely divisible random variable. This is described near the end of Section 2 below. Given this representation and the general spirit of our approach, the question of investigating the tail behavior of infinitely divisible laws by direct probabilistic means rather than by Fourier–analytic ones poses itself. Such an investigation of these tails is reported in paper [26] in the present volume. The representation also leads to a purely probabilistic calculus of infinitely divisible random variables given in [11], which, combined with an observation in connection with one of the main results in [16], results in a purely probabilistic version of the theory of domains of partial attraction. This theory, described in [11] and touched upon at the end of Section 2, contains some improvements over the classical theory and a number of new results. Applications of the new theory providing new insight into the case of Poisson limits are given in [14, 15].

Many of the results on the asymptotic distribution of full, lightly and moderately trimmed sums from [16] and [17], discussed in the next two sections, have been

extended to linear combinations of order statistics in [72] and [73]. These results
are briefly outlined in Section 6. Further results, which together with the results in
the two papers just cited, constitute a rounded–off study of the asymptotic distribution of L–statistics via the quantile–transform – empirical–process technique are
presented in [81] and [74] in this volume.

So far we have discussed exclusively the asymptotic distributional behavior
of sums of order statistics from independent and identically distributed samples.
The quantile–transform–empirical–process approach is also well suited to study the
corresponding asymptotic almost sure behavior. However, though the approach is
again based on the quantile–function representation (1.2) of $X_1, X_2, \ldots$, there is
an essential difference between the treatments of the two types of asymptotics. In
the analysis of strong laws and laws of the iterated logarithm the known optimal
strong approximation results for empirical and quantile processes are less useful
than the weak approximations (1.5) and (1.6) used in the study of the distributional
behavior, because when applied to obtain strong laws for sums of order statistics
they do not lead to results under optimal conditions. Therefore, one has to avoid
approximation techniques when analyzing the strong asymptotic behavior. Instead,
the direct application of almost sure results for uniform empirical processes and
uniform order statistics in combination with classical techniques for proving strong
laws and laws of the iterated logarithm is appropriate. A survey of some of the
results obtainable by this method will be given in Section 5 below. Moreover, in
paper [51] of this volume it is shown in detail how this method is applied to obtain
a law of the iterated logarithm for moderately trimmed sums and a stability result
for lightly trimmed sums from a distribution with a slowly varying upper tail.

We close this introduction by mentioning that necessarily less complete surveys of earlier stages of the work described here were given in [23] and [10]. In the present survey we do not discuss results from [27] and [28] on intermediate sums $T_n(n - \lceil bk_n \rceil, \lceil ak_n \rceil)$, where $0 < a < b$ and $k_n \to \infty$, $k_n/n \to 0$ as $n \to \infty$, even though they also belong to the present theory. These results are also obtained by the quantile–transform method, but use a Wiener–process approximation to the uniform tail empirical process (cf. [70]).

2. Full and lightly trimmed sums [16]

The aim here is to determine all possible limiting distributions of the suitably centered and normalized sequence

$$T_n(m, k) = \sum_{j=m+1}^{n-k} X_{j,n},$$

where $m \geq 0$ and $k \geq 0$ are fixed, along subsequences of $\{n\}$ under the broadest possible conditions.

Choose the integers l and r such that $m \leq l \leq r \leq n - r \leq n - l \leq n - k$, and write

$$T_n(m, k) - \mu_n(m + 1, n - (k + 1)) = \left\{ \sum_{j=m+1}^{l} X_{j,n} - \mu_n(m + 1, l + 1) \right\}$$

$$+ \left\{ \sum_{j=l+1}^{r} X_{j,n} - \mu_n(l + 1, r + 1) \right\}$$

$$+ \left\{ \sum_{j=r+1}^{n-r} X_{j,n} - \mu_n(r + 1, n - (r + 1)) \right\}$$

$$+ \left\{ \sum_{j=n-r+1}^{n-l} X_{j,n} - \mu_n(n - (r + 1), n - (l + 1)) \right\}$$

$$+ \left\{ \sum_{j=n-l+1}^{n-k} X_{j,n} - \mu_n(n-(l+1), n-(k+1)) \right\}$$

$$= v_m^{(1)}(l,n) + \delta_1(l,r,n) + \overline{m}(r,n) + \delta_2(l,r,n) + v_k^{(2)}(l,n).$$

Now if we introduce

$$\phi_n^{(1)}(s) = \begin{cases} \frac{1}{A_n} Q\left(\frac{s}{n}+\right), & 0 < s \le n - n\alpha_n, \\ \frac{1}{A_n} Q((1-\alpha_n)+), & n - n\alpha_n < s < \infty, \end{cases}$$

and

$$\phi_n^{(2)}(s) = \begin{cases} -\frac{1}{A_n} Q\left(1-\frac{s}{n}\right), & 0 < s \le n - n\alpha_n, \\ -\frac{1}{A_n} Q(\alpha_n), & n - n\alpha_n < s < \infty, \end{cases}$$

where $A_n > 0$ is some potential normalizing sequence and $\alpha_n \to 0$ as $n \to \infty$ such that $n\alpha_n \to 0$ (so that $P\{\alpha_n \le U_{1,n} \le U_{n,n} \le 1 - \alpha_n\} \to 1$ as $n \to \infty$), and also

$$Z_{q,n}^{(j)} = \begin{cases} nU_{q,n}, & j = 1, \\ n(1 - U_{n+1-q,n}), & j = 2, \end{cases}$$

then, using (1.2) and integration by parts, for

$$V_h^{(j)}(l,n) = \frac{1}{A_n} v_h^{(j)}(l,n), \quad h = m, k; \quad j = 1, 2,$$

we can write

$$V_h^{(j)}(l,n) \overset{\mathcal{D}}{=} (-1)^{j+1} \left\{ \int_{Z_{h+1,n}^{(j)}}^{Z_{l+1,n}^{(j)}} \left(s - nG_n^{(j)}\left(\frac{s}{n}\right) \right) d\phi_n^{(j)}(s) \right.$$

$$+ \int_{h+1}^{Z_{h+1,n}^{(j)}} (s - (h+1)) \, d\phi_n^{(j)}(s)$$

$$+ \phi_n^{(j)}(Z_{h+1,n}^{(j)})$$

$$+ \int_{Z_{l+1,n}^{(j)}}^{l+1} (s - (l+1)) \, d\phi_n^{(j)}(s)$$

$$\left. - \phi_n^{(j)}(Z_{l+1}^{(j)}) \right\}.$$

If we now assume that there exists a subsequence $\{n'\}$ of the positive integers such that for two non–decreasing, right–continuous functions ϕ_1 and ϕ_2 we have

(2.1)

$$\phi_{n'}^{(j)}(s) \to \phi_j(s), \text{ as } n' \to \infty, \text{ at every continuity point } s \in (0,\infty) \text{ of } \phi_j, \ j = 1, 2,$$

then it turns out that the right–side of the last distributional equality converges in probability to a limit as $n' \to \infty$ for each fixed l, and these limits converge, if we let $l \to \infty$, in probability to

$$V_h^{(j)} = (-1)^{j+1}\Big\{\int_{S_{h+1}^{(j)}}^{\infty} (s - N_j(s))\, d\phi_j(s) + \int_1^{S_{h+1}^{(j)}} s\, d\phi_j(s)$$

$$+ \phi_j(1) - h\phi_j(S_{h+1}^{(j)}) + \int_1^{h+1} \phi_j(s)\, ds\Big\},$$

$h = m, k; \ j = 1, 2$. These limits are well–defined random variables because condition (2.1) implies that

(2.2)
$$\int_{\epsilon}^{\infty} \phi_j^2(s)\, ds < \infty \quad \text{for any} \quad \epsilon > 0, \ j = 1, 2.$$

Also, it turns out that the terms $\delta_j(l, r, n')$, $j = 1, 2$, above only play the role of separating terms in the sense that under (2.1), $\delta_j(l, r, n')/A_{n'}$ converge to some limits in probability as $n' \to \infty$, $j = 1, 2$, and if we let $l \to \infty$ (forcing $r \to \infty$) then both of these limit sequences converge to zero in probability. This fact shows that these two strips do not contribute to the limit and they only separate $V_n^{(1)}$ and $V_m^{(2)}$ from a possibly vanishing normal component in the limit coming from the middle term

$$M(r, n') = \frac{1}{A_{n'}} \overline{m}(r, n'),$$

which component by later appropriate choices of $l = l_{n'}$ and $r = r_{n'}$ and by an application of a result of Rossberg [79] will be independent of the vector $(V_m^{(1)}, V_k^{(2)})$, the two components of the latter being independent by construction.

Finally, using (1.2) and the representation (1.4) with $m = k = r$, and (1.5), it can be shown that for any sequence $r_{n'} \to \infty$, $r_{n'}/n' \to 0$, we have

$$M(r_{n'}, n') \overset{D}{=} \frac{a_{n'}}{A_{n'}} \sigma_{n'} N_{n'}(0, 1) + o_p(1),$$

as $n' \to \infty$, where

$$0 \le \sigma_{n'} = \frac{\sigma((r_{n'} + 1)/n')}{\sigma(1/n')} \le 1$$

and $a_{n'} = \sqrt{n'}\, \sigma(1/n')$, where for $0 < s < 1$,

$$\sigma^2(s) = \int_s^{1-s} \int_s^{1-s} (\min(u, v) - uv)\, dQ(u)\, dQ(v)$$

and where

$$N_{n'}(0, 1) = \int_{(r_{n'}+1)/n'}^{1-(r_{n'}+1)/n'} B_{n'}(s)\, dQ(s) \,/\, \sigma((r_{n'} + 1)/n')$$

is a standard normal ($N(0, 1)$) random variable for each n'.

This is the way we arrive at the direct half (i) of the following result which comprises the essential elements of Theorems 1–5 in [16].

RESULT. *(i) Assume (2.1) and that*

(2.3) $$a_{n'}/A_{n'} \to \delta < \infty,$$

where δ is some non–negative constant. If $\delta > 0$ and $\phi_1 = \phi_2 \equiv 0$, then for all fixed $m \ge 0$ and $k \ge 0$

$$\frac{1}{A_{n'}} \left\{ \sum_{j=m+1}^{n'-k} X_{j,n'} - \mu_{n'}(m + 1, n' - (k + 1)) \right\} \overset{D}{\to} N(0, \delta^2)$$

as $n' \to \infty$. *If* $\delta = 0$, *then*

$$\frac{1}{A_{n'}} \left\{ \sum_{j=m+1}^{n'-k} X_{j,n'} - \mu_{n'}(m+1, n' - (k+1)) \right\} \xrightarrow{D} V_m^{(1)} + V_k^{(2)}$$

as $n' \to \infty$, *where, necessarily,* $\phi_j(s) = 0$ *if* $s \geq 1$, $j = 1, 2$. *If* $\delta > 0$, *then for any subsequence* $\{n''\}$ *of* $\{n'\}$ *for which* $\sigma_{n''} \to \sigma$ *as* $n'' \to \infty$, *where* $0 \leq \sigma \leq 1$, *we have*

$$\frac{1}{A_{n''}} \left\{ \sum_{j=m+1}^{n''-k} X_{j,n''} - \mu_{n''}(m+1, n'' - (k+1)) \right\} \xrightarrow{D} V_m^{(1)} + \delta\sigma N(0,1) + V_k^{(2)}$$

as $n'' \to \infty$, *where the three terms in the limit are independent. In the last two cases* $V_m^{(1)}$ *is non-degenerate if* $\phi_1 \not\equiv 0$ *and* $V_k^{(2)}$ *is non-degenerate if* $\phi_2 \not\equiv 0$.

(ii) *If there exist two sequences of constants* $A_n > 0$ *and* C_n *and a sequence* $\{n'\}$ *of positive integers such that*

(2.4) $$\frac{1}{A_{n'}} \left\{ \sum_{j=m+1}^{n'-k} X_{j,n'} - C_{n'} \right\}$$

converges in distribution to a non-degenerate limit, then there exist a subsequence $\{n''\}$ *of* $\{n'\}$ *and non-decreasing, non-positive, right-continuous functions* ϕ_1 *and* ϕ_2 *defined on* $(0, \infty)$ *satisfying (2.2) and a constant* $0 \leq \delta < \infty$ *such that (2.1) and (2.3) hold true for* $A_{n''}$ *along* $\{n''\}$. *The limiting random variable of the sequence in (2.4) is necessarily of the form* $V_m^{(1)} + \delta\sigma N(0,1) + V_k^{(2)} + d$ *with independent terms, where*

(2.5) $$d = \lim_{n''' \to \infty} d_{n'''} = \lim_{n''' \to \infty} \{\mu_{n'''}(m+1, n''' - (k+1)) - C_{n'''}\}/a_{n'''}$$

for some subsequence $\{n'''\}$ *of* $\{n''\}$. *If* $\delta > 0$ *then either* $\sigma > 0$ *or at least one of* ϕ_1 *and* ϕ_2 *is not identically zero. If* $\delta = 0$ *then* $\phi_j = 0$ *on* $[1, \infty)$, $j = 1, 2$, *but at least one of them is not identically zero.*

In the proof of the converse half (ii), the case when

(2.6)
$$\limsup_{n' \to \infty} \frac{A_{n'}}{a_{n'}} |\phi_{n'}^{(j)}(s)| < \infty, \ 0 < s < \infty; \ j = 1, 2,$$

is trivial, for then by Helly–Bray selection and the convergence of types theorem there exist a subsequence $\{n'''\}$ such that (2.1), (2.3) and (2.4) all hold along it and $\delta > 0$ in (2.3), and we can apply the direct half with $\sigma \geq 0$.

When, contrary to (2.6), there exists $\{n''\} \subset \{n'\}$ such that

(2.7)
$$\lim_{n'' \to \infty} \frac{A_{n''}}{a_{n''}} \phi_{n''}^{(1)}(s) = -\infty$$

for some $s > 0$, for which one can show that necessarily $s < 1$, then the sequence in (2.4) is equal in distribution to

(2.8)
$$\frac{a_{n'}}{A_{n'}} \left\{ R_{n'}^{(1)} + W_{n'} + R_{n'}^{(2)} \right\} + d_{n'}$$

where $R_n^{(1)}$, W_n and $R_n^{(2)}$ result from dividing by $a_{n'}$ the three terms on the right–side of (1.4), respectively. Then again Rossberg's [79] result implies that the two sequences $|R_n^{(1)}|$ and $|R_n^{(2)}|$ are asymptotically independent, and we can show that

$$\lim_{M \to \infty} \liminf_{n \to \infty} P\left\{ |R_n^{(j)}| < M \right\} > 0, \ j = 1, 2,$$

which is somewhat less than the stochastic boundedness of the sequences of $R_n^{(j)}$, $j = 1, 2$. However, it can be shown that these two facts and the stochastic bound-edness of the sequence in (2.8) (which holds since by assumption it has a limiting distribution) already imply that both sequences

(2.9)
$$D_{n'}^{(j)} = H_{n'} |R_{n'}^{(j)}| = a_{n'} |R_{n'}^{(j)}|/\max(a_{n'}, A_{n'}),$$
$$j = 1, 2, \quad \text{are stochastically bounded.}$$

However, on the event $\{U_{m+1,n''} < s/n''\}$ with positive limiting probability $P\{S_{m+1} < s\}$, where s is as in (2.7), we have

$$D_{n''}^{(1)} \geq \left| Q\left(\frac{s}{n''}+\right) \right| / \max(a_{n''}, A_{n''}) = H_{n''} \left| \frac{A_{n''}}{a_{n''}} \phi_n^{(1)}(s) \right|$$

because the integral term in $R_n^{(1)}$ is non–positive for large enough n and $Q(s/n'')$ is non–positive for large enough n'' (otherwise (2.7) could not happen). This fact, together with (2.7) and (2.9) imply that $a_{n''}/A_{n''} \to 0$ as $n'' \to \infty$ and

$$\limsup_{n'' \to \infty} |\phi_{n''}^{(1)}(s)| < \infty, \quad 0 < s < \infty.$$

By repeating this proof if necessary one can choose a further subsequence $\{n'''\} \subset \{n''\}$ to arrive at

$$\limsup_{n''' \to \infty} |\phi_{n'''}^{(2)}(s)| < \infty, \quad 0 < s < \infty,$$

and hence by a final application of a Helly–Bray selection we are done again.

Noting that the integral term in $R_n^{(2)}$ is non–negative for large enough n, the subcase when (2.6) fails for $j = 2$ is treated in an entirely analogous way.

The special case $m = k = 0$ of the result above gives an equivalent version of the classical theory of the asymptotic distribution of independent, identically distributed random variables (see, e.g. [42]) with a condition formulated in terms of the quantile function. In this case, the limiting random variable in the direct half (i) is in general $V_{0,0} = V_0^{(1)} + \rho N(0,1) + V_0^{(2)}$, where $\rho = \delta\sigma \geq 0$ and

$$V_0^{(j)} = (-1)^{j+1} \left\{ \int_{S_1^{(j)}}^{\infty} (s - N_j(s)) \, d\phi_j(s) + \int_1^{S_1^{(j)}} s d\phi_j(s) + \phi_j(1) \right\}$$

$$= (-1)^{j+1} \left\{ \int_1^{\infty} (s - N_j(s)) \, d\phi_j(s) - \int_0^1 N_j(s) \, d\phi_j(s) + \phi_j(1) \right\},$$

for $j = 1, 2$. This is an infinitely divisible random variable with characteristic function

(2.10)
$$Ee^{itV_{0,0}} = \exp\left(it\gamma - \frac{1}{2}\rho^2 t^2 + \int_{-\infty}^{0} \left(e^{itx} - 1 - \frac{itx}{1+x^2}\right) dL(x)\right.$$
$$\left. + \int_{0}^{\infty} \left(e^{itx} - 1 - \frac{itx}{1+x^2}\right) dR(x)\right),$$

$t \in \mathbf{R}$, where $\gamma = \gamma_1 + \gamma_2$ with

$$\gamma_j = (-1)^{j+1} \left\{ \int_0^1 \frac{\phi_j(s)}{1+\phi_j^2(s)} ds - \int_1^{\infty} \frac{\phi_j^3(s)}{1+\phi_j^2(s)} ds \right\}, \ j = 1, 2,$$

and $L(x) = \inf\{s > 0 : \phi_1(s) \geq x\}$, $-\infty < x < 0$, and $R(x) = \inf\{s < 0 : -\phi_2(-s) \geq x\}$, $0 < x < \infty$. In fact, any infinitely divisible random variable can be represented as $V_{0,0}$ plus a constant (Theorem 3 in [16]) by reversing the definitions of the inverse functions if a pair (L, R) of left and right Lévy measures is given. (See Section 7.2 below.) So the result above shows rather directly how these measures arise, while the proof just sketched indicates which portions of the whole sum contribute these extreme parts of the limiting infinitely divisible law.

The direct and converse halfs of the result above are used in [16] to derive necessary and sufficient conditions for full or lightly trimmed sums to be in the domain of attraction of a normal law (the normal convergence criterion) or to be in the domain of partial attraction of a normal law, for full sums to be in the domain of attraction of a non–normal stable law (the stable convergence criterion; stable laws of exponent $0 < \alpha < 2$ arise with the functions $\phi_j(s) = -c_j s^{-1/\alpha}$, $0 < s < \infty$, $j = 1, 2$, where $c_1, c_2 \geq 0$ are constants such that $c_1 + c_2 > 0$) or to be in the domain of partial attraction of a non–normal stable law. The domains of normal attraction of these laws are also characterized. Analogous characterization results are derived for the domain of partial attraction of some infinitely divisible law or its lightly trimmed version $V_{m,k} = V_m^{(1)} + \rho N(0,1) + V_k^{(2)}$, and necessary and sufficient conditions are derived for the stochastic compactness and subsequential compactness of lightly trimmed or full sums, together with a Pruitt–type [77] quantile description of the arising subsequential limiting laws in the compact case. All these results are deduced from (i) and (ii) above, independently of the existing literature, all the obtained necessary and sufficient conditions are expressed in terms of the quantile function and hence are of independent interest. Moreover, most of the results are effectively new as far as light trimming is concerned.

The paper [11] is an organic continuation of [16]. It starts with the observation that for σ in part (i) of the Result above we necessarily have $\underline{\sigma} \leq \sigma \leq \overline{\sigma}$, where $0 \leq \underline{\sigma} = \underline{\sigma}_{\{n'\}} \leq \overline{\sigma} = \overline{\sigma}_{\{n'\}} \leq 1$ and they are defined as

$$\underline{\sigma} = \lim_{h \to \infty} \liminf_{n' \to \infty} \frac{\sigma(h/n')}{\sigma(1/n')} \quad \text{and} \quad \overline{\sigma} = \lim_{h \to \infty} \limsup_{n' \to \infty} \frac{\sigma(h/n')}{\sigma(1/n')}.$$

This fact gives an analytic control on the choice of the variance of a possible limiting normal component that depends only on the underlying distribution and the subsequence $\{n'\}$ along which (2.1) and (2.3) are assumed to be satisfied. In principle, the observation makes it possible to characterize the domain of partial attraction of an arbitrary infinitely divisible law. Several examples are included in [11], some of them improving on those in [16]. Using one of the main results from [11] as a starting point, the domain of partial attraction of a Poisson distribution with mean $\lambda > 0$ is characterized in [14]. It turns out that λ has an interesting role: limiting Poisson distributions arise in two principally different ways depending on whether $\lambda > 1$ or $\lambda < 1$. If

$$\frac{1}{A_{n'}} \left\{ \sum_{j=1}^{n'} X_j - C_{n'} \right\}$$

converges in distributions to a Poisson random variable with mean λ, then $a_{n'}/A_{n'}$ is necessarily bounded away from zero whenever $\lambda > 1$ and $a_{n'}/A_{n'} \to 0$ as $n' \to \infty$ whenever $\lambda < 1$. In the boundary case of $\lambda = 1$ both possibilities may occur. Results of this type do not follow from the work of Groshev [47] who first tackled the problem using classical methods.

The probabilistic calculus of infinitely divisible distributions given in [11] and mentioned in the introduction is derived from the following fact. Using the notation in (2.10), set $V(\phi_1, \phi_2, \rho) = V_{0,0} - \phi_1(1) + \phi_2(1) - \gamma$ and let $V_l(\phi_1, \phi_2, \rho)$, $l = 1, \ldots, r$, be independent copies of $V(\phi_1, \phi_2, \rho)$. Then we have

$$\sum_{l=1}^{r} V_l(\phi_1, \phi_2, \rho) \overset{\mathcal{D}}{=} V(\phi_1^{(r)}, \phi_2^{(r)}, r^{1/2)}\rho),$$

where

$$\phi^{(r)}(s) = \phi\left(\frac{s}{r}\right), \quad s > 0.$$

This relation then leads to purely probabilistic proofs of all the main results of Gnedenko [41] and Doeblin [34] for domains of partial attraction (cf. also [42], pp. 189–190) with a notable improvement. It turns out that if a distribution is in the domain of partial attraction of a non–stable type, then it belongs to the domain of partial attraction of a family of types where the cardinality of this family is that of the *continuum*.

The third feature of [11] is a purely probabilistic proof of an equivalent version of Gnedenko's [40] basic theorem on the convergence of infinitely divisible distributions which is Theorem 2 on pp. 88–92 in [42], one of the core results in that book. The proof in [11] is based on the simple fact that a standard Poisson process has stationary independent increments. This equivalent version then leads to characterization results for infinitely divisible distributions to belong to the domain of partial attraction of given infinitely divisible distributions. The characterizations themselves provide interesting examples, many of which are new. Again, the case of an attracting Poisson distribution is of special interest and is investigated separately in [15].

Finally we note here that extensions of the theory of the asymptotic distributions of lightly trimmed or full sums have been obtained in [29] for vectors of various power sums based on a sequence of independent and identically distributed random variables. Results of this type extend those of Szeidl [84, 85], obtained by the classical characteristic function technique and motivated by the problem of the asymptotic distribution of symmetric polynomials of a fixed degree of such random variables. This problem has recently been exposed by Zolotarev [89, 90].

3. Moderately and heavily trimmed sums [17]

We call the sum

$$T_n = T_n(m_n, k_n) = \sum_{j=m_n+1}^{n-k_n} X_{j,n}$$

moderately trimmed if the integers m_n and k_n are such that, as $n \to \infty$,

(3.1) $m_n \to \infty, \; k_n \to \infty, \; m_n/n \to 0, \; k_n/n \to 0.$

Now, fixing these two sequences $\{m_n\}$ $\{k_n\}$, with

$$\mu_n = \mu_n(m_n, n - k_n) = n \int_{m_n.n}^{1-k_n/n} Q(s)\, ds$$

the equality (1.4) simplifies to

$$\frac{1}{A_n}\{T_n - \mu_n\} \overset{\mathcal{D}}{=} \frac{n}{A_n} \int_{m_n/n}^{U_{m_n,n}} \left(G_n(s) - \frac{m_n}{n}\right) dQ(s)$$

(3.2)
$$+ \frac{n}{A_n} \int_{m_n/n}^{1-k_n/n} (s - G_n(s)) \, dQ(s)$$

$$+ \frac{n}{A_n} \int_{U_{n-k_n,n}}^{1-k_n/n} \left(G_n(s) - \frac{n-k_n}{n}\right) dQ(s)$$

$$= R_{1,n} + Y_n + R_{2,n},$$

where $A_n > 0$ is some potential norming sequence, and $R_{1,n} \leq 0$, $R_{2,n} \geq 0$.

First, using (1.5) it turns out that

$$Y_n = \frac{a_n}{A_n}(Z_n + o_p(1)),$$

where now a_n is defined to be $a_n = \sqrt{n}\sigma(m_n/n, 1 - k_n/n)$, where for $0 \leq s \leq t \leq 1$,

(3.3)
$$\sigma^2(s,t) = \int_s^t \int_s^t (\min(u,v) - uv) \, dQ(u) \, dQ(v),$$

and where

$$Z_n = -\frac{1}{\sigma(m_n/n,\, 1 - k_n/n)} \int_{m_n/n}^{1-k_n/n} B_n(s) \, dQ(s)$$

is a standard normal variable for each n.

Concerning $R_{1,n}$, it is easy to see from (1.6) that

$$\frac{n}{m_n^{1/2}} \left(U_{m_n,n} - \frac{m_n}{n}\right) + Z_{1,n} = O_p(m_n^{-\nu})$$

for any $0 < \nu < 1/4$, where $Z_{1,n} = (n/m_n)^{1/2} B_n(m_n/n)$. Using this in conjunction with (1.5), it can be shown that $R_{1,n}$ behaves asymptotically as

$$\frac{n^{1/2}}{A_n} \int_{m_n/n}^{(m_n - m_n^{1/2} Z_{1,n})/n} \left\{ B_n(s) + n^{1/2}\left(s - \frac{m_n}{n}\right) \right\} dQ(s)$$

$$= \frac{n^{1/2}}{A_n} \int_0^{-Z_{1,n}} \left\{ B_n\left(\frac{m_n}{n} + x\,\frac{m_n^{1/2}}{n}\right) + x\left(\frac{m_n}{n}\right)^{1/2} \right\} dQ\left(\frac{m_n}{n} + x\,\frac{m_n^{1/2}}{n}\right)$$

$$= \int_0^{-Z_{1,n}} \left\{ \left(\frac{n}{m_n}\right)^{1/2} B_n\left(\frac{m_n}{n} + x\,\frac{m_n^{1/2}}{n}\right) + x \right\} d\psi_n^{(1)}(x),$$

which in turn behaves asymptotically as

$$\int_0^{-Z_{1,n}} (Z_{1,n} + x)\, d\psi_n^{(1)}(x) = \int_{-Z_{1,n}}^0 \psi_n^{(1)}(x)\, dx\,,$$

provided the sequence of functions

$$\psi_n^{(1)}(x) = \begin{cases} \psi_n^{(1)}\left(-\frac{m_n^{1/2}}{2}\right), & -\infty < x < -\frac{m_n^{1/2}}{2}, \\ \frac{m_n^{1/2}}{A_n}\left\{Q\left(\frac{m_n}{n} + x\,\frac{m_n^{1/2}}{n}\right) - Q\left(\frac{m_n}{n}\right)\right\}, & |x| \le \frac{m_n^{1/2}}{2}, \\ \psi_n^{(1)}\left(\frac{m_n^{1/2}}{2}\right), & \frac{m_n^{1/2}}{2} < x < \infty, \end{cases}$$

is at least bounded. Similarly, it can be shown that $R_{2,n}$ behaves asymptotically as

$$\int_{-Z_{2,n}}^0 (Z_{2,n} + x)\, d\psi_n^{(2)}(x) = -\int_{-Z_{1,n}}^0 \psi_n^{(2)}(x)\, dx\,,$$

where $Z_{2,n} = (n/k_n)^{1/2}\, B_n(1 - k_n/n)$ and

$$\psi_n^{(2)}(x) = \begin{cases} \psi_n^{(2)}\left(-\frac{k_n^{1/2}}{2}\right), & -\infty < x < -\frac{k_n^{1/2}}{2}, \\ \frac{k_n^{1/2}}{A_n}\left\{Q\left(1 - \frac{k_n}{n} + x\,\frac{k_n^{1/2}}{n}\right) - Q\left(1 - \frac{k_n}{n}\right)\right\}, & |x| \le \frac{k_n^{1/2}}{2}, \\ \psi_n^{(2)}\left(\frac{k_n^{1/2}}{2}\right), & \frac{k_n^{1/2}}{2} < x < \infty. \end{cases}$$

For each n, $(Z_{1,n}, Z_n, Z_{2,n})$ is a trivariate normal vector with covariance matrix

$$\begin{pmatrix} 1 - \frac{m_n}{n} & r_{1,n} & \left(\frac{m_n k_n}{n^2}\right)^{1/2} \\[2mm] r_{1,n} & 1 & r_{2,n} \\[2mm] \left(\frac{m_n k_n}{n^2}\right)^{1/2} & r_{2,n} & 1 - \frac{k_n}{n} \end{pmatrix}$$

where

$$-\left(1 - \frac{m_n}{n}\right)^{1/2} \le r_{1,n} = -\left(\frac{m_n}{n}\right)^{1/2} \int_{m_n/n}^{1-k_n/n} (1-s)\, dQ(s) \,/\, \sigma\left(\frac{m_n}{n}, 1 - \frac{k_n}{n}\right) \le 0$$

and

$$-\left(1-\frac{k_n}{n}\right)^{1/2} \le r_{2,n} = -\left(\frac{k_n}{n}\right)^{1/2} \int_{m_n/n}^{1-k_n/n} s\,dQ(s) \Big/ \sigma\left(\frac{m_n}{n}, 1-\frac{k_n}{n}\right) \le 0.$$

If we break up Z_n as

$$
\begin{aligned}
Z_n &= -\frac{1}{\sigma(m_n/n, 1-k_n/n)} \int_{m_n/n}^{1/2} B_n(s)\,dQ(s) \\
&\quad - \frac{1}{\sigma(m_n/n, 1-k_n/n)} \int_{1/2}^{1-k_n/n} B_n(s)\,dQ(s) \\
&= W_{1,n} + W_{2,n},
\end{aligned}
$$

then

$$EW_{1,n}^2 = \sigma_{1,n}^2 = \sigma^2\left(\frac{m_n}{n}, \frac{1}{2}\right) \Big/ \sigma^2\left(\frac{m_n}{n}, 1-\frac{k_n}{n}\right),$$

$$EW_{2,n}^2 = \sigma_{2,n}^2 = \sigma^2\left(\frac{1}{2}, 1-\frac{k_n}{n}\right) \Big/ \sigma^2\left(\frac{m_n}{n}, 1-\frac{k_n}{n}\right),$$

and it can be shown that if $EX^2 = \infty$ then the three covariances $\mathrm{Cov}(Z_{1,n}, W_{2,n})$, $\mathrm{Cov}(W_{1,n}, W_{2,n})$ and $\mathrm{Cov}(Z_{2,n}, W_{1,n})$ all converge to zero as $n \to \infty$.

This is the way we arrive at the direct half (i) of the following main result of [17], where this result is formulated somewhat differently. The proof of the converse half (ii) goes along the same line as that of the proof of the converse half in the preceding section, the last step being technically different but the same in spirit.

RESULT. (i) *Assume that there exists a subsequence* $\{n'\}$ *of the positive integers such that for two non-decreasing, left-continuous functions* ψ_1 *and* ψ_2 *satisfying* $\psi_j(0) \le 0$, $\psi_j(0+) \ge 0$, $j = 1, 2$, *we have*

(3.4) $\psi_{n'}^{(j)}(x) \to \psi_j(x)$, *at every continuity point* $x \in \mathbf{R}$ *of* ψ_j, $j = 1, 2$,

and that

(3.5) $a_{n'}/A_{n'} \to \delta < \infty$, where $a_n = n^{1/2} \, \sigma(m_n/n, \, 1 - k_n/n)$ and δ is some non-negative constant. If $\delta = 0$, then necessarily $\psi_1(x) = \psi_2(-x) = 0$ for all $x > 0$, and, with $\mu_n = \mu_n(m_n, \, n - k_n)$ given above

$$\frac{1}{A_{n'}} \left\{ \sum_{j=m_{n'}+1}^{n'-k_{n'}} V_{j,n'} - \mu_{n'} \right\} \xrightarrow{\mathcal{D}} V_1 + V_2 \,,$$

where

$$V_j = V_j(\psi_j) = (-1)^{j+1} \int_{-Z_j}^{0} \psi_j(x) \, dx \,, j = 1, 2,$$

where Z_1 and Z_2 are independent standard normal random variables. If $\delta > 0$, then for any subsequence $\{n''\}$ of $\{n'\}$ for which $r_{j,n''} \to r_j$, $j = 1, 2$, where $-1 \le r_1, r_2 \le 0$, we necessarily have $\psi_1(x) \le -r_1$ and $\psi_2(x) \ge r_2$ for all $x \in \mathbf{R}$, and

$$\frac{1}{A_{n''}} \left\{ \sum_{j=m_{n''}+1}^{n''-k_{n''}} X_{j,n''} - \mu_{n''} \right\} \xrightarrow{\mathcal{D}} V_1 + \delta Z + V_2 \,,$$

where, with Z_1 and Z_2 figuring in V_1 and V_2, (Z_1, Z, Z_2) is a trivariate normal random vector with zero mean and covariance matrix

$$\begin{pmatrix} 1 & r_1 & 0 \\ r_1 & 1 & r_2 \\ 0 & r_2 & 1 \end{pmatrix} \,.$$

Moreover, if $\mathrm{Var}(X) = \infty$ and, if necessary, $\{n'''\}$ is a further subsequence of $\{n''\}$ such that for some positive constants σ_1 and σ_2 with $\sigma_1^2 + \sigma_2^2 = 1$, $\sigma_{j,n'''} \to \sigma_j$, $j = 1, 2$, then

$$\frac{1}{A_{n'''}} \left\{ \sum_{j+m_{n'''}+1}^{n'''-k_{n'''}} X_{j,n'''} - \mu_{n'''} \right\} \xrightarrow{\mathcal{D}} V_1 + \delta(W_1 + W_2) + V_2 \,,$$

where (Z_1, W_1, W_2, Z_2) is a quadrivariate normal vector with mean zero and covariance matrix

$$\begin{pmatrix} 1 & r_1 & 0 & 0 \\ r_1 & \sigma_1^2 & 0 & 0 \\ 0 & 0 & \sigma_2^2 & r_2 \\ 0 & 0 & r_2 & 1 \end{pmatrix} \,.$$

(ii) If there exist two sequences of constants $A_n > 0$ and C_n and a sequence $\{n'\}$ of positive integers such that

$$(3.6) \qquad \frac{1}{A_{n'}} \left\{ \sum_{j=m_{n'}+1}^{n'-k_{n'}} X_{j,n'} - C_{n'} \right\}$$

converges in distribution to a non–degenerate limit, then there exist a subsequence $\{n''\}$ of $\{n'\}$ and non–decreasing, left–continuous functions ψ_1 and ψ_2 satisfying $\psi_j(0) \le 0$ and $\psi_j(0+) \ge 0$, $j = 1, 2$, and a constant $0 \le \delta < \infty$ such that (3.4) and (3.5) hold true for $A_{n''}$ along $\{n''\}$, where at least one of ψ_1 and ψ_2 is not identically zero if $\delta = 0$, in which case $\psi_1(x) = \psi_2(-x) = 0$ for all $x > 0$. The limiting random variable of the sequence in (3.6) is necessarily of the form $V_1 + \delta Z + V_2 + d$, with V_1, Z and V_2 described above, where

$$d = \lim_{n''' \to \infty} (\mu_{n'''} - C_{n'''})/A_{n'''}$$

for some subsequence $\{n'''\} \subset \{n''\}$.

This result implies as a corollary (by showing that the possible limits can only be normal if $\psi_1 \equiv 0 \equiv \psi_2$) that the sequence in (3.6) converges in distribution to a standard normal random variable if and only if $\varphi_{n'}^{(j)}(x) \to 0$, as $n' \to \infty$, for all $x \in \mathbf{R}$, $j = 1, 2$. In this case $A_{n'}$ can be chosen to be $a_{n'}$ and $C_{n'}$ to be $\mu_{n'}$. Here $\{n'\}$ is arbitrary and can of course be $\{n\}$.

For a discussion of the conditions (3.4) and (3.5) we refer to the original paper [17] and [51]. Subsequent to [17], Griffin and Pruitt [46] used the more classical characteristic function methodology to prove mathematically equivalent versions of the above results and showed that all possible subsequential limits indeed arise. In another paper, [45], they deal with the analogous problem of moderately trimmed sums when $k_n(k_n \to \infty, k_n/n \to 0)$ of the summands largest in absolute value are discarded at each step (see also [78]), assuming that the underlying distribution is symmetric about zero. A comparison of the two sets of results shows that (perhaps contrary to intuition) even if we assume symmetry and $m_k = k_n$ above, the two trimming problems are wholly different. In Section 8 below we describe a quantile–transform approach to modulus trimming and in paper [19] in this volume we provide proofs of the sufficiency part of the results of Griffin and Pruitt [45], based on our methodology. For further results on various interesting versions of moderate trimming see Kuelbs and Ledoux [60], Hahn, Kuelbs and Samur [54], Hahn and Kuelbs [53], Hahn, Kuelbs and Weiner [55, 56] and Maller [66].

Finally, we turn to heavy trimming assuming instead of (3.1) that

(3.7) $$m_n = [n\alpha] \quad \text{and} \quad k_n = n - [\beta n], \ 0 < \alpha < \beta < 1.$$

This is the case of the classical trimmed sum, for which Stigler [83] completely solved the problem of asymptotic distribution. Suppose that $\sigma(\alpha, \beta) > 0$, where $\sigma(\cdot, \cdot)$ is as in (3.3). The proof sketched above produces the following version of Stigler's theorem (Theorem 5 in [17]), where a_n and μ_n are defined in terms of the present m_n and k_n: For any underlying distribution,

$$\frac{1}{a_n} \left\{ \sum_{j=[n\alpha]+1}^{[n\beta]} X_{j,n} - \mu_n \right\} \overset{D}{\to} V_1(\psi_1) + Z + V_2(\psi_2),$$

as $n \to \infty$, where

$$\psi_1(x) = \begin{cases} 0, & x \le 0, \\ \frac{\sqrt{\alpha}}{\sigma(\alpha,\beta)}(Q(\alpha+) - Q(\alpha)), & x > 0, \end{cases}$$

and

$$\psi_2(x) = \begin{cases} 0, & x \le 0, \\ \frac{\sqrt{1-\beta}}{\sigma(\alpha,\beta)}(Q(\beta+) - Q(\beta)), & x > 0, \end{cases}$$

so that

$$V_1(\psi_1) = \frac{\sqrt{\alpha}}{\sigma(\alpha, \beta)}(Q(\alpha+) - Q(\alpha)) \min(0, Z_1)$$

and

$$V_2(\psi_2) = \frac{\sqrt{1-\beta}}{\sigma(\alpha, \beta)}(Q(\beta+) - Q(\beta)) \max(0, -Z_2),$$

where (Z_1, Z, Z_2) is a trivariate normal random vector with mean zero and covariance matrix

$$\begin{pmatrix} 1-\alpha & r_1 & (\alpha(1-\beta))^{1/2} \\ r_1 & 1 & r_2 \\ (\alpha(1-\beta))^{1/2} & r_2 & \beta \end{pmatrix},$$

with

$$r_1 = -\sqrt{\alpha} \int_\alpha^\beta (1-s) \frac{dQ(s)}{\sigma(\alpha, \beta)} \quad \text{and} \quad r_2 = -\sqrt{1-\beta} \int_\alpha^\beta \frac{s\,dQ(s)}{\sigma(\alpha, \beta)}.$$

This version puts Stigler's theorem (giving asymptotic normality if and only if Q is continuous both at α and β) into a broader picture. Substituting α and β for m_n/n and $1 - k_n/n$ in the arguments of $\psi_n^{(1)}$ and $\psi_n^{(2)}$, the proof also works for more general m_n and k_n sequences, provided $\sqrt{n}(m_n/n - \alpha) \to 0$ and $\sqrt{n}(1 - k_n/n - \beta) \to 0$ as $n \to \infty$. For rates of convergence in Stigler's theorem in the case when the limit is normal, see Bjerve [2] and Egorov and Nevzorov [36], who in [37] also investigate the related problem in the case of magnitude trimming. See also Griffin and Pruitt [45].

4. Extreme sums [18]

Here we are interested in the sums of extreme values

$$E_n = E_n(k_n) = \sum_{j=1}^{k_n} X_{n+1-j,n} = \sum_{j=n-k_n+1}^{n} X_{j,n},$$

where $k_n \to \infty$ and either $k_n/n \to 0$ as $n \to \infty$, or $k_n = [n\alpha]$ with $0 < \alpha < 1$. (We shall refer to the first case when $k_n/n \to 0$ as the case $\alpha = 0$.) Here it is more convenient to work with the function

$$H(s) = -Q((1-s)-), \quad 0 \le s < 1,$$

instead of Q itself, for which we have

$$(X_{1,n}, \dots, X_{n,n}) \overset{\mathcal{D}}{=} (-H(U_{n,n}), \dots, -H(U_{1,n}))$$

instead of (1.2). The natural centering sequence now turns out to be

$$\mu_n = \mu_n(k_n) = -n \int_{1/n}^{k_n/n} H(s)\,ds - H\left(\frac{1}{n}\right).$$

Consider a potential normalizing sequence $A_n > 0$. Now the role of (1.4) or (3.2) is taken over by the decomposition

$$\frac{1}{A_n}\{E_n - \mu_n\} \overset{\mathcal{D}}{=} \Delta_n^{(1)}(m_n) + \Delta_n^{(2)}(m_n, l_n) + \Delta_n^{(3)}(l_n, k_n),$$

where $1 \le m_n \le l_n \le k_n$, and

$$\Delta_n^{(1)}(m_n) = \int_{nU_{1,n}}^{m_n} \left(nG_n\left(\frac{u}{n}\right) - u\right) d\frac{H(u/n) - H(1/n)}{A_n}$$
$$+ \int_{nU_{1,n}}^{1} (u-1)\, d\frac{H(u/n) - H(1/n)}{A_n} - \frac{H(nU_{1,n}/n) - H(1/n)}{A_n},$$

$$\Delta_n^{(2)}(m_n, l_n) = \int_{m_n}^{l_n} \left(nG_n\left(\frac{u}{n}\right) - u\right) d\frac{H(u/n) - H(1/n)}{A_n},$$

and

$$\Delta_n^{(3)}(l_n, k_n) = \int_{l_n/n}^{k_n/n} n(G_n(u) - u)\, d\frac{H(u)}{A_n} + \int_{k_n/n}^{U_{k_n,n}} n(G_n(u) - u)\, d\frac{H(u)}{A_n}.$$

The numbers m_n and l_n (not necessarily integers) will be appropriately chosen such that $m_n \to \infty$, $l_n/m_n \to \infty$ and $k_n/l_n \to \infty$ as $n \to \infty$, and we see that the term $\Delta_n^{(3)}$ presents a "trimmed–sum problem" considered in Section 4 with only one R_n–like term. Therefore, we need both ϕ_n and ψ_n type functions, which are presently defined as

$$\phi_n(s) = \begin{cases} \frac{1}{A_n} \left\{ H\left(\frac{s}{n}\right) - H\left(\frac{1}{n}\right) \right\}, & 0 < s \leq n - n\alpha_n, \\ \frac{1}{A_n} \left\{ H(1 - \alpha_n) - H\left(\frac{1}{n}\right) \right\}, & n - n\alpha_n < s < \infty, \end{cases}$$

where α_n is as in Section 2, and

$$\psi_n(x) = \begin{cases} \psi_n\left(-\frac{k_n^{1/2}}{2}\right), & -\infty < x < -\frac{k_n^{1/2}}{2}, \\ \frac{k_n^{1/2}}{A_n} \left\{ H\left(\frac{k_n}{n} + x\frac{k_n^{1/2}}{n}\right) - H\left(\frac{k_n}{n}\right) \right\}, & |x| \leq \frac{k_n^{1/2}}{2}, \\ \psi_n\left(\frac{k_n^{1/2}}{2}\right), & \frac{k_n^{1/2}}{2} < x < \infty. \end{cases}$$

Again, the middle term $\Delta_n^{(2)}$ turns out to be a separating term converging to zero in probability, and redefining

$$\sigma^2(s, t) = \int_s^t \int_s^t (\min(u, v) - uv) \, dH(u) \, dH(v), \; 0 \leq s \leq t \leq 1,$$

and

$$a_n = \begin{cases} n^{1/2} \, \sigma(1/n, k_n/n), & \text{if } \sigma(1/n, k_n/n) > 0, \\ n^{1/2}, & \text{otherwise,} \end{cases}$$

and introducing

$$0 \leq r_n = \left(\frac{n}{k_n}\right)^{1/2} \frac{1 - k_n/n}{\sigma(l_n/n, k_n/n)} \int_{l_n/n}^{k_n/n} s \, dH(s) \leq \left(1 - \frac{k_n}{n}\right)^{1/2}$$

and

$$N(t) = \sum_{k=1}^{\infty} I(S_k^{(1)} \leq t), \; 0 \leq t < \infty,$$

the right–continuous version of the Poisson process $N_1(\cdot)$ in Sections 1 and 2, the proof of the following main result (Theorems 1 and 2 in [18]) is obtained by an intricate and involved combination of the techniques of [16] and [17], i.e., those of the preceding two sections.

RESULT. *(i) Assume that there exist a subsequence $\{n'\}$ of the positive integers, a left–continuous, non–decreasing function ϕ defined on $(0, \infty)$ with $\phi(1) \leq 0$ and $\phi(1+) \geq 0$, a left–continuous, non–decreasing function ψ defined on $(-\infty, \infty)$ with $\psi(0) \leq 0$ and $\psi(0+) \geq 0$, and a constant $0 \leq \delta < \infty$ such that, as $n' \to \infty$,*

$$(4.1) \qquad \phi_{n'}(s) \to \phi(s) \quad at\, every\, continuity\, point \quad s \in (0, \infty) \quad of \quad \phi,$$

$$(4.2) \qquad \psi_{n'}(x) \to \psi(x) \quad at\, every\, continuity\, point \quad x \in \mathbf{R} \quad of \quad \psi,$$

$$(4.3) \qquad\qquad\qquad\qquad a_{n'}/A_{n'} \to \delta.$$

Then, necessarily, $\phi(s) \leq \delta$ for all $s \in (0, \infty)$,

$$(4.4) \qquad \int_{\epsilon}^{\infty} (\phi(s) - \phi(\infty))^2 \, ds < \infty \quad for\, all \quad \epsilon > 0,$$

and there exist a subsequence $\{n''\} \subset \{n'\}$ and a sequence of positive numbers $l_{n''}$ satisfying $l_{n''} \to \infty$ and $l_{n''}/k_{n''} \to 0$, as $n'' \to \infty$, such that for some $0 \leq b \leq \delta$ and $0 \leq r \leq (1 - \alpha)^{1/2}$, $\sqrt{n''}\, \sigma(l_{n''}/n'', k_{n''}/n'')/A_{n''} \to b$, $r_{n''} \to r$, and

$$\frac{1}{A_{n''}} \left\{ \sum_{j=1}^{k_{n''}} X_{n''+1-j,n''} - \mu_{n''} \right\} \xrightarrow{\mathcal{D}} V(\phi, \psi, b, r, \alpha)$$

as $n'' \to \infty$, where α is zero or positive according to the two cases, ψ necessarily satisfies

$$(4.5) \qquad\qquad \psi(x) \geq -\delta r/(1 - \alpha), \quad -\infty < x < \infty,$$

and

$$V(\phi, \psi, b, r, \alpha) = \int_1^\infty (N(t) - t)\, d\phi(t) + \int_0^1 N(t)\, d\phi(t) + bZ_1 + \int_{-Z(r,\alpha)}^0 \psi(x)\, dx\,,$$

where $Z(r;\alpha) := -rZ_1 + (1 - \alpha - r^2)^{1/2} Z_2$, where Z_1 and Z_2 are standard normal random variables such that Z_1, Z_2 and $N(\cdot)$ are independent. Moreover, if $\phi \equiv 0$ then $b = \delta$, while if $\delta = 0$ then $\phi(s) = 0$ for all $s \geq 1$.

(ii) If there exist a subsequence $\{n'\}$ of the positive integers and two sequences $A_{n'} > 0$ and $C_{n'}$ along it such that

(4.6)
$$\frac{1}{A_{n'}} \left\{ \sum_{j=1}^{k_{n'}} X_{n'+1-j,n'} - C_{n'} \right\}$$

converges in distribution to a non-degenerate limit, then there exists a subsequence $\{n''\} \subset \{n'\}$ such that conditions (4.1), (4.2) and (4.3) hold along the sequence $\{n''\}$ for $A_{n''}$ in (4.6) and for appropriate functions ϕ and ψ with the properties listed above (4.1) and for some constant $0 \leq \delta < \infty$, with ϕ satisfying (4.4) and ψ satisfying (4.5) with an $r \in (0, (1 - \alpha)^{1/2})$ arising along a possible further subsequence. The limiting random variable of the sequence in (4.6) is necessarily of the form $V(\phi, \psi, b, r, \alpha) + d$ for appropriate constants $0 \leq b \leq \delta$, $0 \leq r \leq (1 - \alpha)^{1/2}$ and

$$d = \lim_{n''' \to \infty} (\mu_{n'''} - C_{n'''})/A_{n'''},$$

for some subsequence $\{n'''\} \subset \{n''\}$. Moreover, either $\phi \not\equiv 0$ or $\psi \not\equiv 0$ or $b > 0$.

Just as in the case of full sums in Section 2, it is possible to see the effect on the limiting distribution of deleting a finite number $k \geq 0$ of the largest summands form the extreme sums $E_n(k_n)$. Replacing $E_n(k_n)$ by $\sum_{j=n-k_n+1}^{n-k} X_{j,n}$, μ_n by

$$\mu_n(k) = -n \int_{(k+1)/n}^{k_n/n} H(u)\, du - H\left(\frac{k+1}{n}\right),$$

and the first two integrals in $V(\phi, \psi, b, r, \alpha)$ by

$$\int_{S_{k+1}^{(1)}}^{\infty} (N(t) - t)\, d\phi(t) - \int_{1}^{S_{k+1}^{(1)}} t\, d\phi(t) + k\phi(S_{k+1}^{(1)}) - \int_{1}^{k+1} \phi(t)\, dt,$$

the result above remains true word for word.

The result can be formulated for the sum of lower extremes $\sum_{j=1}^{m_n} X_{j,n}$, where $m_n \to \infty$ and $m_n/n \to 0$ or $m_n = [n\beta]$ with $0 < \beta < 1$. The limiting random variable is of the form $-V(\phi, \psi, b, r, \beta)$ with appropriate ingredients. In fact, if at least one of α and β is zero then the two convergence statements hold jointly with the limiting random variable being independent.

One corollary of the result above is that the sequence in (4.6) converges in distribution to a non-degenerate normal variable if and only if (4.1) and (4.2) are satisfied with $A_{n'} \equiv a_{n'}$, $\phi \equiv 0$ and $\psi \equiv 0$, in which case, choosing $A_{n'} \equiv a_{n'}$ and $C_{n'} \equiv \mu_{n'}$ in (4.6) the limit is standard normal.

Another exhaustive corollary is the convergence in distribution of $(E_n - \mu_n)/a_n$, along the whole sequence $\{n\}$, when the underlying distribution is in the domain of attraction of one of the three possible limiting extremal distributions in the sense of extreme value theory. The details are contained in Corollary 2 in [18], being a common generalization of results from [20], [22] and [62] mentioned in the introduction.

5. Almost sure behavior: Laws of the iterated logarithm and stability theorems.

The quantile-transform – empirical-process approach not only provides an effective method to analyze the distributional limits of sums of order statistics from independent and identically distributed observations, but also to study the corresponding almost sure behavior. It turns out that, in a sense, the strong limit theory is even more complex than its weak counterpart. In particular, this is due to the fact that the law of the iterated logarithm (LIL) behavior of the moderately trimmed sums $T_n(m_n, k_n) = \sum_{j=m_n+1}^{n-k_n} X_{j,n}$ depends as heavily on the growth rate of the trimming sequences m_n and k_n as it does on the tail behavior of the underlying distribution function F. Therefore, in order to obtain nice results these sums have been investigated so far under more restrictive assumptions on the tails of F. Yet, the theory is far from being complete. Presently, a rather complete picture is available only for positive random variables in the domain of attraction of a non–normal stable law, and we will restrict our attention here mainly to a review of these results, indicating possible generalizations only briefly. For this, let F satisfy $F(0-) = 0$ and assume that F belongs to the domain of attraction of a stable law with index $\alpha \in (0,2)$, written $F \in D(\alpha)$. Then the classical domain of attraction criteria can be expressed in terms of the quantile function Q pertaining to F as follows (cf. [5] and [16]): $F \in D(\alpha)$ and $F(0-) = 0$ holds if and only if

(5.1) $Q(1 - s) = s^{-\alpha}L(s),\ 0 < s < 1,\ \text{and}\ \ Q(0+) \geq 0,$

where the function L is slowly varying at zero, i.e., is positive on $(0, u_0)$ for some $u_0 > 0$ and satisfies $L(\lambda u)/L(u) \to 1$ as $u \downarrow 0$ for all $0 < \lambda < \infty$. Since the distribution F has no lower tail, the trimmed sums of interest are now of the form

$$T_n(k) = \sum_{j=1}^{n-k} X_{j,n}$$

with truncated means as centering constants of the form

$$\mu_n(k) = n \int_0^{1-k/n} Q(u)\, du\,.$$

Moreover, now set

$$\sigma^2(s) = \int_0^{1-s} \int_0^{1-s} (\min(u,v) - uv)\, dQ(u)\, dQ(v), \quad 0 < s < 1.$$

According to the results on convergence in distribution of moderately trimmed sums, cf. Section 3, we have

$$n^{-1/2}\sigma(k_n/n)^{-1}\{T_n(k_n) - \mu_n(k_n)\} \xrightarrow{\mathcal{D}} N(0,1)$$

for every sequence $(k_n)_{n\geq 1}$ of positive integers such that $k_n \to \infty$ and $k_n/n \to 0$ as $n \to \infty$. Given this central limit theorem, it is only natural to suppose that the corresponding LIL should also hold, i.e., that

(5.2) $$\limsup_{n\to\infty} \pm (2n\,\log_2 n)^{-1/2}\, \sigma(k_n/n)^{-1}\, \{T_n(k_n) - \mu_n(k_n)\} = 1 \quad a.s.,$$

where $\log_2 n = \log\log n$. In [49] it is shown that (5.2) is true *only* if k_n converges to infinity fast enough (and satisfies some additional monotonicity assumptions which are mainly of technical significance). More precisely, one has

RESULT 1. *Assume $F \in D(\alpha)$ for $0 < \alpha < 2$ and $F(0-) = 0$. Let $(k_n)_{n\geq 1}$ satisfy*

(K) $$k_n \sim \alpha_n \uparrow \infty \quad and \quad k_n/n \sim \beta_n \downarrow 0 \quad as \quad n \to \infty$$

for some sequences of constants $(\alpha_n)_{n\geq 1}$ and $(\beta_n)_{n\geq 1}$.

(i) If $k_n/\log_2 n \to \infty$ as $n \to \infty$, then (5.2) holds.

(ii) If $k_n/\log_2 n \to 0$ as $n \to \infty$, then

$$\limsup_{n\to\infty} (2n\,\log_2 n)^{-1/2}\, \sigma(k_n/n)^{-1}\{T_n(k_n) - \mu_n(k_n)\} = \infty \quad a.s.$$

with the corresponding lim inf *being equal to 0 almost surely.*

For the proof of part (i) one uses the basic quantile–function representation (1.2), which gives

$$\left\{ \sum_{j=1}^{n-k_n} X_{j,n} - \mu_n(k_n) : n \geq 1 \right\} \overset{\mathcal{D}}{=}$$

(5.3)
$$\left\{ n \int_0^{n-k_n} (u - G_n(u))\, dQ(u) + n \int_{U_{n-k_n,n}}^{1-k_n/n} \left(G_n(u) - 1 + \frac{k_n}{n} \right) dQ(u) : n \geq 1 \right\},$$

a representation of the centered sums in terms of integrals involving $G_n, U_{n-k_n,n}$ and Q. Consequently, the asymptotic almost sure behavior of the two sequences in (5.3) is the same. To exploit this representation one now combines results about the asymptotic almost sure behavior of G_n and $U_{n-k_n,n}$ with the analytic form of Q as given by (5.1) which for example entails that

(5.4)
$$\sigma^2(s) \sim \frac{2}{2-\alpha} s^{1-2/\alpha} L^2(s) \quad \text{as} \quad s \downarrow 0.$$

The required results about G_n and $U_{n-k_n,n}$ are

(5.5)
$$\limsup_{n\to\infty} \left(\frac{n}{\log_2 n} \right)^{1/2} \sup_{0 \leq u \leq 1 - (\log_2 n)/n} \frac{|G_n(u) - u|}{(1-u)^{1/2}} = 2 \quad a.s.$$

from Csáki [3] and

(5.6)
$$nU_{n-k_n,n}/k_n \to 1 \quad a.s. \text{ as } \quad n \to \infty,$$

following e.g. from Theorem 4 in Wellner [91]. Relations (5.5) and (5.6) together imply

(5.7)
$$\limsup_{n\to\infty} \frac{n\,|U_{n-k_n,n} - (1 - k_n/n)|}{(k_n \log_2 n)^{1/2}} \leq 2 \quad a.s.,$$

and it is an easy exercise to deduce from (5.1), (5.4), (5.6) and (5.7) that

$$\left(\frac{n}{\log_2 n}\right)^{1/2} \sigma(k_n/n)^{-1} \int_{U_{n-k_n,n}}^{1-k_n/n} \left(G_n(u) - 1 + \frac{k_n}{n}\right) dQ(u) \to 0 \quad a.s. \text{ as } \quad n \to \infty.$$

Consequently, (i) will follow from

(5.8) $\quad \limsup_{n\to\infty} \pm \left(\frac{n}{2\log_2 n}\right)^{1/2} \sigma(k_n/n)^{-1} \int_0^{1-k_n/n} (u - G_n(u)) \, dQ(u) = 1 \quad a.s.$

But

$$n \int_0^{1-k_n/n} (u - G_n(u)) \, dQ(u) = \sum_{j=1}^{n} \int_0^{1-k_n/n} (u - I(U_j \leq u)) \, dQ(u),$$

which means that in (5.8) we are confronted with the sums of a triangular array of row-wise independent random variables, where these random variables are deterministic functions of the independent uniform $(0,1)$ random variables $U_j, j \geq 1$. To obtain (5.8) we can therefore use the classical technique for proving LIL's based on blocking procedures and exponential inequalities. The blocking procedure is a little bit more involved than in the classical situation of sums of independent and identically distributed random variables because we have to take care not only of the n appearing in $\sum_{j=1}^{n}$ (as in the classical case), but also of the n in $\int_0^{1-k_n/n}$. This, however, can be achieved by another application of (5.5), and for this, the monotonicity assumptions (K) are crucial.

Part (ii) of Result 1 is an easy consequence of the analytic form (5.1) of Q and the following two results on uniform order statistics due to Kiefer [59]: Let $(k_n)_{n\geq 1}$ be a sequence of positive integers with $k_n \to \infty$ and $k_n/\log_2 n \to 0$ as $n \to \infty$. Then

(5.9) $\quad\quad\quad\quad\quad\quad \limsup_{n\to\infty} \frac{n(1 - U_{n-k_n,n})}{\log_2 n} = 1 \quad a.s.$

and

$$(5.10) \qquad \liminf_{n \to \infty} \frac{k_n}{\log_2 n} \log \left(\frac{n(1 - U_{n-k_n,n})}{k_n} \right) = -1 \quad a.s.$$

Equations (5.9) and (5.10) should be compared with the LIL (5.7) for $U_{n-k_n,n}$ in the case when $k_n/\log_2 n \to \infty$. They say that in the range of k_n–sequences for which $k_n/\log_2 n \to 0$ holds, the almost sure fluctuations of $U_{n-k_n,n}$ are much larger than in the range of k_n–sequences for which $k_n/\log_2 n \to \infty$ holds. This behavior is in complete accordance with one's intuition. The fluctuations of $U_{n-k_n,n}$ should be larger for smaller k_n, because for smaller and smaller k_n there are fewer and fewer observations between $U_{n-k_n,n}$ and the endpoint 1 of the unit interval to confine the room in which $U_{n-k_n,n}$ can oscillate. Since the uniform order statistics $U_{j,n}$ and the order statistics $X_{j,n}$ from the distribution $F \in D(\alpha)$ are linked by the deterministic quantile function Q of F according to (1.2), it is quite natural that the trimmed sums $T_n(k_n)$ inherit the change in their asymptotic almost sure behavior from the corresponding change in the asymptotic almost sure behavior of the uniform order statistics. Moreover, it is understandable that the borderline case of k_n–sequences where the change takes place is the same in both cases. As is clear from (i) and (ii) in Result 1 and from (5.7), (5.9) and (5.10), this borderline case consists exactly of the sequences $k_n \sim c \log_2 n$ as $n \to \infty$ for some $0 < c < \infty$. In this situation Kiefer [59] (see also Deheuvels [31]) has shown that

$$\limsup_{n \to \infty} \frac{nU_{n-k_n,n}}{\log_2 n} = \gamma_c^+ \quad a.s.$$

and

$$\liminf_{n \to \infty} \frac{nU_{n-k_n,n}}{\log_2 n} = \gamma_c^- \quad a.s.$$

for two constants $0 < \gamma_c^- < \gamma_c^+ < \infty$ depending only on c and determined as roots of certain transcendental equations. Thus it is reasonable to expect a similar result for $T_n(k_n)$ if $k_n \sim c \log_2 n$. In fact, in [48] the following is obtained:

RESULT 2. *Assume* $F \in D(\alpha)$ *for* $0 < \alpha < 2$ *and* $F(0-) = 0$. *Let* $(k_n)_{n\geq 1}$ *satisfy* $k_n \sim c \log_2 n$ *as* $n \to \infty$ *for some* $0 < c < \infty$. *Then*

$$\left.\begin{array}{c} \limsup_{n\to\infty} \\ \liminf_{n\to\infty} \end{array}\right\} (2n \, \log_2 n)^{-1/2} \, \sigma(k_n/n)^{-1} \, \{T_n(k_n) - \mu_n(k_n)\} = \begin{cases} \overline{M}(\alpha,c) \\ \underline{M}(\alpha,c) \end{cases} \quad a.s.$$

where

$$\overline{M}(\alpha,c) = \begin{cases} \frac{(2-\alpha)^{1/2}\alpha}{2(\alpha-1)}c^{1/2}\left(1 + \max\left\{\left(\frac{1}{c\nu}-1\right)e^{\nu/\alpha}\right) : \nu > 0 \right. \\ \text{satisfies } \nu^\alpha e^{-\nu} \int_0^\nu x^{-\alpha+1} e^x \, dx = \frac{1}{c}\right\}, & \text{if } \alpha \neq 1, \\ \frac{c^{1/2}}{2}\left(\int_0^\nu \left(\log \frac{\nu}{x}\right) e^x \, dx + \nu\right) \text{ with } \nu > 0 \text{ satisfying} \\ \nu(1 - e^{-\nu}) = \frac{1}{c}, & \text{if } \alpha = 1, \end{cases}$$

and

$$\underline{M}(\alpha,c) = \begin{cases} \frac{(2-\alpha)^{1/2}\alpha}{2(\alpha-1)}c^{1/2}\left(1 - \left(\frac{1}{c\nu}+1\right)e^{-\nu/\alpha}\right), & \text{if } \alpha \neq 1, \\ \frac{c^{1/2}}{2}\left(\int_0^\nu \left(\log \frac{x}{\nu}\right) e^{-x} \, dx - \nu\right), & \text{if } \alpha = 1, \end{cases}$$

where $\nu > 0$ *is the unique solution of the equation*

$$\nu^\alpha e^\nu \int_0^\nu x^{-\alpha+1} e^x \, dx = \frac{1}{c}.$$

For the proof of this result representation (5.3) is no longer useful. It yields the existence of the constants $\overline{M}(\alpha,c)$ and $\underline{M}(\alpha,c)$ as shown in [49], but does not lead to their analytic description. To obtain this description, one introduces an auxiliary sequence $l_n \sim q \log_2 n$ with $c < q < \infty$ and breaks $T_n(k_n)$ up into two parts:

$$(2n \log_2 n)^{-1/2} \sigma(k_n/n)^{-1} \{T_n(k_n) - \mu_n(k_n)\}$$

$$=(2n \log_2 n)^{-1/2} \sigma(k_n/n)^{-1} \left\{ \sum_{j=1}^{n-l_n} X_{j,n} - n \int_0^{1-l_n/n} Q(u)\,du \right\}$$

$$+ (2n \log_2 n)^{-1/2} \sigma(k_n/n)^{-1} \left\{ \sum_{j=n+1-l_n}^{n-k_n} X_{j,n} - n \int_{1-l_n/n}^{1-k_n/n} Q(u)\,du \right\}$$

$$\equiv \Delta_n(q) + M_n(q).$$

From part (i) of Result 1 it can be deduced that

$$\lim_{q \to \infty} \limsup_{n \to \infty} |\Delta_n(q)| = 0 \quad a.s.$$

holds, which means that $\overline{M}(\alpha, c)$ and $\underline{M}(\alpha, c)$ can be obtained from analytic descriptions of $\overline{M}(\alpha, c, q)$ and $\underline{M}(\alpha, c, q)$ through

$$\limsup_{n \to \infty} M_n(q) = \overline{M}(\alpha, c, q) \quad a.s.$$

and

$$\liminf_{n \to \infty} M_n(q) = \underline{M}(\alpha, c, q) \quad a.s.$$

by taking limits as $q \to \infty$. To determine $\overline{M}(\alpha, c, q)$ and $\underline{M}(\alpha, c, q)$ one uses again elements of the quantile–transform methodology, but the main point now is the observation that the underlying distribution $F \in D(\alpha)$ with $0 < \alpha < 2$ belongs to the domain of attraction (in the sense of extreme value theory) of the extreme value distribution $\Phi_\alpha(x) = \exp(-x^{-\alpha})$, $x > 0$. This fact enables one to use tools from extreme value theory to approximate $\sum_{j=n+1-l_n}^{n-k_n} X_{j,n}$ by heavily trimmed sums with an underlying Pareto distribution of index α divided by appropriate powers of independent sample means from a standard exponential distribution. For these ratios one can infer large deviation theorems from the theory of large derivations of empirical measures, and these large deviation results determine $\overline{M}(\alpha, c, q)$ and $\underline{M}(\alpha, c, q)$ as functions of α, c and q. The analytic description of $\overline{M}(\alpha, c, q)$ and $\underline{M}(\alpha, c, q)$ can be made explicit enough to obtain the above formulas for $\overline{M}(\alpha, c)$ and $\underline{M}(\alpha, c)$ by evaluating the limits as $q \to \infty$.

The asymptotic almost sure behavior of the lightly trimmed sums $T_n(k) = \sum_{j=1}^{n-k} X_{j,n}$, where $k \geq 0$ now is a fixed integer, has been studied in [38], also via the quantile–transform – empirical–process approach. In this case the fluctuations of the uniform order statistics $U_{n-k,n}$ and hence of $X_{n-k,n} \overset{D}{=} Q(U_{n-k,n})$ are so big that it is impossible to center and normalize $T_n(k)$ in such a way that a finite but strictly positive limsup occurs. Precisely, one has the following stability theorem:

RESULT 3. *Assume $F \in D(\alpha)$ for $0 < \alpha < 2$ and $F(0-) = 0$. Let $(b_n)_{n \geq 1}$ be any sequence of non–decreasing constants converging to infinity and set $a_n = b_n Q(1 - 1/n)$. Then the following three statements are equivalent for any fixed integer $k \geq 0$:*

$$(5.11) \qquad \sum_{n=1}^{\infty} n^k (1 - F(a_n))^{k+1} < \infty,$$

$$(5.12) \qquad a_n^{-1} X_{n-k,n} \to 0 \quad a.s.\ as \quad n \to \infty,$$

and there exists a sequence $(c_n)_{n \geq 1}$ of constants such that

$$(5.13) \qquad a_n^{-1} \left\{ \sum_{j=1}^{n-k} X_{j,n} - c_n \right\} \to 0 \quad a.s.\ as \quad n \to \infty.$$

If (5.13) is true, then one can choose

$$c_n = \begin{cases} 0, & \text{if } 0 < \alpha < 1, \\ n \int_0^{1-1/n} Q(u)\, du, & \text{if } \alpha = 1, \\ n \int_0^{1} Q(u)\, du, & \text{if } 1 < \alpha < 2. \end{cases}$$

In addition, the following three statements are equivalent for any fixed integer $k \geq 0$:

$$(5.14) \qquad \sum_{n=1}^{\infty} n^k (1 - F(a_n))^{k+1} = \infty,$$

$$(5.15) \qquad \limsup_{n \to \infty} a_n^{-1} X_{n-k,n} = \infty \quad a.s.$$

and for any sequence $(c_n)_{n \geq 1}$ of constants

$$(5.16) \qquad \limsup_{n \to \infty} a_n^{-1} \left| \sum_{j=1}^{n-k} X_{j,n} - c_n \right| = \infty \quad a.s.$$

Results 1–3 contain a rather complete description of the almost sure behavior of lightly and moderately trimmed sums for positive random variables in the domain of attraction of a non–normal stable law. Of course, one can go beyond this restrictive assumption. Griffin [43], [44], considered the problem of the LIL behavior for modulus trimmed sums, using techniques quite different from the quantile–transform – empirical–process approach. For an underlying distribution F which is stochastically compact in the sense of Feller [39] (see also [16]) he demonstrates in [43] that a bounded LIL with classical normalizing constants holds for all trimming sequences $(k_n)_{n \geq 1}$ with $\liminf_{n \to \infty} k_n / \log_2 n > 0$. In that paper he poses the question whether if one assumes that F is also symmetric about 0 and $k_n / \log_2 n \to \infty$ as $n \to \infty$ holds, is it true the the constant appearing in his LIL result is 1? In [52] it is shown by an application of an appropriate version of the quantile–transform – empirical–process approach that the answer to this question is yes provided that $(k_n)_{n \geq 1}$ satisfies the monotonicity condition (K).

Part (ii) of Result 1 above shows that for k_n–sequences with $k_n / \log_2 n \to 0$ the natural centering constants $\mu_n(k_n)$ and norming constants $(2n \log_2 n)^{1/2} \sigma(k_n/n)$

are not appropriate for the asymptotic description of the almost sure fluctuations of $T_n(k_n)$. In [44], Griffin has determined different centering and norming constants which for modulus trimmed sums from a stochastically compact F, not in the domain of partial attraction of the normal distribution, lead to a finite, strictly positive limsup. For $F \in D(\alpha)$ with $0 < \alpha < 2$ and $k_n/(\log_2 n)^{1/2} \to 0$ as $n \to \infty$, he also specifies the constant in this limsup–statement. Thus, when applied to positive random variables, where modulus and ordinary trimming are the same, his results indicate that the behavior of $T_n(k_n)$ for $k_n \to \infty$ with $k_n/\log_2 n \to 0$ may split up into several subcases. This range of k_n–sequences requires further investigation.

Recently the quantile–transform – empirical–process approach has led to a number of new almost sure results. In particular, for a 'tail–empirical process' approach to some non–standard LIL's, see [32, 33] and for a universal lim inf law of the iterated logarithm for sums of i.i.d. positive random variables refer to [71].

6. L–statistics

Mason and Shorack [72, 73] consider linear combinations of order statistics of the form

$$\sum_{j=m_n+1}^{n-k_n} c_{jn} X_{j,n} \overset{\mathcal{D}}{=} c_{jn} Q(U_{j,n}),$$

or more generally

$$T_n^* = \sum_{j=m_n+1}^{n-k_n} c_{jn} g(U_{j,n}),$$

where g is some function and

$$c_{jn} = n \int_{(j-1)/n}^{j/n} J(t)\, dt, \quad 1 \le j \le n,$$

with some function J regularly varying at 0 and 1. In the moderate trimming case they use a reduction principle showing that the asymptotic distribution problem for T_n^* is the same as that for

$$\overline{T}_n = \sum_{j=m_n+1}^{n-k_n} K(U_{j,n}),$$

where K, as a measure, is defined by $dK = Jdg$. Thus, under certain conditions on g and J, in [73] they obtain results parallel to those in [17] sketched in Section 3 above. Even for fixed (light) trimming (or no trimming), in [72, 74] they still obtain a theory parallel to that in [16], sketched in Section 2 above.

We should mention here that the quantile–transform – empirical–process approach has been used for a long time in the study of L–statistics. See, in particular, the work on the central limit theorem for L–statistics by Shorack [80] and Mason [67], the strong law by Wellner [89], van Zwet [87] and Mason [68] and the law of the iterated logarithm by Wellner [90]. The paper by Shorack [81] in this volume is an up–dated version based on the approximations (1.5) and (1.6) of his 1972 paper [80].

7. Some related results

7.1. Extreme and self–normalized sums in the domain of attraction of a stable law. The paper [8] gives a unified theory of such sums based on the preliminary results in [5]. (A somewhat incomplete such theory was given earlier by LePage, Woodroofe and Zinn [61].) The idea is that properly centered whole sums $\sum_{j=1}^{n} Q(U_j)$ and the individual extremes $Q(U_{1,n}), \ldots, Q(U_{m,n})$, $Q(U_{n-k,n}), \ldots, Q(U_{n,n})$ converge jointly with the same normalizing factor. This

is trivial in our approach, where convergence is, in fact, in probability. Parallel-
ing a result of Hall [57], an approximation of an arbitrary stable law by suitably
centered sums, these being the asymptotic representations of the sum of a finite
number of extremes, is given. (For a generalization, see the next subsection.) The
self–normalized sums are those considered by Logan, Mallows, Rice and Shepp [63]:

$$L_n(p) = \left(\sum_{j=1}^{n} X_i \right) \bigg/ \left(\sum_{j=1}^{n} |X_j|^p \right)^{1/p}.$$ The emphasis in [8] concerning this is the

investigation of the properties of the limiting distribution using a representation
arising out of our approach. The extremes have a definite role in these properties.
In [23], we used our quantile approach to prove half of a conjecture in [63] stat-
ing that if F is in the domain of attraction of a normal law and $EX = 0$, then
$L_n(2) \overset{D}{\to} N(0,1)$. This was proved earlier by Maller [64]. The converse half of the
conjecture is still open. For further results see [56].

**7.2. An "extreme–sum" approximation of infinitely divisible laws
without a normal component.** Given an infinitely divisible characteristic function of the form of the right–side of (2.10) with γ replaced by a general constant θ,
where L and R are left– and right–continuous Lévy measures, respectively, so that
$L(-\infty) = 0 = R(\infty)$ and

$$\int_{-\epsilon}^{0} x^2 \, dL(x) + \int_{0}^{\epsilon} x^2 \, dR(x) < \infty \quad \text{for any} \quad \epsilon > 0,$$

we see that forming $\phi_1(s) = \inf\{x < 0 : L(x) > s\}$, $0 < s < \infty$ and $\phi_2(s) = \inf\{x < 0 : -R(-x) > s\}$, $0 < s < \infty$, so that (2.2) is satisfied, the random variable
$V_{0,0} + \theta - \gamma$ has the given infinitely divisible distribution. From now on suppose that
$\rho = 0$, and let $F_0(\cdot) = F_0(\phi_1, \phi_2, \theta; \cdot)$ be the distribution function of $V_{0,0} + \theta - \gamma$. The

approach sketched in Section 2 implies that under the said conditions the vector

$$\frac{1}{A_n}\left(\left(\sum_{j=1}^{n} X_j - \mu_n(1, n-1)\right), \sum_{j=1}^{m} X_{j,n} + \sum_{j=n-k+1}^{n} X_{j,n}\right)$$

converges in distribution along some subsequence to

$$\left(V_{0,0}, \sum_{j=1}^{m} \phi_1(S_j^{(1)}) - \sum_{j=1}^{k} \phi_2(S_j^{(2)})\right).$$

So the second component here represents the asymptotic contribution of the extremes in the limiting infinitely divisible law of the full sum. Hence it is conceivable that a suitably centered form of this second component can approximate now $V_{0,0}$ if $m, k \to \infty$.

Let $L_{m,k}$ be the Lévy distance between F_0 and the distribution function of

$$\sum_{j=1}^{m} \phi_1(S_j^{(1)}) - \sum_{j=1}^{k} \phi_2(S_j^{(2)}) - \left(\int_1^m \phi_1(s)\, ds - \int_1^k \phi_2(s)\, ds\right) + \theta - \gamma.$$

Then it is shown in [9] that $L_{m,k} \to 0$ as $m, k \to \infty$, and, depending on how fast $\phi_1(s)$ and $\phi_2(s)$ converge up to zero as $s \to \infty$, rates of this convergence are also provided. These rates are sometimes amazingly fast. In the special case of a stable distribution with exponent $0 < \alpha < 2$, given by $\phi_j(s) = -c_j s^{-1/\alpha}$, $s > 0$, $c_1, c_2 \geq 0$, $c_1 + c_2 > 0$, when $L_{m,k}$ can be replaced by the supremum distance $K_{m,k}$, we obtain

$$K_{m,k} = o\left(\max\left(c_1 m^{-\epsilon\left(\frac{1}{\alpha} - \frac{1}{2}\right)}, c_2 k^{-\epsilon\left(\frac{1}{\alpha} - \frac{1}{2}\right)}\right)\right) \quad \text{as} \quad m, k \to \infty,$$

where $0 < \epsilon < 1$ is as close to 1 as we wish. (A slight improvement of this last rate has been recently given in [58].)

8. The quantile–transform approach to modulus trimming

Let, as above, $X, X_1, X_2, \ldots$ be independent random variables with a common non–degenerate distribution function F which is *symmetric about zero*. For $1 \leq j \leq n$, set

$$m_n(j) = \#\{1 \leq i \leq n : |X_i| > |X_j| \quad \text{or} \quad (|X_i| = |X_j| \quad \text{and} \quad i \leq j)\},$$

and let $^{(k)}X_n = X_j$ if $m_n(j) = k$, $1 \leq k \leq n$. Thus $^{(k)}X_n$ is the kth largest random variable in absolute value from amongst $X_1, \ldots, X_n$, with ties broken according to the order in which the sample occurs. In Section 1 we cited a number of papers which investigate the asymptotic distribution of the 'modulus' trimmed sums

$$^{(k)}S_n = {}^{(k+1)}X_n + \cdots + {}^{(n)}X_n$$

as $n \to \infty$, when the largest k terms in absolute value are excluded from the nth partial sums for a fixed $k \geq 1$. Following a preliminary investigation by Pruitt [78], Griffin and Pruitt [45] undertook a detailed study of the asymptotic distribution of $^{(k_n)}S_n$, where $\{k_n\}$ is a sequence of integers such that $0 \leq k_n \leq n-1$ for $n \geq 1$ and both

$$k_n \to \infty \quad \text{and} \quad k_n/n \to 0 \quad \text{as} \quad n \to \infty.$$

Let

$$G(x) = Pr\{|X| \leq x\}, \quad -\infty < x < \infty,$$

be the distribution function of $|X|$ and let

$$K(u) = \inf\{x : G(X) \geq u\}, \quad 0 < u \leq 1, \quad K(0) = K(0+)$$

denote the pertaining quantile function. Also, consider a sequence $U, U_1, U_2, \ldots$ of independent random variables uniformly distributed on (0,1) and, independent of this sequence, a sequence $s, s_1, s_2, \ldots$ of independent and identically distributed random signs, that is, $P(s = 1) = P(s = -1) = 1/2$. Since $K(U) =_{\mathcal{D}} |X|$, it is elementary to show using symmetry that

$$(X, |X|) =_{\mathcal{D}} (sK(U), K(U))$$

from which we have immediately that

(8.1) $\{(X_i, |X_i|) : i \geq 1\} \overset{\mathcal{D}}{=} \{(s_i K(U_i), K(U_i)) : i \geq 1\}.$

For each $n \geq 1$ let $U_{1,n} \leq \cdots \leq U_{n,n}$ denote the order statistics of $U_1, \ldots, U_n$. Let $D_{1,n}, \ldots, D_{n,n}$ be the antiranks pertaining to $U_1, \ldots, U_n$, i.e. $U_{D_{i,n}} = U_{i,n}$ for $1 \leq i \leq n$. Then from (2.1) we easily obtain that for each n,

$$({}^{(n)}X_n, \ldots, {}^{(1)}X_n) \overset{\mathcal{D}}{=} (s_{D_{1,n}} K(U_{1,n}), \ldots, s_{D_{n,n}} K(U_{n,n}))$$

so that

$$({}^{(k_n)}S_n =_{\mathcal{D}} \sum_{i=1}^{n-k_n} s_{D_{i,n}} K(U_{i,n}), \quad n \geq 1.$$

For $n \geq 1$, we set

$$H_n(t) = n^{-1} \sum_{i=1}^{n} s_i \, I(U_i \leq t), \quad 0 \leq t \leq 1.$$

Notice that

$$H_n(t) = \frac{1}{n} \sum_{i=1}^{n} s_{D_{i,n}} I(U_{i,n} \leq t), \, 0 \leq t \leq 1, \quad n \geq 1.$$

Therefore we can write

$$({}^{(k_n)}S_n =_{\mathcal{D}} n \int_0^{U_{n-k_n,n}} K(u) \, dH_n(u).$$

This distributional equality may be viewed as the analogue of (1.4) in the present

context and will be our starting point in [19] in the present volume. In order to

exploit it, we prove the analogue of the approximation (1.5) and (1.6) for processes

of current interest. These are then used to derive all the sufficiency results of Griffin

and Pruitt [45] under certain clean forms of their necessary and sufficient conditions.

References

[1] D. Z. AROV and A. A. BOBROV, The extreme terms of a sample and their role in the sum of independent variables. *Theory Probab. Appl.* **5** (1960), 377–369.

[2] S. BJERVE, Error bounds for linear combinations of order statistics. *Ann. Statist.* **5** (1977), 357–369.

[3] E. CSÁKI, The law of the iterated logarithm for normalized empirical distribution functions. *Z. Wahrsch. Verw. Gebiete* **38** (1977), 147–167.

[4] M. CSÖRGŐ, S. CSÖRGŐ, L. HORVÁTH and D. M. MASON, Weighted empirical and quantile processes. *Ann. Probab.* **14** (1986), 31–85.

[5] M. CSÖRGŐ, S. CSÖRGŐ, L. HORVÁTH and D. M. MASON, Normal and stable convergence of integral functions of the empirical distribution function. *Ann. Probab.* **14** (1986), 86–118.

[6] M. CSÖRGŐ, S. CSÖRGŐ, L. HORVÁTH and D. M. MASON, Sup–norm convergence of the empirical process indexed by functions and applications. *Probab. Math. Statist.* **7** (1986), 13–26.

[7] M. CSÖRGŐ, and L. HORVÁTH, Approximations of weighted empirical and quantile processes. *Statist. Probab. Letters* **4** (1986), 275–280.

[8] S. CSÖRGŐ, Notes on extreme and self–normalized sums from the domain of attraction of a stable law. *J. London Math. Soc.* (2), **39** (1989), 369–384.

[9] S. CSÖRGŐ, An extreme–sum approximation to infinitely divisible laws without a normal component. In: *Probability on Vector Spaces IV* (S. Cambanis and A. Weron, eds.), pp. 47–58. *Lecture Notes in Mathematics* **1391**. Springer, Berlin, 1989.

[10] S. CSÖRGŐ, Limit theorems for sums of order statistics. In: *Sixth Interna-*

tional Summer School in Probability theory and Mathematical Statistics, Varna, 1988, pp. 5–37. Publishing House Bulgarian Acad. Sci., Sofia, 1989.

[11] S. CSÖRGŐ, A probabilistic approach to domains of partial attraction. *Adv. in Appl. Math.* **11** (1990), to appear.

[12] S. CSÖRGŐ, P. DEHEUVELS and D. M. MASON, Kernel estimates of the tail index of a distribution. *Ann. Statist.* **13** (1985), 1050–1077.

[13] S. CSÖRGŐ, and R. DODUNEKOVA, Limit theorems for the Petersburg game. In this volume.

[14] S. CSÖRGŐ, and R. DODUNEKOVA, The domain of partial attraction of a Poisson law. Submitted.

[15] S. CSÖRGŐ, and R. DODUNEKOVA, Infinitely divisible laws partially attracted to a Poisson law. *Math. Balcanica*, to appear.

[16] S. CSÖRGŐ, E. HAEUSLER and D. M. MASON, A probabilistic approach to the asymptotic distribution of sums of independent, identically distributed random variables. *Adv. in Appl. Math.* **9** (1988), 259–333.

[17] S. CSÖRGŐ, E. HAEUSLER and D. M. MASON, The asymptotic distribution of trimmed sums. *Ann. Probab.* **16** (1988), 672–699.

[18] S. CSÖRGŐ, E. HAEUSLER and D. M. MASON, The asymptotic distribution of extreme sums. *Ann. Probab.* **18** (1990), to appear.

[19] S. CSÖRGŐ, E. HAEUSLER and D. M. MASON, The quantile–transform approach to the asymptotic distribution of modulus trimmed sums. In this volume.

[20] S. CSÖRGŐ, L. HORVÁTH and D. M. MASON, What portion of the sample makes a partial sum asymptotically stable or normal? *Probab. Theory Rel. Fields* **72** (1986), 1–16.

[21] S. CSÖRGŐ and D. M. MASON, Central limit theorems for sums of extreme values. *Math. Proc. Cambridge Philos. Soc.* **98** (1985), 547–558.

[22] S. CSÖRGŐ and D. M. MASON, The asymptotic distribution of sums of extreme values from a regularly varying distribution. *Ann. Probab.* **14** (1986), 974–983.

[23] S. CSÖRGŐ and D. M. MASON, Approximations of weighted empirical processes with applications to extreme, trimmed and self–normalized sums. In: *Proc. First World Congress Bernoulli Soc.* Vol. 2, pp. 811–819. VNU Sci. Press, Utrecht, 1987.

[24] S. CSÖRGŐ and D. M. MASON, Simple estimators of the endpoint of a distribution. In: *Extreme Value Theory, Oberwolfach 1987* (J. Hüsler and R.-D. Reiss, eds.), pp. 132–147. *Lecture Notes in Statistics* **51**. Springer, Berlin, 1989.

[25] S. CSÖRGŐ and D. M. MASON, Bootstrapping empirical functions. *Ann. Statist.* **17** (1989), 1447–1471

[26] S. CSÖRGŐ and D. M. MASON, A probabilistic approach to the tails of infinitely divisible laws. In this volume.

[27] S. CSÖRGŐ and D. M. MASON, Intermediate sums and stochastic compactness of maxima. Submitted.

[28] S. CSÖRGŐ and D. M. MASON, Intermediate– and extreme–sum processes. Submitted.

[29] S. CSÖRGŐ and L. VIHAROS, Asymptotic distributions for vectors of power sums. In: *Colloquia Math. Soc. J. Bolyai* 00. *Limit Theorems in Probability and Statistics* (E. Csáki and P. Révész, eds.), pp. 00–00. North–Holland, Amsterdam, 1990. To appear.

[30] D. A. DARLING, The influence of the maximum term in the addition of independent random variables. *Trans. Amer. Math. Soc.* **73** (1952), 95–107.

[31] P. DEHEUVELS, Strong laws for the k–th order statistic when $k \leq c \log_2 n$. *Probab. Theory Rel. Fields* **72** (1986), 133–156.

[32] P. DEHEUVELS and D. M. MASON, Non–standard functional laws of the iterated logarithm for tail empirical and quantile processes. *Ann. Probab.* **18** (1990), to appear.

[33] P. DEHEUVELS and D. M. MASON, A tail empirical process approach to some non–standard laws of the iterated logarithm. *J. Theoret. Probab.*, to appear.

[34] W. DOEBLIN, Sur l'ensemble de puissances d'une loi de probabilité, *Studia Math.* **9** (1940), 71–96.

[35] V. A. EGOROV, The central limit theorem in the absence of extremal absolute order statistics. (In Russian) *Zap. Nauchn. Sem. Leningrad Otdel. Mat. Inst. Steklov (LOMI)* **142** (1985), 59–67.

[36] V. A. EGOROV and V. B. NEVZOROV, Some rates of convergence of sums of order statistics to the normal law. (In Russian) *Zap. Nauchn. Sem. Leningrad Otdel. Mat. Inst. Steklov (LOMI)* **41** (1974), 105–128.

[37] V. A. EGOROV and V. B. NEVZOROV, Summation of order statistics and the normal law. (In Russian) *Vestnik Leningrad Univ. Mat. Mekh. Astronom.*

1974 No. 1, 5–11.

[38] J. H. J. EINMAHL, E. HAEUSLER and D. M. MASON, On the relationship between the almost sure stability of weighted empirical distributions and sums of order statistics. *Probab. Theory Rel. Fields* **79** (1988), 59–74.

[39] W. FELLER, On regular variation and local limit theorems. In: *Proc. Fifth Berkeley Symp. Math. Statist. Probab.* Vol. 2, pp. 373–388. University of California Press, Berkeley, 1967.

[40] B. V. GNEDENKO, On the theory of limit theorems for sums of independent random variables. (In Russian) *Izvestiya Akad. Nauk SSSR, Ser. Mat.* (1939), 181–232, 643–647.

[41] B. V. GNEDENKO, Some theorems on the powers of distribution functions. (In Russian) *Uchen. Zap. Moskov. Gos. Univ. Mat.* **45** (1940), 61–72.

[42] B. V. GNEDENKO and A. N. KOLMOGOROV, *Limit Distributions for Sums of Independent Random Variables.* Addison–Wesley, Reading, Mass., 1954.

[43] P. S. GRIFFIN, The influence of extremes on the law of the iterated logarithm. *Probab. Theory Rel. Fields* **77** (1988), 241–270.

[44] P. S. GRIFFIN, Non–classical law of the iterated logarithm behavior for trimmed sums. *Probab. Theory Rel. Fields* **78** (1988), 293–319.

[45] P. S. GRIFFIN and W. E. PRUITT, The central limit problem for trimmed sums. *Math. Proc. Cambridge Philos. Soc.* **102** (1987), 329–349.

[46] P. S. GRIFFIN and W. E. PRUITT, Asymptotic normality and subsequential limits of trimmed sums. *Ann. Probab.* **17** (1989), 1186–1219.

[47] A. V. GROSHEV, The domain of attraction of the Poisson law. (In Russian) *Izvestiya Akad. Nauk SSSR, Ser. Mat.* **5** (1941), 165–172.

[48] E. HAEUSLER, Laws of the iterated logarithm for sums of order statistics from a distribution with a regularly varying upper tail. Habilitationsschrift, University of Munich, 1988.

[49] E. HAEUSLER and D. M. MASON, Laws of the iterated logarithm for sums of the middle portion of the sample. *Math. Proc. Cambridge Philos. Soc.* **101** (1987), 301–312.

[50] E. HAEUSLER and D. M. MASON, A law of the iterated logarithm for sums of extreme values from a distribution with a regularly varying upper tail. *Ann. Probab.* **15** (1987), 932–953.

[51] E. HAEUSLER and D. M. MASON, On the asymptotic behavior of sums of order statistics from a distribution with a slowly varying upper tail. In this volume.

[52] E. HAEUSLER and D. M. MASON, A law of the iterated logarithm for modulus trimming. In: *Colloquia Math Soc. J. Bolyai 00.* Limit Theorems in Probablity and Statistics (E. Csáki and P. Révész, eds.), pp. 000–000. North-Holland, Amsterdam, 1990. To appear.

[53] M. G. HAHN and J. KUELBS, Universal asymptotic normality for conditionally trimmed sums. *Statist. Probab. Letters* **7** (1989), 9–15.

[54] M. G. HAHN, J. KUELBS and J. D. SAMUR, Asymptotic normality of trimmed sums of Φ–mixing random variables. *Ann. Probab.* **15** (1987), 1395–1418.

[55] M. G. HAHN, J. KUELBS and D. C. WEINER, On the asymptotic distribution of trimmed and winsorized sums. *J. Theoret. Probab.* **3** (1990), 137–168.

[56] M. G. HAHN, J. KUELBS and D. C. WEINER, Self-normalization of censored sums and sums–of–squares in joint estimation of certain center and scale sequences. *Ann. Probab.*, to appear.

[57] P. HALL, On the extreme terms of a sample from the domain of attraction of a stable law. *J. London Math. Soc.* **(2), 18** (1978), 181–191.

[58] A. JANSSEN and D. M. MASON, On the rate of convergence of sums of extremes to a stable law. *Probab. Theory Rel. Fields*, to appear.

[59] J. KIEFER, Iterated logarithm analogues for sample quantiles when $p_n \downarrow 0$. *Proc. Sixth Berkeley Symp. Math. Statist. Probab.* Vol. 1, pp. 227–244. University of California Press, Berkeley, 1972.

[60] J. KUELBS and M. LEDOUX, Extreme values for vector valued random variables and a Gaussian central limit theorem. *Probab. Theory Rel. Fields* **74** (1987), 341–355.

[61] R. LEPAGE, M. WOODROOFE and J. ZINN, Convergence to a stable distribution via order statistics. *Ann. Probab.* **9** (1981), 624–632.

[62] G. S. LO, A note on the asymptotic normality of sums of extreme values. *J. Statist. Planning Inference* **22** (1989), 127–136.

[63] B. F. LOGAN, C. L. MALLOWS, S. O. RICE and L. A. SHEPP, Limit distributions of self-normalized sums. *Ann. Probab.* **1** (1973), 788–809.

[64] R. A. MALLER, A theorem on products of random variables, with application to regression. *Austral. J. Statist.* **23** (1981), 177–185.

[65] R. A. MALLER, Asymptotic normality of lightly trimmed means – a converse. *Math. Proc. Cambridge Philos. Soc.* **92** (1982), 535–545.

[66] R. A. MALLER, Asymptotic normality of trimmed means in higher dimensions. *Ann. Probab.* **16** (1988), 1608–1622.

[67] D. M. MASON, Asymptotic normality of linear combinations of order statistics with a smooth score function. *Ann. Statist.* **9** (1981), 899–904.

[68] D. M. MASON, Some characterizations of strong laws for linear functions of order statistics. *Ann. Probab.* **10** (1982), 1051–1057.

[69] D. M. MASON, A note on weighted approximations to the uniform empirical and quantile processes. In this volume.

[70] D. M. MASON, A strong invariance theorem for the tail empirical process. *Ann. Inst. H. Poincaré Sect. B (N.S.)* **24** (1988), 491–506.

[71] D. M. MASON, A universal one–sided law of the iterated logarithm. *Ann. Probab.*, to appear.

[72] D. M. MASON and G. R. SHORACK, Necessary and sufficient conditions for asymptotic normality of L–statistics. *Ann. Probab.*, to appear.

[73] D. M. MASON and G. R. SHORACK, Necessary and sufficient conditions for the asymptotic normality of trimmed L–statistics. *J. Statist. Planning Inference* **23** (1990), to appear.

[74] D. M. MASON and G. R. SHORACK, Non–normality of a class of random variables. In this volume.

[75] D. M. MASON and W. R. VAN ZWET, A refinement of the KMT inequality for the uniform empirical process. *Ann. Probab.* **15** (1987), 871–884.

[76] T. MORI, On the limit distributions of lightly trimmed sums. *Math. Proc. Cambridge Philos. Soc.* **96** (1984), 507–516.

[77] W. E. PRUITT, The class of limit laws for stochastically compact normed sums. *Ann. Probab.* **11** (1983), 962–969.

[78] W. E. PRUITT, Sums of independent random variables with the extreme terms excluded. In: Probability and Statistics. Essays in Honor of Franklin A. Graybill (J. N. Srivastava, Ed.) pp. 201–216. Elsevier, Amsterdam, 1988.

[79] H. -J. ROSSBERG, Über das asymptotische Verhalten der Rand – und Zentralglieder einer Variationsreihe II. *Publ. Math. Debrecen* **14** (1967), 83–90.

[80] G. R. SHORACK, Functions of order statistics. *Ann. Math. Statist.* **43** (1972), 412–427.

[81] G. R. SHORACK, Limit results for linear combinations. In this volume.

[82] G. R. SHORACK and J. A. WELLNER, *Empirical Processes with Applications to Statistics*. Wiley, New York, 1986.

[83] S. M. STIGLER, The asymptotic distribution of the trimmed mean. *Ann. Statist.* **1** (1973), 472–477.

[84] L. SZEIDL, On the limit distributions of symmetric functions. *Theory Probab. Appl.* **31** (1986), 590–603.

[85] L. SZEIDL, On limit distributions of random symmetric polynomials, *Theory Probab. Appl.* **33** (1988), 248–259.

[86] J. L. TEUGELS, Limit theorems on order statistics. *Ann. Probab.* **9** (1981), 868–880.

[87] W. R. VAN ZWET, A strong law for linear functions of order statistics. *Ann. Probab.* **8** (1980), 986–990.

[88] V. V. VINOGRADOV and V. V. GODOVAN'CHUK, On large deviations of sums of independent random variables without some maximal summands. (In Russian) *Teor. Verojatn. Primenen,* **34** (1989), 569–571.

[89] J. A. WELLNER, A Glivenko–Cantelli theorem and strong laws of large numbers for functions of order statistics. *Ann. Statist.* **5** (1977), 473–480.

[90] J. A. WELLNER, A law of the iterated logarithm for functions of order statistics. *Ann. Statist.* **5** (1977), 481–494.

[91] J. A. WELLNER, Limit theorems for the ratio of the empirical distribution function to the true distribution function. *Z. Wahrsch. Verw. Gebiete* **45** (1978), 73–88.

[92] V. M. ZOLOTAREV, Limit theorems for random symmetric polynomials. *Theory Probab. Appl.* **30** (1985), 636–637.

[93] V. M. ZOLOTAREV, On random symmetric polynomials. (In Russian) In: *Probability Distributions and Mathematical Statistics*, pp. 170–188, FAN, Tashkent, 1986.

Sándor Csörgő
Department of Statistics
University of Michigan
Ann Arbor, Michigan 48109

Erich Haeusler
University of Munich
Therresienstrasse 39
8000 Munich 2
West Germany

David M. Mason
Department of Mathematical Scienc
University of Delaware
Newark, Delaware 19716

A NOTE ON WEIGHTED APPROXIMATIONS TO THE UNIFORM EMPIRICAL AND QUANTILE PROCESSES

David M. Mason[1]

Recently, M. Csörgő, S. Csörgő, Horváth and Mason (1986a) obtained a weighted approximation to the uniform empirical and quantile processes by a sequence of Brownian bridges. The purpose of this note is to give a short and elementary proof of their weighted approximation to the uniform quantile process. Their corresponding weighted approximation to the uniform empirical process follows in a direct fashion from that of the uniform quantile process. The present proof, as was the former, is based on the Komlós, Major and Tusnády (1976) strong approximation to the partial sum process. It is shown, however, that almost the same weighted approximation to the uniform empirical and quantile processes can be derived from the older Skorokhod (1965) embedding. This alternate weighted approximation, obtained via the Skorokhod embedding, is likely to be sufficient for nearly all applications of the weighted approximation methodology and has the advantage that its proof is more suitable for instructional purposes.

Introduction. Let $U_1, U_2 \ldots$, be a sequence of independent uniform $(0, 1)$ random variables, and for each integer $n \geq 1$, let $U_{1,n} \leq \cdots \leq U_{n,n}$ denote the order statistics based on the first n of these uniform $(0, 1)$ random variables. Define the

---
[1] Research partially supported by the Alexander von Humboldt Foundation and partially by NSF Grant #DMS-8803209.

uniform empirical quantile function to be, for each integer $n \geq 1$,

$$U_n(s) = U_{k,n}, \quad (k-1)/n < s \leq k/n, \quad k = 1, \ldots, n,$$

where $U_n(0) = U_{1,n}$, and the uniform quantile process

$$\beta_n(s) = n^{1/2} \{s - U_n(s)\}, \quad 0 \leq s \leq 1.$$

Also let

$$\alpha_n(s) = n^{1/2} \{G_n(s) - s\}, \quad 0 \leq s \leq 1,$$

be the uniform empirical process, where G_n is the right continuous empirical distribution function based on $U_1, \ldots, U_n$.

M. Csörgő, S. Csörgő, Horváth and Mason [Cs–Cs–H–M] (1986a) have obtained a weighted approximation of the uniform empirical and quantile processes by a sequence of Brownian bridges. The essential properties, in terms of applications, of this weighted approximation are contained in the following two theorems.

THEOREM 1. (Theorem 2.1 of Cs–Cs–H–M (1986a)) *There exists a sequence of independent uniform* $(0,1)$ *random variables* $U_1, U_2, \ldots$, *and a sequence of Brownian bridges* $B_1, B_2, \ldots$, *sitting on the same probability space such that for all* $0 \leq \nu_1 < 1/2$

$$(1) \qquad \sup_{1/(n+1) \leq s \leq n/(n+1)} n^{\nu_1} |\beta_n(s) - B_n(s)| (s(1-s))^{-1/2+\nu_1} = O_p(1).$$

THEOREM 2. (Corollary 4.2.2 of Cs–Cs–H–M (1986a)) *On the probability space of Theorem 1, for all* $0 \leq \nu_2 < 1/4$

$$(2) \qquad \sup_{0 \leq s \leq 1} n^{\nu_2} |\alpha_n(s) - \bar{B}_n(s)| (s(1-s))^{-1/2+\nu_2} = O_p(1),$$

where $\bar{B}_n(s) = B_n(s)$ for $1/n \leq s \leq 1 - 1/n$ and zero elsewhere.

Recently, Mason and van Zwet (1987) have constructed a probability space dual to that of Theorems 1 and 2 on which (1) holds for all $0 \leq \nu_1 < 1/4$ and (2) holds for all $0 \leq \nu_2 < 1/2$, i.e. reversing the bounds on ν_1 and ν_2.

Theorems 1 and 2 have proven to be very powerful tools in the solution of a large number of problems in probability and statistics. Besides providing quite simple proofs for the important Chibisov (1964), O'Reilly (1974), Eicker (1979) and Jaeschke (1979) results on the weak convergence and limiting distribution of weighted uniform empirical and quantile processes, cf. Cs–Cs–H–M (1986a), the application of Theorems 1 and 2, which we call the weighted approximation methodology, has led to a number of new results in probability and statistics. The reader is referred to M. Csörgő, S. Csörgő, Horváth and Mason (1986b), S. Csörgő, Horváth and Mason (1986), S. Csörgő, Haeusler and Mason (1988), S. Csörgő and Mason (1986), S. Csörgő, Deheuvels and Mason (1985), S. Csörgő and Mason (1985), M. Csörgő and Mason (1985), S. Csörgő and Mason (1989), and Mason (1985).

The proof of the Cs–Cs–H–M (1986a) version of the weighted approximation given in (1) and (2) is based on a refinement of the M. Csörgő, and Révész (1978) inequality for the strong approximation of the uniform quantile process by a sequence of Brownian bridges, which in turn is based on the Komlós, Major and Tusnády [KMT] (1976) inequality for the strong approximation of the partial sum process by a Wiener process. Whereas, the Mason and van Zwet (1987) version of (1) and (2) is based on a refinement of the KMT (1975) inequality for the strong approximation of the uniform empirical process by a sequence of Brownian bridges. By necessity, the complete proofs of these inequalities entail a substantial amount of technical detail and thus form a formidable barrier to the non–expert to the complete understanding and appreciation of this weighted approximation methodology.

The purpose of this note is to provide a short, simple, and direct proof of Theorem 1, which avoids the necessity of first proving a refined inequality of the sort mentioned above, and thus to make this methodology more readily accessible. The present proof, given in the next section, is based on the KMT (1976) strong approximation to the partial sum process. It will be pointed out, however, in the proof of Theorem 1, that if we were to replace the use of the KMT (1976) approximation given in (3) below by the Skorokhod (1965) embedding, that the same steps in the proof that follow (3) would show that the statement of Theorem 1 holds for all $0 \leq \nu_1 < 1/4$, instead of for all $0 \leq \nu_1 < 1/2$. This version of Theorem 1 based on the Skorokhod embedding is likely to be of some pedagogical importance in making these weighted approximations more easily comprehensible to those not well versed in the KMT (1975, 1976) strong approximations.

Theorem 2 follows from Theorem 1, either for the version based on KMT (1976), where $0 \leq \nu_1 < 1/2$, or for the version based on the Skorokhod embedding, where $0 \leq \nu_1 < 1/4$, since it is a direct consequence of whichever version of Theorem 1 combined with the fact proven in Cs–Cs–H–M (1986a) that for all $0 \leq \nu_2 < 1/4$ and $c > 0$,

$$\sup_{c/n \leq s \leq 1-c/n} n^{\nu_2} |\alpha_n(s) - \beta_n(s)| (s(1-s))^{-1/2+\nu_2} = O_p(1),$$

cf. Corollary 2.3 of Cs–Cs–H–M (1986a), this result holding on any probability space on which sits a sequence $U_1, U_2, \ldots$, of independent uniform $(0,1)$ random variables. Thus we see that in the Skorokhod embedding version of Theorems 1 and 2 the assumption on both ν_1 and ν_2 is that they be less than $1/4$.

Finally, we mention, as an aside, that so far in all the important applications of these weighted approximations the specified upper bounds on ν_1 and ν_2 have played no role in the proofs. The only essential requirement has been that (1) or (2) hold for small enough non–negative ν_1 and ν_2, so that all of the results cited above, which have proofs based on Theorems 1 and 2, could have been proven by means of the Skorokhod embedding version of these theorems. This includes the theorems of Chibisov (1964), O'Reilly (1974), Eicker (1979) and Jaeschke (1979).

2. Proof of Theorem 1. As in Cs–Cs–H–M (1986a), let $\{W^{(i)}(s) : 0 \leq s < \infty\}$ for $i = 1, 2$ be two independent standard Wiener processes sitting on the same probability space. On this probability space construct two independent sequences of independent exponential random variables with expectation equal to one $Y_1^{(i)}, Y_2^{(i)}, \ldots$, as a function of $W^{(i)}(i = 1, 2)$, by means of Theorem 1 of KMT (1976), such that

$$(3) \qquad m^{-1/p} |S_m^{(i)} - m - W^{(i)}(m)| \to 0 \quad \text{a.s. as} \quad m \to \infty$$

for all $0 < p < \infty$, where for $i = 1, 2$

$$S_m^{(i)} = \sum_{j=1}^{m} Y_j^{(i)}, \quad m = 1, 2, \ldots$$

(We remark that if, instead, the Skorokhod embedding technique had been used here to construct for each $i = 1, 2$, the sequence $Y_1^{(i)}, Y_2^{(i)}, \ldots$, as a function of the Wiener process $W^{(i)}$, then (3) would hold for all $0 < p < 4$, cf. Breiman (1967), and the remainder of this proof would then establish that (1) holds for all $0 \leq \nu_1 < 1/4$ on an appropriately constructed probability space.)

For each integer $n \geq 2$, let

$$Y_j(n) = \begin{cases} Y_j^{(1)} & \text{for } j = 1, \ldots, [n/2] \\ Y_{n+2-j}^{(2)} & \text{for } j = [n/2] + 1, \ldots, n+1, \end{cases}$$

where $[x]$ denotes the integer part of x, and set

$$S_m(n) = \sum_{j=1}^{m} Y_j(n) \quad \text{for} \quad m = 1, \ldots, n+1.$$

For notational convenience, we write from now on Y_j and S_m for $Y_j(n)$ and $S_m(n)$. Also for each integer $n \geq 2$, let

$$W_n(s) = \begin{cases} W^{(1)}(s) & \text{for } 0 \leq s \leq [n/2] \\ W^{(1)}([n/2]) + W^{(2)}(n+1-[n/2]) - W^{(2)}(n+1-s) & \text{for } [n/2] < s \leq n+1. \end{cases}$$

It is easily checked that for each integer $n \geq 2$, $\{W_n(s) : 0 \leq s \leq n+1\}$ is a standard Wiener process on $[0, n+1]$.

We will first show that for all $0 \leq \nu < 1/2$

$$(4) \quad C_{n,1} = \max_{1 \leq i \leq n/2} |n\{S_i/S_{n+1} - i/n\} - \{W_n(i) - \frac{i}{n}W_n(n)\}|\, i^{-1/2+\nu} = O_p(1)$$

and

$$(5) \quad C_{n,2} =$$

$$\max_{n/2+1 \leq i \leq n} |n\{S_i/S_{n+1} - i/n\} - \{W_n(i) - \frac{i}{n}W_n(n)\}|(n+1-i)^{-1/2+\nu} = O_p(1).$$

First consider (4). Notice that by the law of large numbers

$$\max_{1 \leq i \leq n/2} n\,|S_i/S_{n+1} - S_i/S_n|\, i^{-1/2+\nu} =$$

$$\max_{1 \leq i \leq n/2} nS_i\, i^{-1/2+\nu}\, Y_{n+1}/(S_n S_{n+1}) = O_p(n^{-1/2+\nu}) = o_p(1).$$

Also, by the law of large numbers, the law of the iterated logarithm, and the central limit theorem

$$\max_{1 \leq i \leq n/2} |n(S_i - i - \tfrac{i}{n}(S_n - n))/S_n - (S_i - i - \tfrac{i}{n}(S_n - n))| \, i^{-1/2+\nu}$$

$$\leq \{ \max_{1 \leq i \leq n/2} |S_i - i| \, i^{-1/2+\nu} + n^{-1/2+\nu} |S_n - n|\} \, |S_n - n|/S_n$$

$$= \{O_p((\log \log n)^{1/2} n^\nu) + O_p(n^\nu)\} \, O_p(n^{-1/2}) = o_p(1).$$

Thus in light of the above two $o_p(1)$ statements and a little algebra, to prove (4) it is enough to show that

$$(6) \quad D_n = \max_{1 \leq i \leq n/2} |S_i - i - \tfrac{i}{n}(S_n - n) - (W_n(i) - \tfrac{i}{n} W_n(n))| \, i^{-1/2+\nu} = O_p(1).$$

Observe that

$$D_n \leq \max_{1 \leq i \leq n/2} |S_i^{(1)} - i - W^{(1)}(i)| \, i^{-1/2+\nu} + n^{-1/2+\nu} |S_{[n/2]}^{(1)} - [n/2] - W^{(1)}([n/2])|$$

$$+ n^{-1/2+\nu} |S_{n+1-[n/2]}^{(2)} - (n+1 - [n/2]) - W^{(2)}(n+1 - [n/2])|$$

$$+ n^{-1/2+\nu} |Y_1^{(2)} - 1| + n^{-1/2+\nu} |W_n(n+1) - W_n(n)|.$$

Since $1/2 - \nu > 0$, we see by (3) that the first term on the right side of this inequality is $O_p(1)$ and the next two terms are $o_p(1)$. The last two terms are obviously $o_p(1)$ random variables. Hence we have established (6) and therefore (4). Assertion (5) is proven in almost the same way using the symmetry of the construction given above.

Next set for each integer $n \geq 2$

$$\tilde{U}_{i,n} = S_i/S_{n+1} \quad \text{for} \quad i = 1, \dots, n,$$

and

$$\tilde{B}_n(s) = n^{-1/2}(s W_n(n) - W_n(sn)) \quad \text{for} \quad 0 \leq s \leq 1.$$

We see that for each integer $n \geq 2$

$$(\tilde{U}_{1,n}, \dots, \tilde{U}_{n,n}) \overset{\mathcal{D}}{=} (U_{1,n}, \dots, U_{n,n}),$$

and $\tilde{B}_n$ is a Brownian bridge. Let $\tilde{\beta}_n$ denote the uniform quantile process based on $\tilde{U}_{1,n}, \ldots, \tilde{U}_{n,n}$. We claim that

$$(7) \qquad E_{n,1} = \sup_{1/(n+1) \le s \le [n/2]/n} n^\nu |\tilde{\beta}_n(s) - \tilde{B}_n(s)| \, s^{-1/2+\nu} = O_p(1)$$

and

$$(8) \qquad E_{n,2} = \sup_{[n/2]/n < s \le n/(n+1)} n^\nu |\tilde{\beta}_n(s) - \tilde{B}_n(s)| \, (1-s)^{-1/2+\nu} = O_p(1).$$

Notice that

$$E_{n,1} \le 2 \max_{1 \le i \le n/2} n^\nu |\tilde{\beta}_n(i/n) - \tilde{B}_n(i/n)| \, (i/n)^{-1/2+\nu}$$

$$(9) \qquad + \max_{1 \le i \le [n/2]-1} \sup_{i/n \le s \le (i+1)/n} n^\nu \left| \tilde{B}_n\left(\frac{i+1}{n}\right) - \tilde{B}_n(s) \right| (i/n)^{-1/2+\nu}$$

$$+ \sup_{1/(n+1) \le s \le 1/n} n^\nu \left| \tilde{B}_n(s) - \tilde{B}_n\left(\frac{1}{n}\right) \right| (1/(n+1))^{-1/2+\nu} + 1.$$

The first term on the right side of inequality (9), we recognize to be $2C_{n,1}$, which has just been proven to be $O_p(1)$. To show that the next two terms on the right side of (9) are $O_p(1)$, we require the following probability inequality: for any $0 < a < 1$, $h \ge 0$, and $0 < u < \infty$

$$(10) \qquad P\left(\sup_{s \in [a-h, a+h] \cap [0,1]} |B(a) - B(s)| \ge u h^{1/2} \right) \le A u^{-1} \exp(-u^2/8),$$

where B denotes a Brownian bridge and A is a suitably chosen universal positive constant, cf. (1.11) of Cs–Cs–H–M (1986a).

Using (10) it is routine to verify that for all $\epsilon > 0$ there exists a $0 < M < \infty$ such that for all integers $n > 2$

$$\sum_{i=1}^{[n/2]-1} P\left(\sup_{i/n \le s \le (i+1)/n} \left| \tilde{B}_n\left(\frac{i+1}{n}\right) - \tilde{B}_n(s) \right| \ge M i^{1/2-\nu} n^{-1/2} \right) < \epsilon,$$

which proves that the second term on the right side of (9) is $O_p(1)$. The fact that the third term on the right of (9) is $o_p(1)$ follows easily from (10). Thus we have established (7). Assertion (8) is proven similarly using (5) and (10).

Combining (7) and (8), we get that

$$(11) \qquad \sup_{1/(n+1)\leq s\leq n/(n+1)} n^{\nu} |\tilde{\beta}_n(s) - \tilde{B}_n(s)| (s(1-s))^{-1/2+\nu} = O_p(1).$$

Up to this stage in the proof, we have constructed a probability space on which sit a sequence $\tilde{\beta}_n$ of *versions* of β_n, i.e. for each integer $n \geq 2$

$$\{\tilde{\beta}_n(s) : 0 \leq s \leq 1\} \overset{D}{=} \{\beta_n(s) : 0 \leq s \leq 1\},$$

and a sequence of Brownian bridges $\tilde{B}_n$ such that (11) holds for all $0 \leq \nu < 1/2$. In order to construct a probability space with a sequence $U_1, U_2, \ldots$, of independent uniform $(0,1)$ random variables and a sequence of Brownian bridges $B_1, B_2, \ldots$, such that (1) holds, we now follow the procedure as given in Lemma 3.1.1 of M. Csörgő (1983). This finishes the proof of Theorem 1.

Acknowledgement. The author would like to thank E. Haeusler for some helpful suggestions and useful discussion while this paper was being written.

Addendum. Except for updating references and correcting typographical errors, the above is a verbatim copy of a University of Munich technical report with the same title dated January 1986. M. Csörgő and Horváth (1986) published a proof of Theorem 1 which is mathematically equivalent to that given above and made the same comment about the substitution of the KMT (1976) approximation by the Skorokhod embedding. They also provided a short proof for Theorem 2 based on a Poissonization and the M. Csörgő and Révész (1979) strong law for the Wiener process used in conjunction with a Wiener process strong approximation to the Poisson process. They emphasized that the advantage of their approach to Theorem 2 is that it is not directly based on complicated inequalities. (Of course some formidable inequalities were used to prove the M. Csörgő and Révész strong law and the strong approximation.)

In this Addendum we give a very short and direct proof for Theorem 2 that does not rely on a strong approximation. The price paid for this is that the main tool in the proof now becomes an admittedly *somewhat complicated* inequality for the oscillation modulus of the uniform empirical process. We should mention that the proof presented here is in the same spirit as the original proof given in Cs–Cs–H–M (1986a). Naturally, whichever proof one prefers is a matter of taste and expediency.

Theorem 2 is an immediate consequence of the following proposition which was derived from Theorems 1 and 2 and stated as Corollary 2.3 in Cs–Cs–H–M (1986a).

Proposition. *For any* $0 < d < 1$ *and* $0 \leq \nu < 1/4$

$$(a) \qquad n^\nu \sup_{d/n \leq s \leq 1-d/n} |\alpha_n(s) - \beta_n(s)|/(s(1-s))^{1/2-\nu} = O_p(1).$$

PROOF. Since trivially

$$(b) \qquad \sup_{0 \leq s \leq 1} |\alpha_n(U_n(s)) - \beta_n(s)| = O(n^{-1/2}) \quad \text{a.s.,}$$

to prove (a) it suffices to show that both

$$(c) \qquad \sup_{d/n \leq s \leq 1} n^\nu \frac{|\alpha_n(U_n(s)) - \alpha_n(s)|}{s^{1/2-\nu}} := M_{n,1}(d, \nu) = O_p(1)$$

and

(d) $$\sup_{0 \leq s \leq 1-d/n} n^{\nu} \frac{|\alpha_n(U_n(s)) - \alpha_n(s)|}{(1-s)^{1/2-\nu}} := M_{n,2}(d,\nu) = O_p(1).$$

Also on account of $M_{n,1}(d,\nu) \overset{\mathcal{D}}{=} M_{n,2}(d,\nu)$ it is enough to prove (c). Assertion (c) will be an easy consequence of the following lemmas.

LEMMA i. For any $0 < d < 1$ and $0 \leq 2\delta < 1/2$

(e) $$\sup_{d/n \leq s \leq 1} n^{2\delta} |U_n(s) - s|/s^{1-2\delta} = O_p(1).$$

PROOF. In Mason (1983), (also see Remark 4.4 of Marcus and Zinn (1984)), it is proven that for any $0 \leq 2\delta < 1/2$

$$\sup_{0 \leq s \leq 1} n^{2\delta} |G_n(s) - s|/s^{1-2\delta} = O_p(1).$$

In addition, from Lemma 2 of Wellner (1978) one has for any $0 < d < 1$

$$\sup_{d/n \leq s \leq 1} U_n(s)/s = O_p(1).$$

These two facts when combined with (b) yield (e). ∎

For any $0 < a \leq 1/2$, $0 \leq b < c \leq 1$ and integer $n \geq 1$ set

$$\omega_n(a,b,c) = \sup\{|\alpha_n(s+h) - \alpha_n(s)| : 0 \leq s+h \leq 1, 0 \leq |h| \leq a, b \leq s \leq c\}.$$

LEMMA ii. For universal positive constants A and B for all $0 < a \leq 1/2$, $0 \leq b < c \leq 1$, $n \geq 1$ and $\lambda > 0$

(f) $$P(\omega_n(a,b,c) > \lambda\sqrt{a}) \leq \{((c-b)a^{-1}) \vee 1\} A \, \exp(-B\lambda^2 \psi(\lambda/\sqrt{na})),$$

where for $x \geq 0$

$$\psi(x) = 2x^{-2}\{(x+1)\log(x+1) - x\}.$$

PROOF. The proof is essentially contained in that of Inequality 1 in Mason, Shorack and Wellner (1983). Also see Inequality 1 of Einmahl and Mason (1988) (the ba^{-1} there should be replaced by $(ba^{-1}) \vee 1$). For future reference we record the fact that for $x \geq 0$

(g) $$\psi(x) \searrow \quad \text{as} \quad x \nearrow.$$

Choose $1/4 > \delta > \nu \geq 0$ and set $\rho = \delta - \nu$. Also for any $\gamma > 0$, integer $m \geq 1$ and $1 \leq i \leq n2^m$, let

$$\Delta_n(i) = \omega_n\left(\frac{\gamma i^{1-2\delta}}{n}, \frac{i}{n2^m}, \frac{i+1}{n2^m}\right).$$

LEMMA iii. For any $\epsilon > 0$, $m \geq 1$ and $1/4 > \delta > \nu \geq 0$ there exists a $\gamma > 0$ such that for all n large enough

(h) $$P\left(\max_{1 \leq i < n2^m} n^\nu \Delta_n(i)/(i/n)^{1/2-\nu} > \gamma\right) < \epsilon.$$

PROOF. For all large enough n inequality (f) is applicable to give uniformly in $1 \leq i < n2^m$, after a little algebra,

$$P(\Delta_n(i) > \gamma n^{-1/2} i^{1/2-\nu}) \leq \left\{\left(\frac{2^m}{\gamma}\right) \vee 1\right\} A \exp\left(-Bi^{2\rho}\gamma\psi\left(\frac{i^{2\delta-\nu-1/2}}{n}\right)\right),$$

which by $2\delta - \nu - 1/2 < 0$ and (g) is

$$\leq \left\{\left(\frac{2^m}{\gamma}\right) \vee 1\right\} A \exp(-Bi^{2\rho}\gamma\psi(1)) := P_i(\gamma).$$

Since $\rho > 0$, for every $\epsilon > 0$ we can choose a $\gamma > 0$ such that

$$\sum_{i=1}^{\infty} P_i(\gamma) < \epsilon. \quad \blacksquare$$

Returning to the proof of (c), by Lemma i for any $m \geq 1$ and $\epsilon > 0$ there exists a $\gamma > 0$ such that for all $n \geq 1$

$$P\left(\sup_{1/(n2^m) \leq s \leq 1} n^{2\delta} |U_n(s) - s|/s^{1-2\delta} \leq 2^{m(1-2\delta)-1}\gamma\right) > 1 - \epsilon.$$

Hence with probability greater than $1 - \epsilon$

$$M_{n,1}(2^{-m}, \nu) \leq$$

$$\max_{1 \leq i < n2^m} \sup\left\{ n^\nu \frac{|\alpha_n(s+h) - \alpha_n(s)|}{s^{1/2-\nu}} : 0 \leq s + h \leq 1, 0 \leq |h| \leq \frac{\gamma i^{1-2\delta}}{n}, \frac{i}{n2^m} \leq s \leq \frac{i+1}{n2^m} \right\},$$

which is obviously

$$\leq 2^{m(1/2-\nu)} \max_{1 \leq i < n2^m} n^\nu \Delta_n(i)/(i/n)^{1/2-\nu}.$$

A straightforward argument based on Lemma iii now verifies that $M_{n,1}(2^{-m}, \nu) = O_p(1)$. $\blacksquare$

References

[1] BREIMAN, L. (1967). On the tail behavior of sums of independent random variables. *Z. Wahrsch. verw. Gebiete* **9**, 20–25.

[2] CHIBISOV, D. (1965). Some theorems on the limiting behavior of empirical distribution functions. *Selected Translations Math. Statist. Probability* **6**, 147–156.

[3] CSÖRGŐ, M. *Quantile Processes with Statistical Applications*. Philadelphia: Regional Conference Series on Appl. Math. SIAM, 1983.

[4] CSÖRGŐ, M., CSÖRGŐ, S., HORVÁTH, L. and MASON, D. (1986a). Weighted empirical and quantile processes. *Ann. Probability* **14**, 31–85.

[5] CSÖRGŐ, M., CSÖRGŐ, S., HORVÁTH, L. and MASON, D. (1986b). Normal and stable convergence of integral functions of the empirical distribution function. *Ann. Probability* **14**, 86–118.

[6] CSÖRGŐ, S. and MASON, D. (1989). Bootstrapping empirical functions. *Ann. Statist.* **17**, 1447–1471.

[7] CSÖRGŐ, M. and MASON, D. (1985). On the asymptotic distribution of weighted uniform empirical and quantile processes in the middle and on the tails. *Stochastic Process. Appl.* **21**, 119–132.

[8] CSÖRGŐ, M. and RÉVÉSZ, P. (1978). Strong approximations of the quantile process. *Ann. Statist.* **6**, 882–894.

[9] CSÖRGŐ, S., DEHEUVELS, P. and MASON, D. (1985). Kernel estimates of the tail index of a distribution. *Ann. Statist.* **13**, 1050–1077.

[10] CSÖRGŐ, S., HAEUSLER, E. and MASON, D. (1988). The asymptotic distribution of trimmed sums. **16**, 672–699.

[11] CSÖRGŐ, S., HORVÁTH, L. and MASON, D. (1986). What portion of the sample makes a partial sum asymptotically stable or normal? *Probab. Theory Related Fields.* **72**, 1–16.

[12] CSÖRGŐ, S. and MASON, D. (1985). Central limit theorems for sums of extreme values. *Math. Proc. Cambridge Phil. Soc.* **98**, 547–558.

[13] CSÖRGŐ, S. and MASON, D. (1986). The asymptotic distribution of sums of extreme values from a regularly varying distribution. *Ann. Probability* **14**, 974–983.

[14] EICKER, F. (1979). The asymptotic distribution of the suprema of the standardized empirical process. *Ann. Statist.* **7**, 116–138.

[15] JAESCHKE, D. (1979). The asymptotic distribution of the supremum of the standardized empirical distribution on subintervals. *Ann. Statist.* **7**, 108–115.

[16] KOMLÓS, J., MAJOR, P. and TUSNÁDY, G. (1975). An approximation of partial sums of rv's and the sample df, I. *Z. Wahrsch, verw. Gebiete* **32**, 111–131.

[17] KOMLÓS, J., MAJOR, P. and TUSNÁDY, G. (1976). An approximation of partial sums of rv's and the sample df, II. *Z. Wahrsch, verw. Gebiete* **34**, 111–131.

[18] MASON, D. (1985). The asymptotic distribution of generalized Rényi statistics. *Acta. Sci. Math. (Szeged)* **48**, 315–323.

[19] MASON, D. and VAN ZWET, W. R. (1987). A refinement of the KMT inequality for the uniform empirical process. *Ann. Probability* **15**, 871–884.

[20] O'REILLY, N. (1974). On the weak convergence of empirical processes in supnorm metrics. *Ann. Probability* **2**, 642–651.

[21] SKOROKHOD, A. V. *Studies in the Theory of Random Processes*. Reading, Mass: Addison–Wesley, 1965.

Additional References For Addendum

[22] CSÖRGŐ, M. and HORVÁTH, L. (1986). Approximations of weighted empirical and quantile processes. *Statistics & Probability Letters* **4**, 275–280.

[23] CSÖRGŐ, M. and RÉVÉSZ, P. (1979). How big are the increments of a Wiener process? *Ann. Probability* **7**, 731–737.

[24] EINMAHL, J. H. J. and MASON, D. (1988). Strong limit theorems for weighted quantile processes. *Ann. Probability* **16**, 1623–1643.

[25] MARCUS, M.B. and ZINN, J. (1984). The bounded law of the iterated logarithm for the weighted empirical distribution process in the non–i.i.d. case. *Probability* **14**, 335–360.

[25] MASON, D. (1983). The asymptotic distribution of weighted empirical distribution functions. *Stochastic Process. Appl.* **15**, 99–109.

[25] MASON, D., SHORACK, G. and WELLNER, J. A. (1983). Strong limit theorems for oscillation moduli of the uniform empirical process. *Z. Wahrsch. verw. Gebiete* **65**, 83–97.

[25] WELLNER, J. A. (1978). Limit theorems for the ratio of the empirical distribution function to the true distribution function. *Z. Wahrsch. verw. Gebiete* **45**, 73–88.

David M. Mason
Department of Mathematical Sciences
University of Delaware
Newark, Delaware 19716

LIMIT THEOREMS FOR THE PETERSBURG GAME

Sándor Csörgő [1] and Rossitza Dodunekova [2]

We determine all possible subsequences $\{n_k\}_{k=1}^{\infty}$ of the positive integers for which the suitably centered and normalized total gain S_{n_k} in n_k Petersburg games has an asymptotic distribution as $k \to \infty$, and identify the corresponding set of limiting distributions. We also solve all the companion problems for lightly, moderately, and heavily trimmed versions of the sum S_{n_k} and for the respective sums of extreme values in S_{n_k}.

1. Introduction. A virtually complete and unified theory of the asymptotic distribution of sums of order statistics has been recently worked out in the three papers [3, 4, 5] with [2] augmenting and rounding off the study in [3]. (See, however [7], and also the survey [6].) Although a number of *ad hoc* examples are scattered in these papers to illustrate various interesting phenomena, we felt that there was a need for a didactic type of a paper that would illustrate all the limit theorems in the above articles on a single, sufficiently complex but still manageable example which, most importantly, is interesting in its own right. Our attention was drawn to the famous Petersburg game by an interesting paper of Martin-Löf [10]. Since the "time-honored" Petersburg paradox, as Feller ([8], X. 4) describes it, has been

---

[1] Partially supported by the Hungarian National Foundation for Scientific Research, Grants 1808/86 and 457/88.
[2] Partially supported by the Ministry of Culture, Science and Education of Bulgaria, Contract No. 16848-F3.

around for 277 years (since the publication of the second edition of Montmort's book in 1713), it is clearly of sufficient interest in itself, and it turns out that it also satisfies the other criteria.

The 'paradox' has been originally posed by Nicolaus Bernoulli to Montmort in 1713, and the English translation of the nice account of it by Daniel Bernoulli in 1738 is given by Martin-Löf [10]. For a recent historical account and references see Shafer [11]. Following Feller and Martin-Löf, and hence doubling the gain of the Bernoullis, the Petersburg game consists in tossing a fair coin until it falls heads; if this occurs at the k-th throw the player receives 2^k ducats. Hence if X is the gain at a single trial we have $P\{X = 2^k\} = 2^{-k}$, $k = 1, 2, \ldots$, and

$$(1.1) \qquad F(x) = P\{X \leq x\} = \begin{cases} 0 & , x < 2, \\ 1 - 2^{-\lfloor \text{Log} x \rfloor} & , x \geq 2, \end{cases}$$

where, and *throughout in this paper* Log *stands for the logarithm to the base 2*, $\lfloor y \rfloor = \max\{j : j \text{ integer, } j \leq y\}$ is the usual integer part function and we shall neeed $\lceil y \rceil = \min\{j : j \text{ integer, } j \geq y\}$, $y \in \mathbf{R}$.

Let $X_1, X_2, \ldots$ be independent copies of X, the gains at the first, second,... trials, so that $S_n = X_1 + \ldots + X_n$ is the total gain in n consecutive Petersburg games. Feller proved that

$$(1.2) \qquad S_n \big/ (n \text{Log} n) \to_P 1 \quad \text{as} \quad n \to \infty,$$

where $\to_P$ denotes convergence in probability and, denoting by $\to_\mathcal{D}$ convergence in distribution, Martin-Löf proved that

$$(1.3) \qquad \frac{S_{2^k}}{2^k} - k \to_\mathcal{D} V \quad \text{as} \quad k \to \infty,$$

where, if $\Delta_r, r = 0, \pm 1, \pm 2, \ldots$, are independent Poisson random variables with $E\Delta_r = 2^{-r}$,

$$V = \sum_{r=0}^{-\infty} (\Delta_r - 2^{-r})2^r + \sum_{r=1}^{\infty} \Delta_r 2^r,$$

and hence, with i denoting the imaginary unit,

$$(1.4) \qquad Ee^{iVt} = \exp \left\{ \sum_{r=0}^{-\infty} 2^{-r}(e^{i2^r t} - 1 - i2^r t) + \sum_{r=1}^{\infty} 2^{-r}(e^{i2^r t} - 1) \right\}$$

for any $t \in \mathbf{R}$. The limit theorem (1.3) then forms a basis for Martin-Löf to develop a premium formula "which clarifies the 'Petersburg paradox'", at least when the game is played in blocks of the size 2^k, $k = 1, 2, \ldots$, and is very favorable for the casino.

At this point two questions arise. The result in (1.3) means that the Petersburg distribution is in the domain of partial attraction of the infinitely divisible distribution of V. Since V is obviously not a stable random variable, it follows from (1.3) and Theorem 10 in [2] that F belongs to the domain of partial attraction of continuum many different types of infinitely divisible distributions and hence there are just as many principle clarifications of the paradox. How different are these from one another? The second question is more practical: What to do if n_k games are played, where $2^{k-1} < n_k < 2^k$? Some qualitative answers to these questions are indicated by the results for the whole sums S_{n_k} in Section 2. It turns out, in particular, that the two questions are more or less the same.

Since the Petersburg game is exciting exactly because of the occasional large single gains in the total gain S_{n_k}, it is interesting to see the influence of these large gains on the limiting distribution. Let $X_{1,n} \leq \cdots \leq X_{n,n}$ be the order statistics pertaining to the sample $X_1, \ldots, X_n$. Together with the full sums S_{n_k}, in Section

2 we also investigate the lightly trimmed sums $S_{n_k}(m) = \sum_{j=1}^{n_k-m} X_{j,n_k}$, where $m \geq 1$ is a fixed integer, while in Section 3 we deal with moderately trimmed sums $S_{n_k}(m_k)$, where $m_k \to \infty$ but $m_k/n_k \to 0$ as $k \to \infty$, and heavily trimmed sums $S_n(n - \lfloor \beta n \rfloor)$, where $1/2 < \beta < 1$. But then it is also interesting to see what happens to the extreme sums $S_{n_k} - S_{n_k}(m_k)$. This is done in Section 4.

The unified method in [3, 4, 5] and [2] is generally based on the quantile function $Q(s) = \inf\{x : F(x) \geq s\}$, $0 < s < 1$, which from (1.1) presently is

(1.5) $Q(s) = 2^r$ if $1 - \dfrac{1}{2^{r-1}} < s \leq 1 - \dfrac{1}{2^r}$, $r = 1, 2, \ldots$,

and we may put $Q(0) = Q(0+) = 2$, or

(1.6) $Q(1-u) = 2^{\lceil \text{Log}(1/u) \rceil}$, $0 < u \leq 1$.

In each of the following three sections we first state all the results of that section, with some discussion when appropriate, and the proofs occupy the remaining part of the section.

2. Full and lightly trimmed sums. The first result restricts the choice of meaningful normalizing and centering sequences. We generally deal with lightly trimmed sums $S_{n_k}(m) = \sum_{j=1}^{n_k-m} X_{j,n_k}$, where $m \geq 0$ is a fixed integer. Since $S_n = \sum_{j=1}^n X_{j,n}$ for each n, we have $S_n(0) = S_n$, so the special case $m = 0$ always corresponds to total gains. Subsequences $\{n_k\}_{k=1}^\infty$ of the positive integers are always assumed to go to infinity.

THEOREM 2.1. *If for some integer $m_0 \geq 0$, a subsequence $\{n_k\}_{k=1}^\infty$ of the positive integers, and some constants $A_{n_k} > 0$ and $C_{n_k} \in \mathbf{R}$ the sequence $(S_{n_k}(m_0) -$*

$C_{n_k})/A_{n_k}$ converges in distribution to a non-degenerate random variable $W(m_0)$, then for each subsequence $\{n'_k\}_{k=1}^{\infty} \subset \{n_k\}_{k=1}^{\infty}$ and each integer $m \geq 0$ there exist a further subsequence $\{n''_k\}_{k=1}^{\infty} \subset \{n'_k\}_{k=1}^{\infty}$ and a constant $a = a_{\{n''\}}$, $0 < a < \infty$, such that $n''_k/A_{n''_k} \to a$ and

$$W_{n''_k}(m) := \frac{S_{n''_k}(m)}{n''_k} - \left(\lceil \text{Log} \frac{n''_k}{m+1} \rceil + 1 - \frac{m+1}{n''_k} 2^{\lceil \text{Log} \frac{n''_k}{m+1} \rceil} \right) \to_{\mathcal{D}} \frac{1}{a} W(m) + c(m)$$

as $k \to \infty$, where $c(m)$ is some constant, $W(m_0)$ is as before, and none of the other random variables $W(m)$ is degenerate.

The theorem shows that using the normalizing sequence $\{n_k\}$ and the indicated centering sequence we can achieve all possible limiting types and hence we can restrict attention to these sequences without loss of generality when answering, in particular, the questions posed in the introduction concerning total gains.

Define $W_{n_k} = W_{n_k}(0)$ as above, replacing n''_k by n_k everywhere. Let us agree that in the dyadic expansion of a number γ, $1/2 < \gamma \leq 1$, given by

$$(2.1) \qquad \gamma = \sum_{j=1}^{\infty} a_j/2^j = \lim_{k \to \infty} s_k, \quad s_k = \sum_{j=1}^{k} a_j/2^j, \quad k \geq 1,$$

where $a_1 = 1$ and $a_j = 0$ or 1 for $j \geq 2$, we exclude all the sequences $(1, a_2, a_3, \ldots)$ which contain only a finite number of non-zero elements. Let $E_1, E_2, \ldots$ be independent exponential random variables with the common expectation 1 and with the Gamma(j) random variables $Y_j = E_1 + \ldots + E_j$ as jump-points, consider the standard left-continuous Poisson process

$$(2.2) \qquad N(s) = \sum_{j=1}^{\infty} I(Y_j < s), \quad s \geq 0,$$

where $I(\cdot)$ is the indicator function. Also, consider the independent increments

(2.3) $\Delta_r(\gamma) = N(\gamma 2^{-r+1}) - N(\gamma 2^{-r}), \; r = 0, \pm 1, \pm 2, \ldots, \quad$ where $\quad \frac{1}{2} \leq \gamma \leq 1$

is a fixed number.

THEOREM 2.2. *For a given subsequence $\{n_k\}_{k=1}^{\infty}$ of the positive integers the sequence W_{n_k} converges in distribution as $k \to \infty$ if and only if*

(2.4) $\lim\limits_{k \to \infty} \dfrac{n_k}{2^{\lceil \mathrm{Log} n_k \rceil}} = \lim\limits_{k \to \infty} 2^{\mathrm{Log} n_k - \lceil \mathrm{Log} n_k \rceil} = \gamma \quad$ *for some* $\quad \frac{1}{2} \leq \gamma \leq 1$.

In this case, for any integer $m \geq 0$,

(2.5) $\dfrac{S_{n_k}(m)}{n_k} - \left(\lceil \mathrm{Log} \dfrac{n_k}{m+1} \rceil - \dfrac{m+1}{n_k} 2^{\lceil \mathrm{Log}(n_k/(m+1)) \rceil} \right) \to_{\mathcal{D}} V_\gamma(m),$

where, with

(2.6) $\varphi_\gamma^*(s) = -2^{-\lfloor \mathrm{Log}(s/\gamma) \rfloor}, \quad s > 0,$

$$V_\gamma(m) = 1 + \frac{1}{\gamma} \left\{ \int_\gamma^\infty (N(s) - s) d\varphi_\gamma^*(s) + \int_{Y_{m+1}}^\gamma N(s) d\varphi_\gamma^*(s) \right.$$
$$\left. + m\varphi_\gamma^*(Y_{m+1}) - \int_1^{m+1} \varphi_\gamma^*(s) ds + 1 \right\}.$$

In particular, if $m = 0$ then under (2.4),

(2.7) $\dfrac{S_{n_k}}{n_k} - \mathrm{Log} n_k \to_{\mathcal{D}} W_\gamma := V_\gamma - \mathrm{Log} \gamma,$

where $V_\gamma = V_\gamma(0) - 1/\gamma$ and we have

(2.8) $\begin{aligned} V_\gamma &= \frac{1}{\gamma} U_\gamma = \frac{1}{\gamma} \left\{ \int_\gamma^\infty (N(s) - s) d\varphi_\gamma^*(s) + \int_0^\gamma N(s) d\varphi_\gamma^*(s) + \gamma \right\} \\ &= \frac{1}{\gamma} \left\{ \sum_{k=1}^\infty (N(\gamma 2^k) - \gamma 2^k) 2^{-k} + \sum_{k=0}^\infty N(\gamma 2^{-k}) 2^k + \gamma \right\} \\ &= \frac{1}{\gamma} \left\{ \sum_{r=0}^{-\infty} (\Delta_r(\gamma) - \gamma 2^{-r}) 2^r + \sum_{r=1}^\infty \Delta_r(\gamma) 2^r \right\}. \end{aligned}$

Furthermore, if $\gamma = 1/2$, then the sequence $n_k = n_k(1/2) = 2^{k-1} + 1$, $k = 1, 2, \ldots$, satisfies (2.4), while if $1/2 < \gamma \leq 1$, then the sequence

$$(2.9) \qquad n_k = n_k(\gamma) = 2^{k-1} + a_2 2^{k-2} + \ldots + a_{k-1} 2 + a_k + 1, \quad k = 1, 2, \ldots,$$

satisfies (2.4), where $a_2, a_3, \ldots$ are the binary digits in the dyadic expansion (2.1).

Martin-Löf's result in (1.3) is obtained by choosing $n_k = n_k(1) = 2^k$ in (2.9) and using the third representation of $V = V_1 = W_1 = U_1$ in (2.8).

It is interesting to note that (2.7) and the general approach in [3], within the framework of which we work in the proofs, at once imply that if $\{n_k^{(l)}\}_{k=1}^{\infty}$, $l = 1, \ldots, r$, are r different sequences satisfying (2.4) with $\gamma = \gamma_l$, $l = 1, \ldots, r$, respectively, then

$$(2.10) \qquad \left(\frac{S_{n_k^{(1)}}}{n_k^{(1)}} - \operatorname{Log} n_k^{(1)}, \ldots, \frac{S_{n_k^{(r)}}}{n_k^{(r)}} - \operatorname{Log} n_k^{(r)} \right) \to_{\mathcal{D}} (W_{\gamma_1}, \ldots, W_{\gamma_r}),$$

where $W_\gamma = (U_\gamma - \gamma \operatorname{Log} \gamma)/\gamma$ for any of $\gamma = \gamma_l$, $l = 1, \ldots, r$, with U_γ given by any of the three representations in (2.8) in terms of the *same* Poisson process. For any $t \in \mathbf{R}$, we have $E \exp(iU_\gamma t) = (E \exp(iVt))^\gamma$ with the expression in (1.4) substituted. We see that U_γ, $1/2 \leq \gamma \leq 1$, determined by (2.10) and explicitly given in (2.8) is a segment of an independent-increment or Lévy process. In fact, U_γ, $1/2 \leq \gamma \leq 1$, is a segment of a special semi-stable process as described by Lévy [9; Section 58] and we will demonstrate below that

$$(2.11) \qquad\qquad\qquad W_{1/2} = W_1.$$

The above results show that the continuum many infinitely divisible types of distributions that partially attract the Petersburg distribution is given by the set

$\{G_\gamma : 1/2 \leq \gamma \leq 1\}$, where $G_\gamma(x) = P\{W_\gamma \leq x\}$, $x \in \mathbf{R}$, and the essential role of the parameter γ is in fact to connect the gaps in the special subsequence $\{2^k\}_{k=1}^\infty$ of Martin-Löf. We plan to return to this problem by providing a general premium formula that makes successive Petersburg games asymptotically fair in a special subsequent note.

In order to get some additional feeling concerning the limiting distribution function G_γ of W_γ in (2.7), for each $n = 1, 2, \ldots$, let $L_n(\gamma)$ be the Lévy distance between G_γ and the approximating distribution function

$$G_\gamma^{(n)}(x) = P\left\{\sum_{j=1}^n \frac{1}{\gamma}2^{-\lfloor \text{Log}(Y_j/\gamma)\rfloor} - \lfloor \text{Log}\frac{n}{\gamma}\rfloor + 1 - \frac{n}{\gamma 2^{\lfloor \text{Log}(n/\gamma)\rfloor}} - \text{Log}\gamma \leq x\right\},$$

$x \in \mathbf{R}$. Note that, in addition to (2.10), the general probabilistic approach in [3] and the details of the proofs below imply that if $\{n_k\}$ satisfies (2.4) then for any integers $l, m \geq 0$,

(2.12)
$$\frac{1}{n_k}\left(S_{n_k}(l) - n_k\lceil \text{Log}\frac{n_k}{l+1}\rceil, X_{n_k-m+1,n_k}, \ldots, X_{n_k,n_k}\right)$$
$$\to_D \left(V_\gamma(l), \frac{1}{\gamma}2^{-\lfloor \text{Log}(Y_m/\gamma)\rfloor}, \ldots, \frac{1}{\gamma}2^{-\lfloor \text{Log}(Y_1/\gamma)\rfloor}\right)$$

as $k \to \infty$, and so the sum

$$\sum_{j=1}^m \frac{1}{\gamma}2^{-\lfloor \text{Log}(Y_j/\gamma)\rfloor}$$

may be looked upon as the asymptotic contribution of the largest m gains to the limiting distribution of the total gain S_{n_k} in (2.5). The following result is a corollary to the Theorem in [1] after some straightforward but lengthier calculations not reproduced here.

THEOREM 2.3. *For any $\gamma > 0$ and any positive number $\rho < 1$, $L_n(\gamma) = o(n^{-\rho/2})$ as $n \to \infty$.*

Now we turn to the proofs of the first two theorems. Throughout the paper, all the asymptotic relations are understood to take place as $k \to \infty$ if not specified otherwise.

Proof of Theorem 2.1. According to the results in [3,2], the limiting behavior of centered and normalized finitely trimmed sums is by and large determined by the asymptotic behavior of the function

$$\sigma^2(s,t) = \int_s^{1-t} \int_s^{1-t} (u \wedge v - uv) dQ(u) dQ(v)$$

(2.13)
$$= sQ^2(s) + tQ^2(1-t) + \int_s^{1-t} Q^2(u) du$$

$$- \left\{ sQ(s) + tQ(1-t) + \int_s^{1-t} Q(u) du \right\}^2$$

as $s, t \downarrow 0$, where $u \wedge v = \min(u,v)$ and this well-known equality holds for any $0 < s < 1 - t < 1$. Presently we need the function $\sigma^2(s) = \sigma^2(s,s)$, $0 < s < 1/2$, for which, substituting (1.5) and (1.6) into (2.13), we obtain by simple calculation that

$$\sigma^2(s) = 2^r 3 - 2 - (r+1)^2, \quad \text{if} \quad \frac{1}{2^r} \le s < \frac{1}{2^{r-1}}, \quad r \ge 2,$$

or, what is the same,

$$\sigma^2(s) = 2^{\lceil \text{Log}(1/s) \rceil} 3 - 2 - (\lceil \text{Log}(1/s) \rceil + 1)^2, \quad 0 < s < 1/2.$$

Hence the asymptotic equality

(2.14)
$$\sigma(s) \sim \sqrt{3}\, 2^{\lceil \text{Log}(1/s) \rceil / 2} \quad \text{as} \quad s \downarrow 0$$

holds, and from the same calculation we also have

(2.15)
$$\int_0^{1-s} Q(u) du = \lceil \text{Log} \frac{1}{s} \rceil + 1 - 2^{\lceil \text{Log}(1/s) \rceil} s,$$

$$\int_0^{1-s} Q^2(u) du = 2^{\lceil \text{Log}(1/s) \rceil} 3 - 2 - 2^{2\lceil \text{Log}(1/s) \rceil} s, \quad 0 < s \le 1.$$

Thus for the sequence

(2.16) $a(n_k) = \sqrt{n_k}\sigma(1/n_k), \quad k = 1, 2, \ldots,$

we have by (2.14),

(2.17) $a(n_k) \sim \sqrt{3}\sqrt{n_k}2^{\lceil \text{Log}n_k \rceil/2} = \left(3\dfrac{2^{\lceil \text{Log}n_k \rceil}}{n_k}\right)^{1/2} n_k.$

Introducing

(2.18) $\alpha_k = \alpha_{n_k} = \lceil \text{Log}n_k \rceil - \text{Log}n_k \quad \text{and} \quad \beta_s = \text{Log}s - \lfloor \text{Log}s \rfloor,$

where, obviously, $0 \le \alpha_k, \beta_s < 1$ for any $k \ge 1$ and $s > 0$, we have

$$\lceil \text{Log}(n_k/s) \rceil = \begin{cases} \lceil \text{Log}n_k \rceil - \lfloor \text{Log}s \rfloor & , \quad \text{if} \quad 0 \le \alpha_k + \beta_s < 1, \\ \lceil \text{Log}n_k \rceil - \lfloor \text{Log}s \rfloor - 1 & , \quad \text{if} \quad 1 \le \alpha_k + \beta_s < 2. \end{cases}$$

Whence and from (1.6) and (2.17),

(2.19) $\varphi_{n_k}(s) := -\dfrac{Q\left(1 - \frac{s}{n_k}\right)}{a(n_k)} \sim \begin{cases} -\dfrac{2^{-\lfloor \text{Log}s \rfloor}}{\sqrt{3}}\sqrt{\dfrac{2^{\lceil \text{Log}n_k \rceil}}{n_k}} & , 0 \le \alpha_k + \beta_s < 1, \\ -\dfrac{2^{-\lfloor \text{Log}s \rfloor}}{2\sqrt{3}}\sqrt{\dfrac{2^{\lceil \text{Log}n_k \rceil}}{n_k}} & , 1 \le \alpha_k + \beta_s < 2, \end{cases}$

for any fixed $s > 0$, where this function plays the role of $\psi_2(n_k, s)$ in [3, 2].

   The assumption of the theorem and Theorem 5 in [3] imply that for each subsequence $\{n'_k\} \subset \{n_k\}$ there is a further subsequence $\{n'''_k\{\subset \{n'_k\}$ such that both sequences $Q(s/n'''_k)/A^{n'''_k}$ and $a(n'''_k)\varphi_{n'''_k}(s)/A_{n'''_k}$ of functions converge weakly (denoted in what follows by $\Rightarrow$) on $(0, \infty)$ to some finite functions φ_1 and φ_2, respectively, and $a(n'''_k)/A_{n'''_k} \to \delta$, where $0 \le \delta < \infty$. Since, necessarily, $A_{n_k} \to \infty$, it follows from (1.5) that $\varphi_1 \equiv 0$. But, since

(2.20) $1 \le 2^{\lceil \text{Log}s \rceil}/s < 2, \quad s > 0,$

it is clear that if δ were zero then we would have $\varphi_2 \equiv 0$, and then by the same Theorem 5 in [3] the limiting random variable $W(m_0)$ would be degenerate. Hence $\delta > 0$ and, necessarily, $\varphi_2 \not\equiv 0$. Therefore, $\varphi_{n_k'''}(\cdot) \Rightarrow \delta^{-1}\varphi_2(\cdot)$, and thus by (2.20) there is a further subsequence $\{n_k^{IV}\} \subset \{n_k'''\}$ such that $a(n_k^{IV})\varphi_{n_k^{IV}}(\cdot)/n_k^{IV} \Rightarrow a\varphi_2(\cdot)$ for some $a > 0$. Taking into account (2.15) and the fact that $\int_0^{1/n_k} Q(u)du = 2/n_k \to 0$, it follows now from Theorem 4 in [3] that for each $m \geq 0$ we can choose a further subsequence $\{n_k''\} \subset \{n_k^{IV}\}$ and a constant $c(m)$, both depending on m, such that $W_{n_k''}(m)$ converges as stated. ∎

We are a little bit sloppy in the definition of $\varphi_{n_k}(\cdot)$ in (2.19). Let $0 < \rho_{n_k} < 1$ be such that $\rho_{n_k} \downarrow 0$ and $n_k\rho_{n_k} \to 0$, and define $\varphi_{n_k}(s)$ by (2.19) if $0 < s \leq n_k - n_k\rho_{n_k}$, while if $s > n_k - n_k\rho_{n_k}$ then set $\varphi_{n_k}(s) = -Q(\rho_{n_k})/a(n_k)$. Now $\varphi_{n_k}(\cdot)$ is a negative, non-decreasing function on the whole $(0, \infty)$ with $\varphi_{n_k}(0) = \varphi_{n_k}(0+) = -\infty$.

LEMMA 2.4. *For a given subsequence $\{n_k\}_{k=1}^{\infty}$ the functions $\varphi_{n_k}(\cdot)$ converge weakly to some finite function given on $(0, \infty)$ if and only if condition (2.4) holds. In this case,*

$$(2.21) \qquad \varphi_{n_k}(\cdot) \Rightarrow \varphi_\gamma(\cdot) := \frac{1}{\sqrt{3\gamma}}\varphi_\gamma^*(\cdot) = -\frac{1}{\sqrt{3\gamma}}2^{-\lfloor \mathrm{Log}(\cdot/\gamma) \rfloor}$$

with the right-continuous function $\varphi_\gamma^*(\cdot)$ taken from (2.6).

Proof. In the notation of (2.18), condition (2.4) is equivalent to

$$(2.22) \qquad \alpha_k = \alpha_{n_k} \to \alpha \quad \text{for some} \quad 0 \leq \alpha \leq 1,$$

in which case

$$(2.23) \qquad \gamma = 2^{-\alpha}.$$

First we prove the necessity of (2.22). Suppose that weak convergence takes place but (2.22) is not true. Then for two subsequences $\{n'_k\}$ and $\{n''_k\}$ we have $\underline{\alpha}_k = \alpha_{n'_k} \to \underline{\alpha} = \liminf \alpha_k$ and $\bar{\alpha}_k = \alpha_{n''_k} \to \bar{\alpha} = \limsup \alpha_k$, where $0 \le \underline{\alpha} < \bar{\alpha} \le 1$. Clearly, we can find a continuity point $s > 0$ of the limiting function such that for β_s defined in (2.18) we have $\underline{\alpha} < 1 - \beta_s < \bar{\alpha}$, and hence for all k large enough $\underline{\alpha}_k < 1 - \beta_s < \bar{\alpha}_k$ or, what is the same, $\underline{\alpha}_k + \beta_s < 1$ and $\bar{\alpha}_k + \beta_s > 1$. Thus from (2.19),

$$\varphi_{n'_k}(s) \to -\sqrt{\frac{2\underline{\alpha}}{3}} 2^{-\lfloor \text{Log} s \rfloor} \quad \text{and} \quad \varphi_{n''_k}(s) \to -\frac{1}{2}\sqrt{\frac{2\bar{\alpha}}{3}} 2^{-\lfloor \text{Log} s \rfloor}.$$

Equating the two limits, we obtain $\bar{\alpha} = 2 + \underline{\alpha}$, which is impossible.

Now we consider sufficiency. Using (2.4), (2.22), and (2.23), from (2.19) we obtain that if $0 < \alpha < 1$, or equivalently, $1/2 < \gamma < 1$, then

$$\varphi_{n_k}(s) \to \varphi_\gamma(s) = \begin{cases} -\dfrac{2^{-\lfloor \text{Log} s \rfloor}}{\sqrt{3\gamma}}, & \text{if } \beta_s < 1 + \text{Log}\gamma, \\ -\dfrac{2^{-\lfloor \text{Log} s \rfloor}}{2\sqrt{3\gamma}}, & \text{if } \beta_s > 1 + \text{Log}\gamma, \end{cases}$$

if $\gamma = 1$, then

$$\varphi_{n_k}(s) \to \varphi_1(s) = -\frac{2^{-\lfloor \text{Log} s \rfloor}}{\sqrt{3}}, \quad s > 0,$$

and if $\gamma = 1/2$, then

$$\varphi_{n_k}(s) \to \varphi_{1/2}(s) = \begin{cases} -\dfrac{2^{-\lfloor \text{Log} s \rfloor}}{\sqrt{3/2}}, & \text{if } \beta_s = 0, \\ -\dfrac{2^{-\lfloor \text{Log} s \rfloor}}{2\sqrt{3/2}}, & \text{if } \beta_s > 0. \end{cases}$$

When $\gamma = 1$ or $\gamma = 1/2$, it is obvious that $\varphi_\gamma(\cdot)$ can be written in the form claimed in (2.21). An elementary argument shows that this is also true when $1/2 < \gamma < 1$. (Of course, if $1/2 < \gamma < 1$ then there is no convergence in general at those $s > 0$ for which $\beta_s = 1 + \text{Log}\gamma$. These are the jump-points $s = \gamma 2^k$, $k = 0, \pm 1, \pm 2, \ldots$. See also (2.27) below.) ∎

Proof of Theorem 2.2. Before starting the actual proof, we note that for any sequence $\{n_k\}$ by (2.14) and (2.20) we have

$$\frac{\sigma^2(h/n_k)}{\sigma(1/n_k)} \sim \frac{1}{h} \frac{n_k}{2^{\lceil \text{Log} n_k \rceil}} \frac{2^{\lceil \text{Log}(n_k/h) \rceil}}{n_k/h} < \frac{2}{h}, \quad h > 0,$$

and hence

(2.24) $$\lim_{h \to \infty} \limsup_{k \to \infty} \sigma(h/n_k)/\sigma(1/n_k) = 0.$$

Also, whatever is the sequence $\{n_k\}$ like, it follows from (2.20) that any subsequence $\{n'_k\} \subset \{n_k\}$ contains a further subsequence $\{n''_k\} \subset \{n'_k\}$ such that (2.4) is satisfied along $\{n''_k\}$ and with some $\gamma = \gamma_{\{n''\}}$, $1/2 \leq \gamma \leq 1$. Then by (2.17), $a(n''_k)/n''_k \to \sqrt{3/\gamma}$ and hence by Lemma 2.4, for $\varphi^*_{n_k}(\cdot) = a(n_k)\varphi_{n_k}(\cdot)/n_k$ we have

$$\varphi^*_{n''_k}(\cdot) \Rightarrow \frac{1}{\gamma}\varphi^*_\gamma(\cdot).$$

Since $Q((s/n_k)+)/n_k \to 0$ for all $s > 0$, this last weak convergence, (2.24) and (2.15) imply by an application of Theorem 1* in [2] (which is an augmented form of Theorem 1 in [3]) that for any $m \geq 0$ in $W_{n''_k}(m)$ of Theorem 2.1 we have

$$W_{n''_k}(m) \to_D \frac{1}{\gamma}V_{0,m}(0, \varphi^*_\gamma, 0)$$

$$:= \frac{1}{\gamma}\left\{ \int_{Y_{m+1}}^\infty (N(s) - s)d\varphi^*_\gamma(s) - \int_1^{Y_{m+1}} sd\varphi^*_\gamma(s) \right.$$

$$\left. + m\varphi^*_\gamma(Y_{m+1}) - \int_1^{m+1} \varphi^*_\gamma(s)ds - \varphi^*_\gamma(1) \right\}.$$

Now if $W_{n_k} = W_{n_k}(0)$ converges in distribution, then its subsequential limits are all of the form $V_{0,0}(0, \varphi^*_\gamma, 0)/\gamma$ having an infinitely divisible distribution. Since this representation, being a special case of the general representation in Theorem 3 in [3], is unique, it follows that γ must be the same for all the above subsequences $\{n_k''\}$, and hence (2.4) is necessarily satisfied.

Suppose now (2.4). Then for any $m \geq 0$, the above convergence of $W_{n_k}(m)$ takes place along the original $\{n_k\}$. It follows then that the left side of (2.5) converges in distribution to

$$(2.25) \qquad \frac{1}{\gamma}V_{0,m}(0, \varphi_\gamma^*, 0) + 1,$$

and since by easy manipulation one can see that

$$\int_{Y_{m+1}}^{\infty} (N(s) - s)d\varphi_\gamma^*(s) - \int_1^{Y_{m+1}} sd\varphi_\gamma^*(s)$$

$$= \int_\gamma^\infty (N(s) - s)d\varphi_\gamma^*(s) + \int_{Y_{m+1}}^\gamma N(s)d\varphi_\gamma^*(s) + \int_\gamma^1 sd\varphi_\gamma^*(s)$$

(in fact, regardless of a meaningful function φ replacing the present φ_γ^*) and, separating the cases $\gamma = 1/2$ and $1/2 < \gamma \leq 1$, that

$$\int_\gamma^1 sd\varphi_\gamma^*(s) - \varphi_\gamma^*(1) = 1, \quad 1/2 \leq \gamma \leq 1$$

(see (2.27) below), it is clear that the limit in (2.25) is the same as $V_\gamma(m)$ in (2.5). Thus (2.5) holds, and since under (2.4),

$$(2.26) \qquad \lceil \text{Log}n_k \rceil - \text{Log}n_k \to -\text{Log}\gamma,$$

the convergence in (2.7) and the first representation in (2.8) also follow. Furthermore, since the function φ_γ^* in (2.6) can be written as

$$(2.27) \qquad \varphi_\gamma^*(s) = -2^{-k}, \quad \gamma 2^k \leq s < \gamma 2^{k+1}, \quad k = 0, \pm 1, \pm 2, \ldots,$$

the second representation in (2.8) follows from the first by carrying out the integration.

To prove the last equation in (2.8), note that by simple computation,

$$\sum_{k=1}^{j}(N(\gamma 2^k) - \gamma 2^k)2^{-k} + \sum_{k=0}^{j}N(\gamma 2^{-k})2^k + \gamma$$

$$= \sum_{k=0}^{j-1}\left([N(\gamma 2^{k+1}) - N(\gamma 2^k)] - \gamma 2^k\right)2^{-k} + [N(\gamma 2^j) - \gamma 2^j]2^{-j}$$

$$+ \sum_{k=1}^{j}[N(\gamma 2^{-k+1}) - N(\gamma 2^{-k})]2^k + N(\gamma 2^{-j})2^{j+1}$$

for each $j \geq 1$. Since the second term goes to zero almost surely by the law of large numbers as $j \to \infty$ and the fourth term is almost surely zero for all j large enough, the last line of (2.8) follows upon letting $j \to \infty$.

Finally, it is trivial that $n_k(1/2) = 2^{k-1} + 1$ satisfies (2.4) with $\gamma = 1/2$. Also, if $1/2 < \gamma \leq 1$, then it is plain that $\lceil Log n_k(\gamma)\rceil = k$ for the sequence $n_k(\gamma)$ in (2.9) and hence, in the notation of (2.1),

$$\frac{n_k(\gamma)}{2^{\lceil Log n_k(\gamma)\rceil}} = \frac{n_k(\gamma)}{2^k} = s_k + 2^{-k} \to \gamma.$$

The theorem is completely proved. ■

Proof of (2.11). From the third representation in (2.8) we obtain

$$V_{1/2} = \sum_{k=0}^{\infty}\left([N(2^k) - N(2^{k-1})] - 2^{k-1}\right)\frac{1}{2^{k-1}} + \sum_{k=1}^{\infty}\left[N\left(\frac{1}{2^k}\right) - N\left(\frac{1}{2^{k+1}}\right)\right]2^{k+1}$$

$$= \sum_{r=0}^{\infty}\left([N(2^{r+1}) - N(2^r)] - 2^r\right)\frac{1}{2^r} + \sum_{r=1}^{\infty}\left[N\left(\frac{1}{2^{r-1}}\right) - N\left(\frac{1}{2^r}\right)\right]2^r$$

$$+ \left([N(1) - N(\tfrac{1}{2})] - \frac{1}{2}\right)2 - [N(1) - N(\tfrac{1}{2})]2$$

$$= V_1 - 1,$$

which is the same as (2.11). ■

3. Moderately and heavily trimmed sums. Let $\{n_k\}_{k=1}^{\infty}$ and $\{m_{n_k}\}_{k=1}^{\infty}$ be two subsequences of the positive integers such that

(3.1) $m_k \to \infty$ and $m_{n_k}/n_k \to 0$,

and consider the moderately trimmed sums

$$S_{n_k}(m_{n_k}) = \sum_{j=1}^{n_k - m_{n_k}} X_{j,n_k}.$$

The first result is an analogue of Theorem 2.1 in the present setting.

THEOREM 3.1. *If (3.1) is satisfied and for some constants $A_{n_k} > 0$ and $C_{n_k} \in \mathbf{R}$ the sequence $(S_{n_k}(m_{n_k}) - C_{n_k})/A_{n_k}$ converges in distribution to a non-degenerate random variable W, then for each subsequence $\{n'_k\}_{k=1}^{\infty} \subset \{n_k\}_{k=1}^{\infty}$ there exist a further subsequence $\{n''_k\}_{k=1}^{\infty} \subset \{n'_k\}_{k=1}^{\infty}$ and a constant $a = a_{\{n''\}}$, $0 < a < \infty$, such that $(n''_k / \sqrt{m_{n''_k}})/A_{n''_k} \to a$ and*

$$W_{n''_k} := \sqrt{m_{n''_k}} \left\{ \frac{S_{n''_k}(m_{n''_k})}{n''_k} - \left(\left\lceil \operatorname{Log} \frac{n''_k}{m_{n''_k}} \right\rceil + 1 - \frac{m_{n''_k}}{n''_k} 2^{\lceil \operatorname{Log}(n''_k/m_{n''_k}) \rceil} \right) \right\}$$

$$\to_{\mathcal{D}} \frac{1}{a} W + c,$$

as $k \to \infty$, where c is some constant.

Again we see that it is enough to deal with centering and norming sequences of the given special form. Define now W_{n_k} as in the above theorem, replacing n''_k by n_k. Set $\varepsilon_{n_k} = \lceil \operatorname{Log}(n_k/m_{n_k}) \rceil - \operatorname{Log}(n_k/m_{n_k})$.

THEOREM 3.2. *For a given subsequence $\{n_k\}_{k=1}^{\infty}$ of the positive integers and a sequence $\{m_{n_k}\}_{k=1}^{\infty}$ satisfying (3.1) the sequence W_{n_k} converges in distribution as*

$k \to \infty$ if and only if one of the following three mutually exclusive conditions holds:

$$(3.2) \qquad \sqrt{m_{n_k}}(1 - 2^{-\varepsilon_{n_k}}) \to \infty \quad \text{and} \quad \sqrt{m_{n_k}}(1 - 2^{1-\varepsilon_{n_k}}) \to -\infty,$$

$$(3.3) \qquad \sqrt{m_{n_k}}(1 - 2^{1-\varepsilon_{n_k}}) \to v \quad \text{for some} \quad -\infty < v \le 0,$$

$$(3.4) \qquad \sqrt{m_{n_k}}(1 - 2^{-\varepsilon_{n_k}}) \to u \quad \text{for some} \quad 0 \le u < \infty,$$

as $k \to \infty$.

If (3.2) holds then

$$(3.5) \qquad \frac{\sqrt{m_{n_k}}}{B(n_k)} \left\{ \frac{S_{n_k}(m_{n_k})}{n_k} - \left(\lceil \mathrm{Log} \frac{n_k}{m_{n_k}} \rceil + 1 - B^2(n_k) \right) \right\} \to_D \sqrt{3}Z$$

as $k \to \infty$, where Z is a standard normal random variable and

$$B(n_k) = \left(\frac{2^{\lceil \mathrm{Log}(n_k/m_{n_k}) \rceil}}{n_k/m_{n_k}} \right)^{1/2}.$$

If (3.3) holds then, as $k \to \infty$,

$$(3.6) \qquad \sqrt{m_{n_k}} \left\{ \frac{S_{n_k}(m_{n_k})}{n_k} - \mathrm{Log} \frac{n_k}{m_{n_k}} \right\} \to_D \sqrt{6}Z + \max(0, v + Z_2),$$

where (Z, Z_2) is a bivariate normal vector with mean vector zero, $EZ^2 = EZ_2^2 = 1$, and $EZZ_2 = -\sqrt{2/3}$.

If (3.4) holds then, as $k \to \infty$,

$$(3.7) \qquad \sqrt{m_{n_k}} \left\{ \frac{S_{n_k}(m_{n_k})}{n_k} - \mathrm{Log} \frac{n_k}{m_{n_k}} \right\} \to_D \sqrt{3}Z + \max(0, -u - Z_2),$$

where (Z, Z_2) is a bivariate normal vector with mean vector zero, $EZ^2 = EZ_2^2 = 1$, and $EZZ_2 = -1/\sqrt{3}$.

Furthermore, the pair of sequences (n_k, m_{n_k}) where m_{n_k} is an arbitrary sequence of positive integers such that $m_{n_k} \to \infty$, as $k \to \infty$, and

$$n_k = m_{n_k}(2^{k-1} + a_2 2^{k-2} + \ldots + a_{k-1} 2 + a_k), \quad k = 1, 2, \ldots,$$

where $a_2, a_3, \ldots$ are the binary digits in the dyadic expansion (2.1) of any fixed
number $1/2 < \gamma < 1$, satisfies (3.2).

The pair (n_k, m_{n_k}) given for any $k = 1, 2, \ldots$ by

$$n_k = 2^{2q_k - 1} + 2^{2q_k r_k - 1}\left\{ \lfloor -v/\sqrt{2} \rfloor + \sum_{j=1}^{2q_k - r_k - 1} v_j 2^{-j} \right\}, \quad m_{n_k} = 2^{2r_k + 1} - 1,$$

where q_k and r_k are arbitrary positive integers such that

(3.8) $q_k \to \infty, \quad r_k \to \infty, \quad and \quad q_k - r_k \to \infty \quad as \quad k \to \infty$

and $v_1, v_2, \ldots$ are the binary digits in the dyadic expansion of $-v/\sqrt{2} - \lfloor -v/\sqrt{2} \rfloor$
satisfies (3.3) with $v < 0$, and the pair (n_k, m_{n_k}) given by

$$n_k = 2^{3q_k - 1} + 2^{3q_k - 2r_k}, \quad m_{n_k} = 2^{3r_k + 1} - 1, \quad k = 1, 2, \ldots,$$

with q_k and r_k satisfying (3.8), satisfies (3.3) with $v = 0$.

Finally, with q_k and r_k satisfying (3.8), the pair (n_k, m_{n_k}) given for any $k = 1, 2, \ldots$ by

$$n_k = 2^{2q_k} - 2^{2q_k - r_k}\left\{ \lfloor u \rfloor + \sum_{j=1}^{2q_k - r_k} u_j 2^{-j} \right\}, \quad m_{n_k} = 2^{2r_k} + 1,$$

where $u_1, u_2, \ldots$ are the binary digits in the dyadic expansion of $u - \lfloor u \rfloor$, satisfies
(3.4) with $u > 0$, while the pair (n_k, m_{n_k}) given for any $k = 1, 2, \ldots$ by

$$n_k = 2^{q_k - 1}, \quad m_{n_k} = 2^{r_k} + 1$$

satisfies (3.4) with $u = 0$.

We note that when talking about a binary expansion of a number in $(0, 1)$ in the above theorem we always use the convention following (2.1), but we also assume that all the binary digits of zero are zero.

It is interesting to point out the fact that the sequence $B(n_k)$ in (3.5), for which by (2.20) we have

$$(3.9) \qquad\qquad 1 \leq B(n_k) < \sqrt{2},$$

does not in general converge under (3.2). In fact, if

$$0 < \liminf_{k \to \infty} \varepsilon_{n_k} \leq \limsup_{k \to \infty} \varepsilon_{n_k} < 1$$

then we always have (3.2).

The last result of this section is for heavily trimmed sums

$$S_n(n - \lfloor \beta n \rfloor) = \sum_{j=1}^{\lfloor \beta n \rfloor} X_{j,n}, \quad \text{where} \quad \frac{1}{2} < \beta < 1.$$

It is a special case of a half-sided version of Theorem 5 in [4] easily stated for an underlying distribution that is concentrated on the positive half line.

THEOREM 3.3. (i) If $-\mathrm{Log}(1 - \beta)$ is not an integer, then

$$\sqrt{n} \left\{ \frac{\sum_{j=1}^{\lfloor \beta n \rfloor} X_{j,n}}{n} - \mu(\beta) \right\} \to_{\mathcal{D}} \sigma_1(0, \beta) Z$$

as $n \to \infty$, where Z is a standard normal random variable,

$$\mu(\beta) = 2 - \lceil \mathrm{Log}(1 - \beta) \rceil - 2(1 - \beta) 2^{-\lceil \mathrm{Log}(1-\beta) \rceil} > 0$$

and

$$\sigma_1(0,\beta) = \left(6\{2^{-\lceil \text{Log}(1-\beta)\rceil} - 1\} + 2\lceil \text{Log}(1-\beta)\rceil - \lceil \text{Log}(1-\beta)\rceil^2\right)^{1/2}.$$

(ii) If $-\text{Log}(1-\beta)$ *is an integer, then*

$$\sqrt{n}\left\{\frac{\sum_{j=1}^{\lfloor \beta n\rfloor} X_{j,n}}{n} - \text{Log}\frac{1}{1-\beta}\right\} \to_{\mathcal{D}} \sigma_2(0,\beta)Z + (1-\beta)^{-1/2}\max(0,-Z_2),$$

where (Z, Z_2) *is a bivariate normal vector with mean vector zero,* $EZ^2 = 1$, $EZ_2^2 = \beta$, *and*

$$EZZ_2 = -\{2\beta + (1-\beta)\text{Log}(1-\beta)\}/\{1 + 2\beta - (1-\beta)(1-\text{Log}(1-\beta))^2\}^{1/2}$$

and where $\sigma_2(0,\beta) = (\{1 + 2\beta - (1-\beta)(1-\text{Log}(1-\beta))^2\}/(1-\beta))^{1/2}.$

Part (i) of this theorem leads to an interesting modification of the Petersburg game. Suppose the casino and the player agree to play n consecutive Petersburg games so that the largest $n - \lfloor \beta n\rfloor$ principal gains will not be paid to the player, where $-\text{Log}(1-\beta)$ is not an integer. Since we have

$$P\left\{\sum_{j=1}^{\lfloor \beta n\rfloor} X_{j,n} > n\mu(\beta)\right\} \to \frac{1}{2} \quad \text{and} \quad P\left\{\sum_{j=1}^{\lfloor \beta n\rfloor} X_{j,n} < n\mu(\beta)\right\} \to \frac{1}{2}$$

as $n \to \infty$, the fair premium for the player to pay to the casino for playing this sequence of n games is $n\mu(\beta)$ if n is large enough. Of course the casino can also determine a different premium formula from part (i) of the theorem to raise unfairly its fair chance $1/2$.

Proof of Theorem 3.1. Presently we need the norming sequence

$$a_{n_k}(m_{n_k}) = \sqrt{n_k}\sigma(m_{n_k}/n_k),$$

instead of the one in (2.16), for which by (2.14),

$$(3.10) \qquad a_{n_k}(m_{n_k}) \sim \sqrt{3} \sqrt{n_k} 2^{\lceil \text{Log}(n_k/m_{n_k}) \rceil/2} = \left(3 \frac{2^{\lceil \text{Log}(n_k/m_{n_k}) \rceil}}{n_k/m_{n_k}} \right)^{1/2} \frac{n_k}{\sqrt{m_{n_k}}}.$$

Also, we use $\varepsilon_k = \varepsilon_{n_k}$ in the formulation of Theorem 3.2, that is

$$(3.11) \qquad \varepsilon_k = \varepsilon_{n_k} = \lceil \text{Log}(n_k/m_{n_k}) \rceil - \text{Log}(n_k/m_{n_k}),$$

and introduce on $\mathbf{R}$ the functions, that play the role of ψ_{2,n_k} in [4],

$$\psi_{n_k}(x) = \frac{\sqrt{m_{n_k}}}{a_{n_k}(m_{n_k})} \left\{ Q\left(1 - \frac{m_{n_k}}{n_k} + x \frac{\sqrt{m_{n_k}}}{n_k}\right) - Q\left(1 - \frac{m_{n_k}}{n_k}\right) \right\}, \quad |x| \le \frac{\sqrt{m_{n_k}}}{2},$$

by setting $\psi_{n_k}(x) = \psi_{n_k}(-\sqrt{m_{n_k}}/2)$ if $x < -\sqrt{m_{n_k}}/2$ and $\psi_{n_k}(x) = \psi_{n_k}(\sqrt{m_{n_k}}/2)$ if $x > \sqrt{m_{n_k}}/2$. We can separate the values of these functions according to the three alternatives $d_k(x) \ge 1$, $0 \le d_k(x) < 1$, and $d_k(x) < 0$, where, with ε_k from (3.11), $d_k(x) = \varepsilon_k + \text{Log}(1 - x/\sqrt{m_{n_k}})$, and after showing that

$$d_k(x) \ge 1 \quad \text{if and only if} \quad x < 0 \quad \text{and} \quad \sqrt{m_{n_k}}(1 - 2^{1-\varepsilon_k}) \ge x,$$

$$0 \le d_k(x) < 1 \quad \text{if and only if} \quad \sqrt{m_{n_k}}(1 - 2^{1-\varepsilon_k}) < x \le \sqrt{m_{n_k}}(1 - 2^{-\varepsilon_k}),$$

$$d_k(x) < 0 \quad \text{if and only if} \quad x > 0 \quad \text{and} \quad x > \sqrt{m_{n_k}}(1 - 2^{-\varepsilon_k}),$$

we obtain by (1.6) and (2.14) that for any fixed $x \in \mathbf{R}$,

$$(3.12) \quad \psi_{n_k}(x) \sim \begin{cases} -\frac{1}{\sqrt{3}} \frac{1}{2} B(n_k) & , \text{if } x < 0 \quad \text{and} \quad \sqrt{m_{n_k}}(1 - 2^{1-\varepsilon_k}) \ge x, \\ 0 & , \text{if } \sqrt{m_{n_k}}(1 - 2^{1-\varepsilon_k}) < x \le \sqrt{m_{n_k}}(1 - 2^{-\varepsilon_k}), \\ \frac{1}{\sqrt{3}} B(n_k) & , \text{if } x > 0 \quad \text{and} \quad \sqrt{m_{n_k}}(1 - 2^{-\varepsilon_k}) < x, \end{cases}$$

where $B(n_k)$ is as in (3.5), satisfying (3.9).

Using the assumption of the theorem and the converse half of Theorem 3 in [4] (as formulated in Section 3 of [6]; see also the end of Section 1 in [5]), we see that for each $\{n_k'\} \subset \{n_k\}$ there is a further $\{n_k'''\} \subset \{n_k'\}$ such that $\psi_{n_k'''}(\cdot) \Rightarrow \psi(\cdot)$ for some finite function $\psi(\cdot)$ on $\mathbf{R}$ and

$$a_{n_k'''}(m_{n_k'''})/A_{n_k'''} \to \delta, \quad 0 < \delta < \infty.$$

(The case $\delta = 0$ can be ruled out again because then the limit would be degenerate.) Thus, using also (3.10), (3.9), (2.15), Theorem 1 in [4], and the convergence of types theorem, it is now routine to see that there exist a subsequence $\{n_k''\} \subset \{n_k'''\}$ and constants $0 < a < \infty$ and $c \in \mathbf{R}$ such that all the statements of the theorem hold true. ∎

Again, before the proof of Theorem 3.2 we need a technical lemma, parallel to Lemma 2.4. The proof of this lemma uses ideas very similar to those in the proof of Lemma 2.4 but, of necessity, is lengthier and more complicated. In the interest of saving space, it is omitted here.

LEMMA 3.4. *For a given $\{n_k\}_{k=1}^{\infty}$ and $\{m_{n_k}\}_{k=1}^{\infty}$ satisfying (3.1) the functions ψ_{n_k} converge weakly to some non-decreasing left-continuous function ψ on $\mathbf{R}$, satisfying $\psi(0) \leq 0$ and $\psi(0+) \geq 0$ if and only if one of the conditions (3.2), (3.3) or (3.4) holds. If (3.2) holds then*

$$(3.13) \qquad\qquad \psi_{n_k}(x) \to 0 \quad \text{for every} \quad x \in \mathbf{R}.$$

If (3.3) holds then

$$(3.14) \qquad \varepsilon_{n_k} \to 1, \quad \psi_{n_k}(\cdot) \Rightarrow \psi_v(\cdot) \quad \text{on} \quad \mathbf{R}, \quad \text{and} \quad B(n_k) \to \sqrt{2},$$

where

$$\psi_v(x) = \begin{cases} -1/\sqrt{6} &, \quad x \leq v, \\ 0 &, \quad x > v. \end{cases}$$

If (3.4) holds then

(3.15) $\varepsilon_{n_k} \to 0, \quad \psi_{n_k}(\cdot) \Rightarrow \psi^u(\cdot) \quad$ on $\quad \mathbf{R}, \quad$ and $\quad B(n_k) \to 1,$

where

$$\psi^u(x) = \begin{cases} 0 & , \quad x \le u, \\ 1/\sqrt{3} & , \quad x > u. \end{cases}$$

Proof of Theorem 3.2. First we consider the three sufficiency statements. If (3.2) holds, then the statement follows by a direct application of Theorem 1 in [4] on account of (3.13) of Lemma 3.4, using (3.10).

If (3.3) holds then by Lemma 3.4 we have (3.14). Using the norming factor $a_{n_k}^* = n_k/\sqrt{m_{n_k}}$ instead of $a_{n_k}(m_{n_k}) \sim \sqrt{6}\, n_k/\sqrt{m_{n_k}}$, the latter obtained from (3.10) and (3.14), the left side of (3.6) converges in distribution to

$$\sqrt{6}\left\{ Z + \int_0^{-Z_2} \psi_v(x)dx \right\} = \sqrt{6}Z + \max(0, v + Z_2)$$

by Theorem 1 in [4], where the covariance $EZZ_2 = -\sqrt{2/3}$ is obtained by the fact that presently

(3.16) $r_{2,n_k} := -\dfrac{\sqrt{m_{n_k}}}{a_{n_k}(m_{n_k})} \displaystyle\int_0^{1-m_{n_k}/n_k} s\, dQ(s) \sim -\dfrac{B^2(n_k)}{\sqrt{6}},$

which follows by integrating by parts and using formulae (1.6) and (2.15), and hence by (3.14) we obtain $-\sqrt{2/3}$ in the limit.

Finally, if (3.4) holds then the proof is exactly the same as above, replacing the use of (3.14) by that of (3.15) and noting that presently we use $a_{n_k}^*$ instead of $a_{n_k}(m_{n_k}) \sim \sqrt{3}\, n_k/\sqrt{m_{n_k}}$, that the limit of the left side of (3.7) is, again by Theorem 1 in [4], now

$$\sqrt{3}\left\{ Z + \int_0^{-Z_2} \psi^u(x)dx \right\} = \sqrt{3}Z + \max(0, -u - Z_2),$$

and that (3.16) is still true with $\sqrt{6}$ replaced by $\sqrt{3}$, and hence by (3.15) we now obtain $r_{2,n_k} \to EZZ_2 = -1/\sqrt{3}$.

Now we turn to necessity. Using (3.10), (3.9), and the converse result referred to in the proof of Theorem 3.1, it follows that if W_{n_k} converges in distribution then for any subsequence $\{n'_k\} \subset \{n_k\}$ there is a further subsequence $\{n''_k\} \subset \{n'_k\}$ such that for ψ_{n_k} in (3.12) we have $\psi_{n''_k}(\cdot) \Rightarrow \psi(\cdot)$ on $\mathbf{R}$ for some appropriate limiting function ψ. But then by Lemma 3.4 we must have exactly one of (3.2), (3.3), and (3.4) along $\{n''_k\}$. Applying now the already proved sufficency results along $\{n''_k\}$, we get one of the three possible limiting behavior. Since, obviously, no two of the three limiting random variables in (3.5), (3.6), and (3.7) can be equal in distribution for any choice of $v \le 0$ and $u \ge 0$, we see that these subsequential limits must be the same, exactly one of those in (3.5), (3.6), or (3.7). Since the subsequence $\{n'_k\} \subset \{n_k\}$ was arbitrary, the necessity statement follows.

The proof of the last statements of the theorem, that is, that the stated constructions for (3.2), (3.3), and (3.4) to hold are indeed valid is elementary but all together extremely tedious and lengthy. We challenge the interested reader to check what we state. ∎

4. Extreme sums. Again and throughout this last section, $\{n_k\}_{k=1}^\infty$ and $\{m_{n_k}\}_{k=1}^\infty$ will be two subsequences of the positive integers satisfying (3.1), and we are interested in the exreme sums

$$E_{n_k}(m_{n_k}) = S_{n_k} - S_{n_k}(m_{n_k}) = \sum_{j=n_k-m_{n_k}+1}^{n_k} X_{j,n_k} = \sum_{j=1}^{m_{n_k}} X_{n_k+1-j,n_k},$$

the sums of the largest m_{n_k} gains. Following the pattern of the preceding two sections, first we state the following.

THEOREM 4.1. *If for some constants $A_{n_k} > 0$ and $C_{n_k} \in \mathbf{R}$ the sequence $(E_{n_k}(m_{n_k}) - C_{n_k})/A_{n_k}$ converges in distribution to a non-degenerate random variable R, then for each subsequence $\{n'_k\}_{k=1}^{\infty} \subset \{n_k\}_{k=1}^{\infty}$ there exist a further subsequence $\{n''_k\}_{k=1}^{\infty} \subset \{n'_k\}_{k=1}^{\infty}$ and a constant $a = a_{\{n''\}}$, $0 < a < \infty$, such that $n''_k/A_{n''_k} \to a$ and*

$$R_{n''_k} := \frac{E_{n''_k}(m_{n''_k})}{n''_k} - r_{n''_k}(m_{n''_k}) \to_D \frac{1}{a}R + c$$

as $k \to \infty$, where

$$r_{n_k}(m_{n_k}) = \left(2^{2\lceil \mathrm{Log} n_k \rceil - \lfloor \mathrm{Log} n_k \rfloor - 1} - 2^{\lceil \mathrm{Log} n_k \rceil}\right)\frac{1}{n_k} + \lceil \mathrm{Log} n_k \rceil - \lceil \mathrm{Log}(n_k/m_{n_k}) \rceil$$
$$+ 2^{\lceil \mathrm{Log}(n_k/m_{n_k}) \rceil}\frac{m_{n_k}}{n_k}$$

and c is some constant.

Again, let R_{n_k} be as above, replacing n''_k by n_k. Our last result is the following.

THEOREM 4.2. *For a given subsequence $\{n_k\}_{k=1}^{\infty}$ of positive integers and an arbitrary subsequence $\{m_{n_k}\}_{k=1}^{\infty}$ of positive integers satisfying (3.1) the sequence R_{n_k} converges in distribution if and only if the sequence $\{n_k\}_{k=1}^{\infty}$ satisfies condition (2.4) of Theorem 2.2 with some $1/2 \le \gamma \le 1$. In this case,*

$$\frac{E_{n_k}(m_{n_k})}{n_k} - \left(\mathrm{Log} m_{n_k} + \left\{2^{\lceil \mathrm{Log}(n_k/m_{n_k}) \rceil}\frac{m_{n_k}}{n_k} + \mathrm{Log}\frac{n_k}{m_{n_k}} - \lceil \mathrm{Log}\frac{n_k}{m_{n_k}} \rceil\right\}\right) \to_D W_{\gamma},$$

where W_γ is the limiting random variable appearing in (2.7).

Of course, the special constructions of $\{n_k\}$ in and above (2.9) for condition (2.4) are still valid, and we emphasize that $\{m_{n_k}\}$ satisfying (3.1) is completely arbitrary. The normalizing sequence $\{n_k\}$ is the same for full sums and extreme sums. This is of course entirely natural having (2.12). The results in Section 2 show that the Petersburg distribution function F is stochastically compact (cf. [3]) and we also see that the largest gain $X_{n,n}$ is also stochastically compact, with all possible limiting distributions given in (2.12). So the Petersburg game exhibits the phenomenon discussed in Corollary 12 in [3].

It is interesting to observe the generally non-convergent oscillatory term in the centering sequence in Theorem 4.2, for which, if log stands for the natural logarithm, it is easy to see that

$$\frac{1}{\log 2} - \operatorname{Log}\frac{1}{\log 2} \leq \left\{ 2^{\lceil \operatorname{Log}(n_k/m_{n_k}) \rceil}\frac{m_{n_k}}{n_k} + \operatorname{Log}\frac{n_k}{m_{n_k}} - \lceil \operatorname{Log}\frac{n_k}{m_{n_k}} \rceil \right\} \leq 1.$$

Proof of Theorem 4.1. Recalling the notation in (2.13), we now need to know the asymptotic behavior of $\sigma^2(1 - m_{n_k}/n_k, 1/n_k)$. Using (1.5), (1.6), (2.13), and both formulae in (2.15) yield

(4.1) $$\sigma^2(1 - m_{n_k}/n_k, 1/n_k) \sim 2^{\lceil \operatorname{Log} n_k \rceil}3$$

for *any* m_{n_k} satisfying (3.1). Hence, instead of the special norming sequence $\sqrt{n_k}\sigma(1 - m_{n_k}/n_k, 1/n_k)$ designed for extreme sums in [5], we can use $a(n_k)$ belonging to whole sums, since by (2.17),

(4.2) $$\sqrt{n_k}\sigma(1 - m_{n_k}/n_k, 1/n_k) \sim \left(3\frac{2^{\lceil \operatorname{Log} n_k \rceil}}{n_k}\right)^{1/2} n_k \sim a(n_k).$$

According to [5], here we need the following variants of the functions φ_{n_k} and ψ_{n_k} in the proofs of Theorems 2.1 and 3.1:

$$\overline{\varphi}_{n_k}(s) := \begin{cases} \dfrac{-Q((1-\frac{s}{n_k})-)+Q((1-\frac{1}{n_k})-)}{a(n_k)} & , \quad 0 < s \leq n_k - n_k\rho_{n_k}, \\[3mm] \dfrac{-Q(\rho_{n_k}-)+Q((1-\frac{1}{n_k})-)}{a(n_k)} & , \quad n_k - n_k\rho_{n_k} < s < \infty, \end{cases}$$

where ρ_{n_k} is the sequence introduced before Lemma 2.4, and

$$\overline{\psi}_{n_k}(x) = \begin{cases} \dfrac{\sqrt{m_{n_k}}}{a(n_k)}\left\{Q((1-\frac{m_{n_k}}{n_k})-) - Q((1-\frac{m_{n_k}}{n_k} - x\frac{\sqrt{m_{n_k}}}{n_k})-)\right\} & , \quad |x| \leq \frac{\sqrt{m_{n_k}}}{2}, \\[2mm] \overline{\psi}_{n_k}(-\sqrt{m_{n_k}}/2) & , \quad x < -\frac{\sqrt{m_{n_k}}}{2}, \\[2mm] \overline{\psi}_{n_k}(\sqrt{m_{n_k}}/2) & , \quad \frac{\sqrt{m_{n_k}}}{2} < x. \end{cases}$$

By (2.17), (2.20), (3.10), (3.12), and (4.2) it is obvious that there is a positive constant $C > 0$ such that $\psi_{n_k}(x) \leq C/\sqrt{m_{n_k}}$ for any $x \in \mathbf{R}$, and hence by (3.1),

$$(4.3) \qquad\qquad \overline{\psi}_{n_k}(x) \to 0, \quad x \in \mathbf{R}.$$

On the other hand, with φ_{n_k} given in (2.19), we have

$$(4.4) \qquad\qquad \overline{\varphi}_{n_k}(s) = \varphi_{n_k}(s-) - \varphi_{n_k}(1-), \quad s > 0.$$

Also, by a computation similar to that leading to (4.1) we obtain that for *any* positive numbers l_{n_k} such that $l_{n_k} \to \infty$ and $l_{n_k}/m_{n_k} \to 0$ we have

$$(4.5) \qquad\qquad \sigma^2(1 - l_{n_k}/n_k, m_{n_k}/n_k) \sim 2^{\lceil \mathrm{Log}(n_k/l_{n_k})\rceil}3,$$

and from (2.15) and (1.6),

$$(4.6) \qquad\qquad \frac{1}{n_k}Q\left((1-\frac{1}{n_k})-\right) + \int_{1-m_{n_k}/n_k}^{1-1/n_k} Q(u)du = r_{n_k}(m_{n_k}).$$

After all these preliminaries we are ready now to prove the theorem. Using the assumption of it, (4.2), (4.3), (4.4), (2.19), and (2.20), an application of Theorem 2 in [5] shows that for any $\{n'_k\} \subset \{n_k\}$ there is a further subsequence $\{n'''_k\} \subset \{n'_k\}$ such that

$$\overline{\psi}_{n'''_k}(x) \to 0, \quad x \in \mathbf{R}, \quad \overline{\varphi}_{n'''_k}(\cdot) \Rightarrow \overline{\varphi}(\cdot) \quad \text{on} \quad (0,\infty) \quad \text{and} \quad a(n'''_k)/A_{n'''_k} \to \delta,$$

where $\overline{\varphi}$ is a finite function and $\delta > 0$. Hence an application of Theorem 1 in [5], and the convergence of types theorem provide a subsequence $\{n''_k\} \subset \{n'''_k\}$ along which the statements hold. ∎

Proof of Theorem 4.2. First we note that by (4.1) and (4.5),

$$(4.7) \qquad \sigma(1 - l_{n_k}/n_k, m_{n_k}/n_k)/\sigma(1 - m_{n_k}/n_k, 1/n_k) \to 0$$

for any sequences $\{n_k\}$ and $\{m_{n_k}\}$ satisfying (3.1) and for any sequence $\{l_{n_k}\}$ of positive numbers such that $l_{n_k} \to \infty$ and $l_{n_k}/m_{n_k} \to 0$.

Also, for each fixed $s > 0$, by (4.4) and (2.19) we have

$$\overline{\varphi}_{n_k}(s) \sim \begin{cases} -\dfrac{2^{-\lfloor \log s - \rfloor - 2}}{\sqrt{3}} \sqrt{\dfrac{2^{\lceil \log n_k \rceil}}{n_k}} & , \quad 0 \le \alpha_k + \beta_s < 1, \\[3mm] -\dfrac{2^{-\lfloor \log s - \rfloor - 2}}{2\sqrt{3}} \sqrt{\dfrac{2^{\lceil \log n_k \rceil}}{n_k}} & , \quad 1 \le \alpha_k + \beta_s < 2, \end{cases}$$

where α_k and β_s are as in (2.19), and clearly the proof of Lemma 2.4 implies that $\overline{\varphi}_{n_k}(\cdot)$ converges weakly on $(0,\infty)$ to some finite function if and only if condition (2.4) holds. In particular,

$$(4.8) \qquad \overline{\varphi}_{n_k}(\cdot) \Rightarrow \frac{1}{\sqrt{3\gamma}}(\overline{\varphi}^*_\gamma(\cdot) - \overline{\varphi}^*_\gamma(1)) \quad \text{on} \quad (0,\infty) \quad \text{if and only if (2.4) holds,}$$

where $\overline{\varphi}^*_\gamma$ is the left-continuous version of φ^*_γ given in (2.6).

Assume now that R_{n_k} converges in distribution along the given $\{n_k\}$. Consider an arbitrary subsequence $\{n'_k\} \subset \{n_k\}$. Then, obviously, there is a further subsequence $\{n''_k\} \subset \{n'_k\}$ such that condition (2.4) is satisfied for some $1/2 \le \gamma \le 1$

along this $\{n_k''\}$. But then, using (4.3), (4.7), and (4.8) along $\{n_k''\}$, it follows from Theorem 1 in [5] that

(4.9)
$$R_{n_k''} \to_D \frac{1}{\gamma} \overline{V}_{0,0}(0, \overline{\varphi}_\gamma^*, 0) := \frac{1}{\gamma} \left\{ \int_{Y_1}^\infty (\overline{N}(s) - s) d\overline{\varphi}_\gamma^*(s) - \int_1^{Y_1} s d\overline{\varphi}_\gamma^*(s) - \overline{\varphi}_\gamma^*(1) \right\}$$

where

$$\overline{N}(s) = \sum_{j=1}^\infty I(Y_j \le s), \quad s \le 0,$$

is the right-continuous version of the left-continuous Poisson process given in (2.2) and where we use the convention that the symbol $\int_x^y$ means $\int_{[x,y)}$ whenever we integrate with respect to a left-continuous function, just as the convention that $\int_x^y$ means $\int_{(x,y]}$ whenever we integrate with respect to a righ-continuous function has been tacitly used throughout the paper. Using these conventions it is easy to see that

$$\overline{V}_{0,0}(0, \overline{\varphi}_\gamma^*, 0) = V_{0,0}(0, \varphi_\gamma^*, 0) = \int_{Y_1}^\infty (N(s) - s) d\varphi_\gamma^*(s) - \int_1^{Y_1} s d\varphi_\gamma^*(s) - \varphi_\gamma^*(1),$$

and hence, using the special case $m = 0$ of the formula right below (2.25), we see that (4.9) is the same as

$$R_{n_k''} \to_D V_\gamma,$$

where V_γ is as in (2.7) and (2.8). At the same time, just as in the proof of Theorem 2.2, we see that γ must be unique for all these subsequences $\{n_k''\}$, and since $\{n_k'\}$ was an arbitrary subsequence of $\{n_k\}$, condition (2.4) must hold along $\{n_k\}$, and hence we also have

(4.10)
$$R_{n_k} = \frac{E_{n_k}(m_{n_k})}{n_k} - r_{n_k}(m_{n_k}) \to_D V_\gamma.$$

Conversely, if condition (2.4) is satisfied then, by (4.3), (4.7), and (4.8), Theorem 1 in [5] implies (4.10). But under (2.4) the first term of $r_{n_k}(m_{n_k})$, given in Thorem 4.1, converges to zero, and hence by (2.26) we obtain

$$\frac{E_{n_k}(m_{n_k})}{n_k} - \left(\text{Log}n_k - \lceil\text{Log}(n_k/m_{n_k})\rceil + 2^{\lceil\text{Log}(n_k/m_{n_k})\rceil}\frac{m_{n_k}}{n_k}\right) \to_{\mathcal{D}} W_\gamma,$$

which by adding and subtracting $\text{Log}m_{n_k}$ in the centering sequence is clearly equivalent to the stated convergence in the theorem. ∎

Acknowledgement. We are grateful to Professor Gordon Simons of the University of North Carolina at Chapel Hill for pointing out an oversight in the first version of the manuscript.

REFERENCES

[1] S. CSÖRGŐ, An extreme-sum approximation to infinitely divisible laws without a normal component. In: *Probability on Vector Spaces IV.* (S. Cambanis and A. Weron, eds.) pp. 47–58. Lecture Notes in Mathematics **1391**, Springer, Berlin, 1989.

[2] S. CSÖRGŐ, A probabilistic approach to domains of partial attraction. *Adv. in Appl. Math.* **11** (1990), to appear.

[3] S. CSÖRGŐ, E. HAEUSLER, and D. M. MASON, A probabilistic approach to the asymptotic distribution of sums of independent, identically distributed random variables. *Adv. in Appl. Math.* **9**(1988), 259-333.

[4] S. CSÖRGŐ, E. HAEUSLER, and D. M. MASON, The asymptotic distribution of trimmed sums. *Ann. Probab.* **16**(1988), 672-699.

[5] S. CSÖRGŐ, E. HAEUSLER, and D. M. MASON, The asymptotic distribution of extreme sums. *Ann. Probab.* **19** (1991), to appear.

[6] S. CSÖRGŐ, E. HAEUSLER, and D. M. MASON, The quantile - transform – empirical - process approach to limit theorems for sums of order statistics. In: *Sums, Trimmed Sums, and Extremes* (M. G. Hahn, D. M. Mason, and D. C. Weiner, eds.). Birkhäuser, Basel, 1990.

[7] S. CSÖRGŐ and D. M. MASON, Intermediate sums and stochastic compactness of maxima. Submitted.

[8] W. FELLER, *An Introduction to Probability Theory and its Applications, Vol.1.* Wiley, New York, 1950.

[9] P. LÉVY, *Théorie de l'addition des variables aléatoires,* 2nd ed. Gauthier - Villars, Paris, 1954.

[10] A. MARTIN - LÖF, A limit theorem which clarifies the 'Petersburg paradox'. *J. Appl. Probab.* **22**(1985), 634-643.

[11] G. SHAFER, The St. Petersburg paradox. In: *Encyclopedia of Statistical Sciences, Vol. 8* (S. Kotz, N. L. Johnson, and C. B. Read, eds.), pp. 865-870. Wiley, New-York, 1988.

Sándor Csörgő
Department of Statistics
The University of Michigan
1444 Mason Hall
Ann Arbor, MI 48109-1027
U. S. A.

Rossitza Dodunekova
Department of Probability and Statistics
University of Sofia
P. O. Box 373
1090 Sofia
Bulgaria

A PROBABILISTIC APPROACH
TO THE TAILS OF INFINITELY DIVISIBLE LAWS

Sándor Csörgő[1] and David M. Mason[2]

1. Introduction. Consider an arbitrary non–degenerate infinitely divisible real random variable X. Usually this is described by the Lévy formula for its characteristic function. For any t in the real line $I\!R$, we have

$$C(t) = E \exp(iXt)$$

$$= \exp\left\{ i\theta t - \frac{\sigma^2}{2}t^2 \right.$$

$$\left. + \int_{-\infty}^{0} \left(e^{ixt} - 1 - \frac{ixt}{1+x^2} \right) dL(x) + \int_{0}^{\infty} \left(e^{ixt} - 1 - \frac{ixt}{1+x^2} \right) dR(x) \right\},$$

where $\theta \in I\!R$ and $\sigma \geq 0$ are uniquely determined constants and L and R are uniquely determined left–continuous and right–continuous functions (the so–called Lévy measures) on $(-\infty, 0)$ and $(0, \infty)$, respectively, that is $L(\cdot)$ and $R(\cdot)$ are non–decreasing functions such that $L(-\infty) = 0 = R(\infty)$ and

(1.1)
$$\int_{-\epsilon}^{0} x^2 \, dL(x) + \int_{0}^{\epsilon} x^2 \, dR(x) < \infty \quad \text{for any} \quad \epsilon > 0.$$

Many authors have investigated the tail behavior of the distribution of X (see [1], [3], [7–11], [13–21], for instance). Of necessity, the methods normally used have a Fourier–analytic character and the results are usually formulated in terms of

---
[1] Partially supported by the Hungarian National Foundation for Scientific Research, Grants 1808/86 and 457/88.
[2] Partially supported by the Alexander von Humboldt Foundation and NSF Grant #DMS–8803209.

conditions on the Lévy measures, conditions that do not always allow an immediate probabilistic insight.

As an integral part of a probabilistic approach to the asymptotic distribution of sums of independent, identically distributed random variables and of the corresponding lightly trimmed sums, a representation of X is given in Theorem 3 of [5]; cf. also the end of Section 2 of [6] in the present volume. Introduce the non–decreasing, non–positive, right–continuous inverse functions

$$\phi_1(s) = \inf\{x < 0 : L(x) > s\}, \ \phi_2(s) = \inf\{x < 0 : -R(-x) > s\}, \ 0 < s < \infty,$$

and consider two standard (intensity one) left–continuous Poisson processes $N_1(s)$ and $N_2(s)$, $s \geq 0$, and a standard normal random variable Z such that $N_1(\cdot)$, Z, and $N_2(\cdot)$ are independent. Since (1.1) implies that

$$(1.2) \qquad \int_\epsilon^\infty \phi_1^2(s)\, ds + \int_\epsilon^\infty \phi_2^2(s)\, ds < \infty \quad \text{for any} \quad \epsilon > 0,$$

the random variables

$$(1.3) \qquad \begin{aligned} V_j &= \int_{Y_j}^\infty (N_j(s) - s)\, d\phi_j(s) - \int_1^{Y_j} s d\phi_j(s) \\ &= \int_1^\infty (N_j(s) - s)\, d\phi_j(s) + \int_0^1 N_j(s)\, d\phi_j(s), \end{aligned}$$

where Y_j is the first jump–point of $N_j(\cdot)$, are well defined for $j = 1, 2$ (the first integrals exist as improper Riemann integrals with probability one), and the constants

$$\theta_j = -\phi_j(1) + \int_0^1 \frac{\phi_j(s)}{1 + \phi_j^2(s)}\, ds - \int_1^\infty \frac{\phi_j^3(s)}{1 + \phi_j^2(s)}\, ds, \ j = 1, 2,$$

are finite, and we have the distributional equality

(1.4) $$X \overset{\mathcal{D}}{=} V_{1,2}(\sigma, \theta) := -V_1 + \sigma Z + V_2 + \theta - \theta_1 + \theta_2,$$

that is $E \exp(iV_{1,2}(\sigma, \theta)t) = C(t)$, $t \in \mathbb{R}$.

The aim of the present paper is to introduce a direct probabilistic approach to the investigation of the tails of X based on the representation in (1.4) and on the use of elementary calculations and inequalities for the Poisson distribution. We derive several earlier results by this approach, sometimes in a polished form, under *natural* new versions of the conditions, including Sato's [15] bounds. The fact that Sato's theorem can be obtained by this approach is particularly interesting to us. This is because we used it in the proof of Corollary 1 in [5] when proving the necessity half of the normal convergence criterion in the framework of our general probabilistic theory of convergence, also joint with Erich Haeusler. With the present proof of Sato's theorem our entire theory becomes independent of the existing literature and uses Fourier analysis only to ensure the uniqueness of ϕ_1, ϕ_2, σ, and θ on the right side of (1.4). The results are stated in Section 2, all the proofs are in Section 3.

We close the present introduction by pointing out that it is very easy to see directly that the right side of (1.4) is infinitely divisible. Indeed, for any integer $n \geq 1$, let $Z(k, n)$, $1 \leq k \leq n$, be Normal $(0, 1/n)$ random variables and $N_j^{(k)}$, $1 \leq k \leq n$, $j = 1, 2$, be standard Poisson processes such that $Z(1, n), \ldots, Z(n, n)$, $N_1^{(1)}, \ldots, N_1^{(n)}$, $N_2^{(1)}, \ldots N_2^{(n)}$ are independent, and form

$$V_j(k, n) = \int_{1/n}^{\infty} (N_j^{(k)}(u) - u) \, d\phi_j(nu) + \int_0^{1/n} N_j^{(k)}(u) \, d\phi_j(nu).$$

Then the random variables

$$V_{1,2}^{(k,n)}(\sigma) = -V_1(k, n) + \sigma Z(k, n) + V_2(k, n), \quad k = 1, \ldots, n,$$

are independent and identically distributed and, interchanging integration and summation and substituting $nu = s$, we see that

$$\sum_{k=1}^{n} V_{1,2}^{(k,n)}(\sigma) \overset{\mathcal{D}}{=} -V_1 + \sigma Z + V_2.$$

It would be interesting to prove the converse statement (that if X is infinitely divisible then with uniquely determined ϕ_1, ϕ_2, σ, and θ we necessarily have (1.4)) *without* a recourse to the Lévy or Lévy–Hinchin formula for the characteristic function. This problem seems to be difficult.

2. Results. The first set of results is formulated for the extreme pieces V_j in (1.4), defined in (1.3). It is worthwhile to separate these because the corresponding results for the general X will follow from these special ones in an extremely elementary fashion.

Introduce the non–negative quantities

$$A_j = -\phi_j(0+) \leq \infty, \; j = 1, 2,$$

and let δ denote any finite constant. Unless otherwise stated, we do not necessarily assume that $A_j < \infty$. If $A_j = \infty$, then $1/A_j$ is interpreted as zero. In Theorem 1 below, j can be 1 or 2, and we only consider the non–degenerate case when $\phi_j \not\equiv 0$. This is equivalent to assuming that $A_j > 0$.

THEOREM 1. (i) *For every $\alpha > 1/A_j$ there exists an $x_0 = x_0(\alpha, \phi_j, \delta) > 0$ such that for all $x \geq x_0$,*

$$P\{V_j + \delta > x\} \geq \exp(-\alpha x \log x).$$

(ii) *If $A_j < \infty$, then for every $\alpha < 1/A_j$ there exists an $x_0 = x_0(\alpha, \phi_j, \delta) > 0$ such that for all $x \geq x_0$,*

$$P\{V_j + \delta \geq x\} \leq \exp(-\alpha x \log x).$$

(iii) *For every $\tau > 0$ there exists an $x_0 = x_0(\tau, \phi_j, \delta) > 0$ such that for all $x \geq x_0$,*

$$P\{V_j + \delta \leq -x\} \leq \exp(-\tau x^2).$$

(iv) *The essential infimum of $V_j + \delta$ is $-\int_1^\infty s d\phi_j(s) + \delta$.*

(v) *The random variable $V_j + \delta$ is bounded from below (with probability one) if and only if $\int_1^\infty s d\phi_j(s) < \infty$, which in turn is equivalent to $-\int_1^\infty \phi_j(s)ds < \infty$.*

Now we consider the general infinitely divisible random variable $X = X(L, R, \sigma, \theta)$ with characteristic function $C(\cdot)$ or, what is the same, $X = X(\phi_1, \phi_2, \sigma, \theta)$ on the left side of (1.4). We exclude the trivial case when X degenerates at θ, which happens if and only if $A_1 = A_2 = \sigma = 0$ (cf. Theorem 4 in [5]). So at least one of A_1, A_2, and σ is assumed to be positive. As usual, Φ will denote the standard normal distribution function.

THEOREM 2. (i) *If $0 < A_2 \leq \infty$, then for every $\alpha > 1/A_2$ there exists an $x_0 > 0$ such that for all $x \geq x_0$,*

$$(2.1) \qquad\qquad P\{X > x\} \geq \exp(-\alpha x \log x).$$

If $0 < A_2 < \infty$, then for every $0 < \alpha < 1/A_2$ there exists an $x_0 > 0$ such that for all $x \geq x_0$,

$$(2.2) \qquad\qquad P\{X \geq x\} \leq \exp(-\alpha x \log x).$$

If $A_2 = 0, A_1 > 0$, and $\sigma > 0$, then for every $0 < \epsilon < 1$ there exists an $x_0 > 0$ such that for all $x \geq x_0$,

$$(2.3) \qquad\qquad P\{X \geq x\} \leq 1 - \Phi\left((1-\epsilon)\frac{x}{\sigma}\right).$$

(ii) If $0 < A_1 \leq \infty$, then for every $\beta > 1/A_1$ there exists an $x_0 > 0$ such that for all $x \geq x_0$,

$$(2.4) \qquad\qquad P\{X < -x\} \geq \exp(-\beta x \log x).$$

If $0 < A_1 < \infty$, then for every $0 < \beta < 1/A_1$ there exists an $x_0 > 0$ such that for all $x \geq x_0$,

$$(2.5) \qquad\qquad P\{X \leq -x\} \leq \exp(-\beta x \log x).$$

If $A_1 = 0, A_2 > 0$, and $\sigma > 0$, then for every $0 < \epsilon < 1$ there exists an $x_0 > 0$ such that for all $x \geq x_0$,

$$(2.6) \qquad\qquad P\{X \leq -x\} \leq \Phi\left(-(1-\epsilon)\frac{x}{\sigma}\right).$$

(iii) *The random variable X is almost surely bounded from below if and only if* $A_1 = 0, \sigma = 0$, *and* $\int_1^\infty s \, d\phi_2(s) < \infty$, *or, equivalently,* $-\int_1^\infty \phi_2(s) ds < \infty$, *in which case*

$$-\infty < \operatorname{ess\ inf} X = \theta + \int_0^\infty \frac{\phi_2(s)}{1 + \phi_2^2(s)} \, ds = \theta - \int_0^\infty \frac{x}{1+x^2} \, dR(x).$$

(iv) *The random variable X is almost surely bounded from above if and only if* $A_2 = 0, \sigma = 0$, *and* $\int_1^\infty s \, d\phi_1(s) < \infty$, *or, equivalently,* $-\int_1^\infty \phi_1(s) ds < \infty$, *in which case*

$$\operatorname{ess\ sup} X = \theta - \int_0^\infty \frac{\phi_1(s)}{1 + \phi_1^2(s)} ds = \theta - \int_{-\infty}^0 \frac{x}{1+x^2} dL(x) < \infty.$$

(v) *The random variable X is almost surely unbounded.*

Results of the type of the first two statements of parts (i) and (ii) of Theorem 2 have been proved by Kruglov [11], Ruegg [13,14], Horn [10], Steutel [16] and Elliott and Erdős [7]. Related is the work of Wolfe [18]. The principally final result has been achieved by Sato [15], who formulates it in terms of the support of the combined Lévy measure $L(-x) - R(x), x > 0$. Sato's result in fact covers the multidimensional case, and hence the univariate special case of his statement is by nature two–sided. However, the univariate special case of his proof allowed Bingham, Goldie, and Teugels [2; page 342] to formulate the more precise one–sided results. Our results in (2.1), (2.2) and (2.4), (2.5) of parts (i) and (ii) of Theorem 2 above are equivalent forms of these one–sided variants of Sato's theorem. An equivalent form of Sato's two–sided univariate result has been proved later but independently by Esseen [9].

We are not aware of a precise formulation of the complementary results in part (iii) of Theorem 1 and in (2.3) and (2.6) of Theorem 2 in the literature.

Parts (iii) and (iv) of Theorem 2 are equivalent forms of results achieved jointly by the two papers of Baxter and Shapiro [1] and Tucker [17]. Independently of Tucker they have been also obtained by Esseen [8]. These two parts trivially imply part (v) of Theorem 2, which has been obtained first by Chatterjee and Pakshirajan [3] by an independent Fourier–analytic proof.

There are, of course, other works that deal with the tail behavior of $X(L, R, \sigma, \theta)$. For example, Yakymiv [19] has recently proved rather precise results for a special class of infinitely divisible variables, extending greatly the main theorem of Zolotarev [20]. It would be interesting to see if these problems can also be approached by our more direct probabilistic methods.

3. Proofs. For the sake of simplicity in the notation, we drop all the subscripts j in the proof of Theorem 1. Thus V_j, ϕ_j, N_j, Y_j, and A_j become V, ϕ, N, Y, and A in this proof.

The main trick in the proof of (i) and (iii) is to introduce the functions

$$\phi_x(s) = \begin{cases} \phi(s), & 0 < s < x^2, \\ \phi(x^2), & x^2 \leq s < \infty, \end{cases} \qquad \bar{\phi}_x(s) = \begin{cases} \phi(x^2), & 0 < s < x^2, \\ \phi(s), & x^2 \leq s < \infty, \end{cases}$$

where $x > 0$, and to write (cf. (1.3))

$$(3.1) \qquad V = \int_Y^\infty (N(s) - s) d\phi(s) - \int_1^Y s \, d\phi(s) = V(x) + \bar{V}(x),$$

where

$$(3.2) \qquad \begin{aligned} V(x) &= \int_Y^\infty (N(s) - s) d\phi_x(s) - \int_1^Y s \, d\phi_x(x) \\ &= \int_Y^\infty N(x) d\phi_x(s) - \int_1^\infty s \, d\phi_x(s) \end{aligned}$$

and

$$(3.3) \qquad \bar{V}(x) = \int_Y^\infty (N(s) - s) d\bar{\phi}_x(s) - \int_1^Y s \, d\bar{\phi}_x(s).$$

Note that for $x > 1$,

$$\int_1^\infty s \, d\phi_x(s) = -\phi(1) + x^2 \phi(x^2) - \int_1^{x^2} \phi(s) \, ds.$$

Clearly, (1.2) implies that $x^2 \phi(x^2) = o(x)$ as $x \to \infty$, and by an elementary argument based on the Cauchy–Schwarz inequality we also have

$$\int_1^{x^2} \phi(s) ds = o(x) \quad \text{as} \quad x \to \infty.$$

Hence

(3.4)
$$\int_1^\infty s d\phi_x(s) = o(x) \quad \text{as} \quad x \to \infty.$$

The following inequality for the Poisson distribution, stated as Inequality 2 in [12], is an easy consequence of Stirling's formula.

LEMMA 1. *For every* $\lambda > 0$ *there exists a constant* $0 < K(\lambda) < \infty$ *such that for all* $y \geq 1$,

$$P\{N(\lambda) \geq y\} \geq K(\lambda) y^{-1/2} \exp\{-y(\log y - 1 - \log \lambda)\}.$$

LEMMA 2. *For all* $x \geq 1, \gamma > 0,$ *and* $\gamma/\phi(x^2) \leq t < \infty,$

$$E \exp(t\bar{V}(x)) \leq \exp\left(\frac{t^2}{2} f(x) e^\gamma\right),$$

where

$$f(x) = x^2 \phi^2(x^2) + \int_{x^2}^\infty \phi^2(s) ds \to 0 \quad \text{as} \quad x \to \infty.$$

Proof. Setting

$$\theta(x) = -\bar{\phi}_x(1) + \int_0^1 \frac{\bar{\phi}_x(s)}{1 + \bar{\phi}_x^2(s)} ds - \int_1^\infty \frac{\bar{\phi}_x^3(s)}{1 + \bar{\phi}_x^2(s)} ds$$

$$= -\frac{\phi^3(x^2)}{1 + \phi^2(x^2)} - \int_1^\infty \frac{\bar{\phi}_x^3(s)}{1 + \bar{\phi}_x^2(s)} ds,$$

the argument used for the computation of the corresponding characteristic function in the proof of Theorem 3 in [5] gives

$$\log E \exp(t\bar{V}(x)) = \int_0^\infty \left\{ \exp(t\bar{\phi}_x(s)) - 1 - t\frac{\bar{\phi}_x(s)}{1 + \bar{\phi}_x^2(s)} \right\} ds + t\theta(x)$$

$$\leq \int_0^\infty \left\{ t\bar{\phi}_x(s) + \frac{t^2}{2} \bar{\phi}_x^2(s) e^\gamma - t\frac{\bar{\phi}_x(s)}{1 + \bar{\phi}_x^2(s)} \right\} ds + t\theta(x)$$

$$= \frac{t^2}{2} f(x) e^\gamma,$$

where we used the inequality

$$e^y - 1 \leq y + \frac{y^2}{2} e^\gamma, \quad -\infty < y \leq \gamma. \quad \blacksquare$$

The next two lemmas together show that the right tial of V is determined by the term $V(x)$ in (1.3).

LEMMA 3. *For all $\alpha > 1/A$ and all sufficiently small $\rho > 1$,*

$$\lim_{x \to \infty} \exp(\alpha x \, \log x) \, P\{V(x) > \rho x\} = \infty.$$

Proof. Using (3.4), we see that for any $0 < \epsilon < 1$ and all x large enough,

$$
\begin{aligned}
P\{V(x) > \rho x\} &\geq P\left\{ \int_Y^\infty N(s)\phi_x(s) > (1+\epsilon)\rho x \right\} \\
&\geq P\{N(\epsilon)(\phi(x^2) - \phi(\epsilon)) > (1+\epsilon)\rho x\} \\
&= P\{N(\epsilon) > \beta x\} \\
&\geq K(\epsilon)(\beta x)^{-1/2} \exp\{-\beta x(\log x + \log(\beta/\epsilon) - 1)\}
\end{aligned}
$$

by Lemma 1 in the last step, where $\epsilon > 0$ and $\rho > 1$ are chosen so small and x so large that

$$\frac{1}{A} < \beta = \beta(x) = \frac{(1+\epsilon)\rho}{\phi(x^2) - \phi(\epsilon)} < \alpha.$$

This inequality clearly implies the lemma. $\blacksquare$

LEMMA 4. *For all $\alpha > 0$ and $\rho > 0$,*

$$\lim_{x \to \infty} \exp(\alpha x^2) \, P\{|\bar{V}(x)| \geq \rho x\} = 0.$$

Proof. For any $0 < t < \infty$, the Markov inequality and Lemma 2 give that

$$P\{\bar{V}(x) \geq \rho x\} \leq \exp\left(-\rho t x + \frac{t^2}{2} f(x) e^\rho \right).$$

Upon choosing $t = \rho e^{-\rho} x / f(x) > 0$, since $f(x) \to 0$ as $x \to \infty$, we obtain

$$\lim_{x \to \infty} \exp(\alpha x^2) P\{\bar{V}(x) \geq \rho x\} \leq \lim_{x \to \infty} \exp\left\{ x^2 \left(\alpha - \frac{\rho^2}{2e^\rho} \frac{1}{f(x)} \right) \right\} = 0.$$

On the other hand, for any $t > 0$ such that $\rho / \phi(x^2) \leq -t < 0$,

$$P\{-\bar{V}(x) \geq \rho x\} \leq \exp\left(-\rho t x + \frac{t^2}{2} f(x) e^\rho \right).$$

It is easy to see that presently the choice $t = \rho(x^2/(e^\rho f(x)))^{1/2}$ is permissible, whence we get

$$\lim_{x \to \infty} \exp(\alpha x^2) P\{-\bar{V}(x) \geq \rho x\} \leq \lim_{x \to \infty} \exp\left\{ x^2 \left(\alpha + \frac{\rho^2}{2} - \frac{\rho^2}{\sqrt{e^\rho}} \frac{1}{\sqrt{f(x)}} \right) \right\} = 0$$

again since $f(x) \to 0$ as $x \to \infty$. The two limit relations prove the lemma. ■

Proof of Theorem 1. (i) Choose $\alpha > 1/A$ and let $\epsilon > 0$ be small enough so that Lemma 3 holds for $\rho = 1 + \epsilon$. Then by (3.1),

$$P\{V > x\} \geq P\{V(x) > (1 + \epsilon)x, \ |\bar{V}(x)| < \epsilon x\}$$

$$\geq P\{V(x) > (1 + \epsilon)x\} - P\{|\bar{V}(x)| \geq \epsilon x\}.$$

Now Lemmas 3 and 4 imply that

$$\lim_{x \to \infty} \exp(\alpha x \log x) P\{V > x\} = \infty,$$

from which the statement follows. ■

(iii) For any $0 < \epsilon < 1$ and $x > 0$, by (3.1) we have

$$P\{V \leq -x\} \leq P\{V(x) \leq -(1 - \epsilon)x\} + P\{-\bar{V}(x) \geq \epsilon x\}.$$

Since by (3.2) and (3.4),

$$V(x) = \int_Y^{x^2} N(s) d\phi(s) - o(x) \quad \text{as} \quad x \to \infty,$$

we notice that the first probability in this bound becomes zero for all x large enough. Hence the statement follows from Lemma 4. ∎

(ii) The proof of this statement is more analytical. It goes as the proof of Lemma 1 of Sato [15] and hence uses arguments from the proof of Theorem 1 of Zolotarev [21]. These calculations performed with the moment generating function actually go back to Cramér [4]. Our presentation is self–contained.

First of all note that the condition $A < \infty$ implies that

(3.5)
$$\int_0^\epsilon s d\phi(s) < \infty \quad \text{for all} \quad \epsilon > 0.$$

Introduce the random variable

$$W = \int_0^\infty (N(s) - s) d\phi(s)$$

and note that by (1.3),

(3.6)
$$W = V - \int_0^1 s d\phi(s).$$

Again, arguing as in the proof of Theorem 3 in [5], we find that

(3.7)
$$E \exp(tW) = \exp(\xi(t)), \quad t \in \mathbb{R},$$

where

$$\xi(t) = \int_0^\infty \{\exp(-t\phi(s)) - 1 + t\phi(s)\} \, ds.$$

Since $|\exp(-v) - 1 + v| \leq v^2 \exp(|v|)/2$ for all $v \in \mathbb{R}$, we see that

$$|\xi(t)| \leq \left(\frac{1}{2} \int_0^\infty \phi^2(s) ds\right) t^2 e^{A|t|}, \quad t \in \mathbb{R}.$$

By (1.2) this says that, due to the condition $A < \infty$, $E \exp(tW)$ is finite for all $t \in \mathbb{R}$.

Differentiating ξ we obtain

$$\xi'(t) = -\int_0^\infty \phi(s)\{\exp(-t\phi(s)) - 1\}\, ds, \quad t \in \mathbb{R},$$

and differentiating once more,

$$\xi''(t) = \int_0^\infty \phi^2(s)\exp(-t\phi(s))ds > 0, \quad t \in \mathbb{R},$$

from which we see that

$$\xi'(t) \downarrow \mu := \int_0^\infty \phi(s)ds \quad \text{as} \quad t \downarrow -\infty,$$

where $-\infty \leq \mu < 0$, $\xi'(0) = 0$, and $\xi'(t) \uparrow \infty$ as $t \uparrow \infty$. For any $\mu < x < \infty$, introduce the inverse to ξ':

(3.8) $\xi'(\eta(x)) = x.$

The function η is well defined on (μ, ∞) since ξ' is strictly increasing and continuous. Furthermore, by the inverse function theorem we have

(3.9) $\xi''(\eta(x))\eta'(x) = 1, \quad \mu < x < \infty,$

and we know from the above that $\eta(x) > 0$ if and only if $x > 0$.

Now by (3.7) and (3.8), for any $x > 0$,

$$P\{W \geq x\} = P\{\eta(x)W \geq \eta(x)x\} \leq \exp\{\xi(\eta(x)) - \eta(x)\xi'(\eta(x))\}.$$

Observe that

$$\xi(\eta(x)) - \eta(x)\xi'(\eta(x)) = \int_0^{\eta(x)} \xi'(s)ds - \eta(x)\xi'(\eta(x))$$

$$= -\int_0^{\eta(x)} s\xi''(s)ds$$

$$= -\int_0^x \eta(t)\xi''(\eta(t))\eta'(t)dt$$

$$= -\int_0^x \eta(t)dt$$

by (3.9). Thus for all $x > 0$,

$$(3.10) \qquad\qquad P\{W \geq x\} \leq \exp\left(-\int_0^x \eta(t)dt\right).$$

Since $\exp(v) - 1 \leq v \exp(v)$ for all $v > 0$,

$$\xi'(t) \leq \left(\int_0^\infty \phi^2(s)ds\right) t \exp(At) =: Bt \exp(At), \ t > 0,$$

from which it follows by (3.8) that $t \leq B\eta(t) \exp(A\eta(t))$, $t > 0$. But $\eta(t) \uparrow \infty$ as $t \uparrow \infty$, and thus the last inequality implies that for any $0 < \alpha < 1/A$ there exists an $x_0 \geq e$ such that $t \leq \exp(\eta(t)/\alpha)$ for all $t \geq x_0$. This says that $\alpha \log t \leq \eta(t)$ for all $t \geq x_0$. Substituting this bound into (3.10), we obtain for all $x \geq x_0$ that

$$P\{W \geq x\} \leq \exp\left\{-\int_0^{x_0} \eta(t)dt - \alpha \int_{x_0}^x \log t \, dt\right\}$$

$$\leq \exp\left\{-\alpha \int_{x_0}^x \log t \, dt\right\}$$

$$\leq \exp(\alpha x \log x).$$

By (3.6) this leads immdiately to the statement. ∎

(iv) Starting out from (1.3) again, we now use a different decomposition of V : for any $x > 1$ we have $V = U_1(x) + U_2(x)$, where

$$U_1(x) = \int_0^x N(s)d\phi(s) - \phi(x)N(x) + \phi(x)x - \int_1^x sd\phi(s)$$

and

$$U_2(x) = \int_x^\infty (N(s) - s)d\phi(s) + \phi(x)(N(x) - x)$$

$$= \int_x^\infty \{(N(s) - N(x)) - (s - x)\}d\phi(s).$$

Since (1.2) implies that

$$E\left(\int_x^\infty (N(s) - s)d\phi(s)\right)^2 = \int_x^\infty \int_x^\infty \min(s,t)d\phi(s)d\phi(t) \to 0$$

and

$$E(\phi(x)(N(x) - x))^2 = \phi^2(x)x \to 0$$

as $x \to \infty$, we have

(3.11) $$U_2(x) \to_P 0 \quad \text{as} \quad x \to \infty.$$

On the other hand, since N has independent increments, $U_1(x)$ and $U_2(x)$ are independent for each $x > 1$. Therefore, for any $\epsilon > 0$,

$$P\{V < -\int_1^x sd\phi(s) + \phi(x)x + \epsilon\}$$

$$\geq P\{U_1(x) = -\int_1^x sd\phi(s) + \phi(x)x, \ |U_2(x)| < \epsilon\}$$

$$= P\{N(x) = 0\} P\{|U_2(x)| < \epsilon\}$$

$$= \exp(-x) P\{|U_2(x)| < \epsilon\} > 0$$

for all x large enough.

If $\int_1^\infty sd\phi(s) = \infty$, then this inequality obviously implies that V is unbounded from below.

If $\int_1^\infty sd\phi(s) < \infty$, then

(3.12) $$\phi(x)x \to 0 \quad \text{as} \quad x \to \infty,$$

and hence the above inequality implies that for any $\beta > 0$,

$$P\left\{V < -\int_1^\infty sd\phi(s) + \beta\right\} > 0.$$

At the same time, it follows form (3.11), (3.12), and the law of large numbers that

$$P\left\{V \geq -\int_1^\infty sd\phi(s)\right\} = P\left\{\int_0^\infty N(s)d\phi(s) \geq 0\right\} = 1,$$

and hence the statement. ∎

(v) This was in fact proved above, and formally follows from (iv). ∎

Proof of Theorem 2. (i) Let $\alpha > 1/A_2$ be given, fix a $\gamma \in (0,1)$, and choose $\epsilon > 0$ so small that $\alpha > \alpha/(1+\epsilon) > \alpha' > 1/A_2$ for some $\alpha > \alpha' > 1/A_2$. Then, applying (i) of Theorem 1 to V_2, for all $x > 0$ large enough we have by (1.4) that

$$P\{X > x\} \geq P\{-V_1 + \sigma Z + \theta - \theta_1 + \theta_2 > -\epsilon x\}\, P\{V_2 > (1+\epsilon)x\}$$

$$\geq (1-\gamma)\exp\{-\alpha'(1+\epsilon)x\,\log((1+\epsilon)x)\}.$$

Multiplying this by $\exp(\alpha x \log x)$, we see that the resulting lower bound goes to infinity as $x \to \infty$, and hence (2.1) follows.

Next, suppose $0 < A_2 < \infty$, let $0 < \alpha < 1/A_2$ be given, and choose $0 < \epsilon < 1$ so small that $0 < \alpha < \alpha/(1-\epsilon) < \alpha'' < 1/A_2$ for some $\alpha < \alpha'' < 1/A_2$. Then, applying (ii) and (iii) of Theorem 1 to V_2 and V_1, respectively, for any $\tau > 0$ and for all $x > 0$ large enough,

$$P\{X \geq x\} \leq P\{V_2 + \theta - \theta_1 + \theta_2 \geq (1-\epsilon)x\} + P\left\{-V_1 \geq \frac{\epsilon}{2}x\right\} + P\left\{\sigma Z \geq \frac{\epsilon}{2}x\right\}$$

$$\leq \exp\{-\alpha''(1-\epsilon)x\,\log((1-\epsilon)x)\} + \exp(-\tau x^2) + \left(1 - \Phi\left(\frac{\epsilon}{2\sigma}x\right)\right).$$

Multiplying this by $\exp(\alpha x \log x)$, the resulting upper bound goes to zero as $x \to \infty$, and hence (2.2) follows.

Finally, assume that $A_2 = 0$ and $A_1, \sigma > 0$ and fix $0 < \epsilon < 1$. Let $0 < \epsilon' < \epsilon$ and choose τ so that $\tau > (2\sigma^2)^{-1}$. Then, applying (iii) of Theorem 1 to V_1,

$$P\{X \geq x\} = P\{-V_1 + \sigma Z + \theta - \theta_1 \geq x\}$$
$$\leq P\{-V_1 + \theta - \theta_1 \geq \epsilon'x\} + P\{\sigma Z \geq (1 - \epsilon')x\}$$
$$\leq \exp(-\tau x^2) + 1 - \Phi((1 - \epsilon')x/\sigma).$$

Multiplying this inequality by $(1 - \Phi((1 - \epsilon)x/\sigma))^{-1}$ and using standard upper and lower estimates for $1 - \Phi(\cdot)$, we see that the resulting upper bound converges to zero as $x \to \infty$. This proves (2.3). ∎

(ii) This is completely analogous to (i). ∎

(iii) Sufficiency follows from part (v) of Theorem 1. Conversely, suppose that X is bounded from below. Then (2.4) implies that $A_1 = 0$. Obviously there exists a $y > 0$ such that $P\{V_2 < y\} > 0$. thus, for any $x > 0$,

$$P\{X < -x\} = P\{\sigma Z + V_2 + \theta + \theta_2 < -x\}$$
$$\geq P\{\sigma Z + \theta + \theta_2 < -x - y\}\, P\{V_2 < y\} > 0,$$

unless $\sigma = 0$. Hence X is distributed as $V_2 + \theta + \theta_2$, and by (v) and (iv) of Theorem 1, we have $\int_1^\infty s d\phi_2(s) < \infty$ and

$$ess\ inf\ X = -\int_1^\infty s d\phi_2(s) + \theta + \theta_2 = \theta + \int_0^\infty \frac{\phi_2(s)}{1 + \phi_2^2(s)}\, ds.\ \ ∎$$

(iv) This is completely analogous to (iii). ∎

References

[1] BAXTER, G. and SHAPIRO, J. M. (1960). On bounded infinitely divisible random variables. *Sankhyā* **22**, 253–260.

[2] BINGHAM, N. H., GOLDIE, C. M. and TEUGELS, J. L. (1987). *Regular Variation.* Cambridge Univ. Press, Cambridge.

[3] CHATTERJEE, S. D. and PAKSHIRAJAN, R. P. (1956). On the unboundedness of infinitely divisible laws. *Sankhyā* **17**, 349–350.

[4] CRAMÉR, H. (1938). Sur un nouveau théorème–limite de la théorie des probabilités. *Actualités Sci. et Indust.* No. **736**, Paris.

[5] CSÖRGŐ, S., HAEUSLER, E. and MASON, D. M. (1988). A probabilistic approach to the asymptotic distribution of sums of independent, identically distributed random variables. *Adv. in Appl. Math.* **9**, 259–333.

[6] CSÖRGŐ, S., HAEUSLER, E. and MASON, D. M. (1989).

[7] ELLIOTT, P. D. T. A. and ERDŐS, P. (1979). The tails of infinitely divisible laws and a problem in number theory. *J. Number Theory* **11**, 542–551.

[8] ESSEEN, C. -G. (1965). On infinitely divisible one–sided distributions. *Math. Scand.* **17**, 65–76.

[9] ESSEEN, C. -G. (1981). On the tails of a class of infinitely divisible distributions. In: *Contributions to Probability.* A Collection of Papers Dedicated to Eugene Lukacs (J. Gani and V. K. Rohatgi, eds.), pp. 115–122. Academic Press, New York.

[10] HORN, R. A. (1972). On necessary and sufficient conditions for an infinitely divisible distribution to be normal or degenerate. *Z. Wahrsch. Verw. Gebiete* **21**, 179–187.

[11] KRUGLOV, V. M. (1970). A note on infinitely divisible distributions. *Theory Probab. Appl.* **15**, 319–324.

[12] MASON, D. M. and SHORACK, G. R., Non–normality of a class of random variables. In this volume.

[13] RUEGG, A. F. (1970). A characterization of certain infinitely divisible laws. *Ann. Math. Statist.* **41**, 1354–1356.

[14] RUEGG, A. F. (1971). A necessary condition on the infinite divisibility of probability distributions. *Ann. Math. Statist.* **42**, 1681–1685.

[15] SATO, K. (1973). A note on infinitely divisible distributions and their Lévy measures. *Sci. Rep. Tokyo Kyoiku Daigaku Sect. A* **12**, 101–109.

[16] STEUTEL, F. W. (1974). On the tails of infinitely divisible distributions. *Z. Wahrsch. Verw. Gebiete* **28**, 273–276.

[17] TUCKER, H. G. (1961). Best one–sided bounds for infinitely divisible random variables. *Sankhyā* **23**, 387–396.

[18] WOLFE, S. J. (1971). On moments of infinitely divisible distribution functions. *Ann. Math. Statist.* **42**, 2036–2043.

[19] YAKYMIV, A. L. (1987). The asymptotic behavior of a class of infinitely divisible distributions. *Theory Probab. Appl.* **32**, 628–639.

[20] ZOLOTAREV, V. M. (1961). On the asymptotic behavior of a class of infinitely divisible distribution laws. *Theory Probab. Appl.* **6**, 304–307.

[21] ZOLOTAREV, V. M. (1965). Asymptotic behavior of the distributions of processes with independent increments. *Theory Probab. Appl.* **10**, 28–44.

Sándor Csörgő
Department of Statistics
University of Michigan
Ann Arbor, Michigan 48109

David M. Mason
Department of Mathematical Sciences
University of Delaware
Newark, Delaware 19716

THE QUANTILE–TRANSFORM – EMPIRICAL–PROCESS
APPROACH TO LIMIT THEOREMS
FOR SUMS OF ORDER STATISTICS

Sándor Csörgő,[1] Erich Haeusler and David M. Mason[2]

1. Introduction

Let $X, X_1, X_2, \ldots$ be independent, real–valued non–degenerate random variables with the common distribution function $F(x) = P\{X \leq x\}$, $x \in \mathbf{R}$, and introduce the inverse or quantile function Q of F defined as

$$Q(s) = \inf\{x : F(x) \geq s\}, \quad 0 < s \leq 1, \quad Q(0) = Q(0+) .$$

Our motivating point of departure in the study of the asymptotic distribution of sums of various ordered portions of the sample $X_1, \ldots, X_n$ is the following simple fact: If $U_1, U_2 \ldots$ are independent random variables on a probability space $(\Omega, \mathcal{F}, P)$ uniformly distributed in $(0, 1)$, then for each n the distributional equality

$$(1.1) \qquad \sum_{j=1}^{n} X_j \overset{\mathcal{D}}{=} \sum_{j=1}^{n} Q(U_j) = n \int_0^1 Q(s) \, dG_n(s)$$

holds, where $G_n(s) = n^{-1} \#\{1 \leq j \leq n : U_j \leq s\}$ is the uniform empirical distribution function on $(0, 1)$. More generally, if $X_{1,n} \leq \cdots \leq X_{n,n}$ are the order statistics

[1] Partially supported by the Hungarian National Foundation for Scientific Research, Grants 1808/86 and 457/88.
[2] Partially supported by the Alexander von Humboldt Foundation and NSF Grant #DMS-8803209.

Let

$$G(x) = P\{|X| \le x\}, \ -\infty < x < \infty,$$

be the distribution function of $|X|$, with corresponding quantile function

$$K(s) = \inf\{x : G(x) \ge s\}, \ 0 < s \le 1, \ K(0) = K(0+).$$

Moreover, for each integer $n \ge 1$ and $-\infty < x < \infty$ set

$$\tau_n^2(x) = n \int_0^{u_n(x)} K^2(u)du,$$

where $u_n(x) = \{(n - k_n + xk_n^{\frac{1}{2}})n^{-1} \vee 0\} \wedge 1$ with $\vee$ and $\wedge$ standing for maximum and minimum.

The following theorem yields the direct halves of Theorems 3.2, 3.4, 3.5, 3.7, 3.9 and 3.10 of Griffin and Pruitt [4] under the cleaner forms of their conditions, which correspond to those given in their paper when the underlying distribution function is continuous. In what follows, the symbols $=_D$ and $\to_D$ will mean equality and convergence in distribution, respectively.

THEOREM. *Let $0 \le k_n \le n - 1$ for $n \ge 1$ be a sequence of positive integers satisfying (1.1). Assume that for a non-degenerate distribution function F, symmetric about zero, and a subsequence $\{n'\} \subset \{n\}$ there exist $A_{n'} > 0$ and $B_{n'}$ such that for some $-\infty < C < \infty$,*

(1.2) $-B_{n'}/A_{n'} \to C \ as \ n' \to \infty$

and for all $-\infty < x < \infty$

(1.3) $v_{n'}^2(x) := \tau_{n'}^2(x)/A_{n'}^2 \to v^2(x) \ as \ n' \to \infty,$

for a (necessarily non-decreasing convex) function v^2 which is not identically equal to zero. Then

(1.4) $A_{n'}^{-1}\{{}^{(k_{n'})}S_{n'} - B_{n'}\} \to_D V$ as $n' \to \infty$,

where the limiting random variable V is the variance mixture of normals with characteristic function

$$E\{\exp(itV)\} = \exp(iCt) \int_{-\infty}^{\infty} \exp(-\frac{v^2(x)t^2}{2}) d\Phi(x), \quad -\infty < t < \infty,$$

where Φ is the standard normal distribution function.

The proof, postponed until the next section, is based on a weighted approximation to a 'signed' empirical process, which is likely to be of separate interest. See the Proposition in Section 2.

From our theorem it is easy to infer that a sufficient condition for the asymptotic normality of ${}^{(k_{n'})}S_{n'}$ is that for all $-\infty < x < \infty$,

(1.5) $\tau_{n'}^2(x)/\tau_{n'}^2(0) \to 1$ as $n' \to \infty$,

implying that

$${}^{(k_{n'})}S_{n'}/\tau_{n'}(0) \to_D V,$$

where V has the characteristic function for $-\infty < t < \infty$,

$$E\{\exp(itV)\} = \int_{-\infty}^{\infty} \exp(-\frac{t^2}{2}) d\Phi(x) = e^{\frac{-t^2}{2}},$$

saying that V is standard normal. (Arguing as in Griffin and Pruitt [4], it can be shown that (1.5) is also necessary for the asymptotic non-degenerate normality of ${}^{(k_{n'})}S_{n'}$.)

It is elementary to show that (1.5) is equivalent to

(1.6) $$k_{n'}^{1/2} K^2(u_{n'}(x))/\tau_{n'}^2(0) \to 0 \quad \text{as} \quad n' \to \infty$$

for all $-\infty < x < \infty$. From this form of the normality condition it can be seen that stochastic compactness of the whole sums S_n, which in the present symmetric case is equivalent to

(1.7) $$\limsup_{s \downarrow 0} s K^2(1-s) \Big/ \int_0^{1-s} K^2(u) du < \infty,$$

implies that $^{(k_n)}S_n/\tau_n(0)$ converges in distribution to a standard normal random variable for all sequences k_n satisfying (1.1) as $n \to \infty$. (Refer to Corollary 10 and (3.13) in [1].) That stochastic compactness of the whole sums S_n implies asymptotic normality of $^{(k_n)}S_n$ for all sequences k_n as in (1.1) was first pointed out by Pruitt [7].

2.   Proof. In the proof of the theorem we work on a probability space that carries a sequence $U, U_1, U_2, \ldots$, of independent random variables uniformly distributed on (0,1) and, independent of this sequence, a sequence $s, s_1, s_2, \ldots$, of independent and identically distributed random signs, that is, $P\{s = 1\} = P\{s = -1\} = \frac{1}{2}$. Since $K(U) =_{\mathcal{D}} |X|$, it is elementary to show using symmetry that

$$(X, |X|) =_{\mathcal{D}} (sK(U), K(U)),$$

from which we have immediately that

$$\{(X_i, |X_i|) : i \geq 1\} =_{\mathcal{D}} \{(s_i K(U_i), K(U_i)) : i \geq 1\}.$$

For each $n \geq 1$, let $U_{1,n} \leq \ldots \leq U_{n,n}$ denote the order statistics of $U_1, \ldots, U_n$. Let $D_{1,n}, \ldots, D_{n,n}$ denote the antiranks of $U_1, \ldots, U_n$, i.e. $U_{D_{i,n}} = U_{i,n}$ for $1 \leq i \leq n$.

Then from the above distributional equality we easily obtain that for each fixed $n \geq 1$,

$$({}^{(n)}X_n, \ldots, {}^{(1)}X_n) =_{\mathcal{D}} (s_{D_{1,n}}K(U_{1,n}), \ldots, s_{D_{n,n}}K(U_{n,n})),$$

so that

$$({}^{(k_n)}S_n =_{\mathcal{D}} \sum_{i=1}^{n-k_n} s_{D_{i,n}}K(U_{i,n}).$$

For each $n \geq 1$, we define the 'signed' empirical process

$$H_n(t) = n^{-1} \sum_{i=1}^{n} s_i I(\{U_i \leq t\}), \ 0 \leq t \leq 1,$$

where $I(A)$ denotes the indicator of a set A. Notice that almost surely for $n \geq 1$,

$$H_n(t) = n^{-1} \sum_{i=1}^{n} s_{D_{i,n}} I(\{U_{i,n} \leq t\}), \ 0 \leq t \leq 1.$$

Therefore we can write for each integer $n \geq 1$,

(2.1) $$({}^{(k_n)}S_n =_{\mathcal{D}} n \int_0^{U_{n-k_n,n}} K(u)dH_n(u).$$

We now introduce the uniform empirical distribution function

$$G_n(s) = n^{-1}\#\{1 \leq i \leq n : U_i \leq s\}, \ 0 \leq s \leq 1,$$

and the empirical quantile function

$$U_n(s) = \begin{cases} U_{k,n}, & \frac{k-1}{n} < s \leq \frac{k}{n}, \ k = 1, \ldots, n, \\ U_{1,n}, & s = 0. \end{cases}$$

The following proposition provides a joint weighted approximation of H_n, G_n and U_n, which will be essential to our approach to the asymptotic distribution of $({}^{(k_n)}S_n - B_n)/A_n$.

PROPOSITION. *On a rich enough probability space there exists a sequence of probabilistically equivalent versions* $(\tilde{H}_n, \tilde{G}_n, \tilde{U}_n)$ *of* (H_n, G_n, U_n), *meaning that for each integer* $n \geq 1$,

$$(\tilde{H}_n, \tilde{G}_n, \tilde{U}_n) =_{\mathcal{D}} (H_n, G_n, U_n),$$

such that for a Brownian bridge B *and a standard Wiener process* W, *with* B *and* W *independent, for all* $0 < \nu < \frac{1}{2}$,

$$(2.2) \qquad \sup_{0 \leq t \leq 1} |n^{\frac{1}{2}} \{\tilde{G}_n(t) - t\} - B(t)| / (1-t)^{\frac{1}{2} - \nu} = \mathcal{O}_P(n^{-\nu}),$$

$$(2.3) \qquad \sup_{0 \leq t \leq 1 - n^{-1}} |n^{\frac{1}{2}} \tilde{H}_n(t) - W(t)| / (1-t)^{\frac{1}{2} - \nu} = \mathcal{O}_P(n^{-\nu}),$$

and for all $0 < \delta < \frac{1}{4}$,

$$(2.4) \qquad \sup_{0 \leq t \leq 1 - n^{-1}} |n^{\frac{1}{2}} \{t - \tilde{U}_n(t)\} - B(t)| / (1-t)^{\frac{1}{2} - \delta} = \mathcal{O}_P(n^{-\delta}).$$

Proof. The proof will be a consequence of the following lemmas. Our first lemma can be derived easily from results and a starting element of the proof in Mason and van Zwet [6].

LEMMA 1. *On a rich enough probability space there exist independent Brownian bridges* B_1, B_2 *and* B_3 *and a triangular array of row-wise independent uniform* $(0,1)$ *random variables* $\xi_1^{(n)}(1), \ldots, \xi_n^{(n)}(1), \xi_1^{(n)}(2), \ldots, \xi_n^{(n)}(2), \xi_1^{(n)}(3), \ldots, \xi_n^{(n)}(3)$, $n \geq 1$, *such that for all* $0 < \nu < \frac{1}{2}$,

$$(2.5) \qquad \sup_{0 \leq t \leq 1} |n^{\frac{1}{2}} \{G_n^{(i)}(t) - t\} - B_i(t)| / (1-t)^{\frac{1}{2} - \nu} = \mathcal{O}_P(n^{-\nu})$$

and

(2.6) $$\left| n^{\frac{1}{2}}\{G_n^{(3)}(\tfrac{1}{2}) - \tfrac{1}{2}\} - B_3(\tfrac{1}{2}) \right| = \mathcal{O}_P(n^{-\frac{1}{2}}),$$

where for $i = 1, 2, 3$, $n \geq 1$,

$$G_n^{(i)}(t) = n^{-1} \sum_{j=1}^{n} I(\{\xi_j^{(n)}(i) \leq t\}), \; 0 \leq t \leq 1.$$

Set now for $n \geq 1$,

$$N_n = \sum_{j=1}^{n} I(\{\xi_j^{(n)}(3) \leq \tfrac{1}{2}\}) \quad \text{and} \quad M_n = n - N_n.$$

Let for $n \geq 1$ and $0 \leq t \leq 1$,

$$n\tilde{G}_n(t) = N_n G_{N_n}^{(1)}(t) + M_n G_{M_n}^{(2)}(t)$$

and

$$n\tilde{H}_n(t) = N_n G_{N_n}^{(1)}(t) - M_n G_{M_n}^{(2)}(t).$$

(We define $G_0^{(i)} \equiv 0$ for $i = 1, 2$.) Also define

$$B(t) = 2^{-\frac{1}{2}}\{B_1(t) + B_2(t)\}, \; 0 \leq t \leq 1,$$

and

$$W(t) = 2^{-\frac{1}{2}}\{B_1(t) - B_2(t)\} + 2tB_3(\tfrac{1}{2}), \; 0 \leq t \leq 1.$$

It is simple to verify that B and W are independent processes with B being a Brownian bridge and W a standard Wiener process.

LEMMA 2. *On the probability space of Lemma 1, for all $0 < \nu < \tfrac{1}{2}$,*

(2.7) $$\sup_{0 \leq t \leq 1} |n^{\frac{1}{2}}\{\tilde{G}_n(t) - t\} - B(t)|/(1-t)^{\frac{1}{2}-\nu} = \mathcal{O}_P(n^{-\nu})$$

and

$$(2.8) \qquad \sup_{0 \le t \le 1 - n^{-1}} |n^{\frac{1}{2}} \tilde{H}_n(t) - W(t)|/(1-t)^{\frac{1}{2}-\nu} = \mathcal{O}_P(n^{-\nu}).$$

Proof. Obviously, since both

$$n^{-1} N_n \to_P \frac{1}{2} \quad \text{and} \quad n^{-1} M_n \to_P \frac{1}{2} \quad \text{as } n \to \infty,$$

we have by (2.5), for all $0 < \nu < \frac{1}{2}$,

$$(2.9) \qquad \sup_{0 \le t \le 1} |N_n^{\frac{1}{2}} \{G_{N_n}^{(1)}(t) - t\} - B_1(t)|/(1-t)^{\frac{1}{2}-\nu} = \mathcal{O}_P(n^{-\nu})$$

and

$$(2.10) \qquad \sup_{0 \le t \le 1} |M_n^{\frac{1}{2}} \{G_{M_n}^{(2)}(t) - t\} - B_2(t)|/(1-t)^{\frac{1}{2}-\nu} = \mathcal{O}_P(n^{-\nu}).$$

Also, by (2.6),

$$(2.11) \qquad \sup_{0 \le t \le 1 - n^{-1}} \frac{t}{(1-t)^{\frac{1}{2}-\nu}} |\frac{2N_n - n}{n^{\frac{1}{2}}} - 2B_3(\frac{1}{2})| = \mathcal{O}_P(n^{-\nu}).$$

Moreover, it is easily checked that

$$2^{-\frac{1}{2}} - (N_n/n)^{\frac{1}{2}} = \mathcal{O}_P(n^{-\frac{1}{2}}) \quad \text{and} \quad 2^{-\frac{1}{2}} - (M_n/n)^{\frac{1}{2}} = \mathcal{O}_P(n^{-\frac{1}{2}}).$$

In addition, by well-known properties of the Brownian bridge, for $i = 1, 2$,

$$\sup_{0 \le t \le 1} |B_i(t)|/(1-t)^{\frac{1}{2}-\nu} = \mathcal{O}_P(1).$$

Thus,

$$(2.12) \qquad \sup_{0 \le t \le 1} |(\frac{N_n}{n})^{\frac{1}{2}} B_1(t) - 2^{-\frac{1}{2}} B_1(t)|/(1-t)^{\frac{1}{2}-\nu} = \mathcal{O}_P(n^{-\frac{1}{2}})$$

and

$$(2.13) \qquad \sup_{0 \le t \le 1} |(\frac{M_n}{n})^{\frac{1}{2}} B_2(t) - 2^{-\frac{1}{2}} B_2(t)|/(1-t)^{\frac{1}{2}-\nu} = \mathcal{O}_P(n^{-\frac{1}{2}}).$$

Using (2.9), (2.10), (2.11), (2.12) and (2.13), a little algebra now gives (2.7) and (2.8). $\square$

For each integer $n \ge 1$, let $\tilde{U}_{1,n} \le \cdots \le \tilde{U}_{n,n}$ be the order statistics of the random variables $\xi_1^{(N_n)}(1), \ldots, \xi_{N_n}^{(N_n)}(1), \xi_1^{(M_n)}(2), \ldots, \xi_{M_n}^{(M_n)}(2)$. Define the empirical quantile function $\tilde{U}_n(t)$ based on these order statistics in the usual way. The following lemma is a consequence of the fact that for all $0 < \delta < \frac{1}{4}$,

$$\sup_{0 \le t \le 1-n^{-1}} n^\delta |\tilde{G}_n(t) + \tilde{U}_n(t) - 2t|/(1-t)^{\frac{1}{2}-\delta} = \mathcal{O}_P(1),$$

which can be readily inferred from Corollary 2.3 in [2] (also see the Proposition in Mason [5]).

LEMMA 3. *On the probability space of Lemma 1, for all $0 < \delta < \frac{1}{4}$,*

$$(2.14) \qquad \sup_{0 \le t \le 1-n^{-1}} |n^{\frac{1}{2}} \{t - \tilde{U}_n(t)\} - B(t)|/(1-t)^{\frac{1}{2}-\delta} = \mathcal{O}_P(n^{-\delta}).$$

Since it is routine to verify that for each integer $n \ge 1$, $(\tilde{H}_n, \tilde{G}_n, \tilde{U}_n) =_D (H_n, G_n, U_n)$, assertions (2.2), (2.3) and (2.4) follow from (2.7), (2.8) and (2.14). This completes the proof of the Proposition. $\square$

Returning to the proof of the Theorem, from now on, since we are only concerned with distributional results, on account of (2.1) for the sake of the proof we can and do identify $(\tilde{H}_n, \tilde{G}_n, \tilde{U}_n)$ with (H_n, G_n, U_n); that is, we work on the probability space of the Proposition. Also for notational convenience we drop the prime on the n'. The proof of the Theorem requires a number of lemmas.

LEMMA 4. *Under (1.3), for each* $0 < M < \infty$,

$$(2.15) \qquad\qquad \limsup_{n\to\infty} k_n^{\frac{1}{4}} K(u_n(M))/A_n < \infty.$$

Proof. Fix any finite $\overline{M} > M$. Then for each $n \geq 1$,

$$A_n^{-2}\tau_n^2(\overline{M}) = A_n^{-2}\tau_n^2(M) + nA_n^{-2}\int_{u_n(M)}^{u_n(\overline{M})} K^2(u)du,$$

from which by letting $n \to \infty$ and using monotonicity of K we get

$$v^2(\overline{M}) - v^2(M) = \lim_{n\to\infty} nA_n^{-2}\int_{u_n(M)}^{u_n(\overline{M})} K^2(u)\,du$$
$$\geq \limsup_{n\to\infty} nA_n^{-2}K^2(u_n(M))(\overline{M} - M)k_n^{1/2}n^{-1},$$

implying (2.15). $\square$

For each $n \geq 1$, set

$$D_n = \{0 < u_n(-Z_n) < 1\},$$

where

$$Z_n = (\frac{n}{k_n})^{\frac{1}{2}}B(1 - \frac{k_n}{k}) =_{\mathcal{D}} N(0, 1 - \frac{k_n}{n}).$$

Clearly, we have

$$(2.16) \qquad\qquad P\{D_n\} \to 1 \text{ as } n \to \infty.$$

LEMMA 5. *Under (1.3) we have as* $n \to \infty$,

$$(2.17) \quad nA_n^{-1} \int_0^{U_{n-k_n,n}} K(u)dH_n(u) = nA_n^{-1} \int_0^{u_n(-Z_n)} K(u)dH_n(u)I(D_n) + o_P(1).$$

Proof. Because of (2.16) the assertion will follow from

$$(2.18) \quad nA_n^{-1} \int_{u_n(-Z_n)}^{U_{n-k_n,n}} K(u)dH_n(u)I(D_n) \to_P 0 \quad \text{as} \quad n \to \infty.$$

Towards a proof of (2.18) notice that from the weighted approximation (2.4) for $U_n(\cdot)$, for any $0 < \delta < \frac{1}{4}$,

$$(2.19) \quad U_{n-k_n,n} - u_n(-Z_n) = \mathcal{O}_P(k_n^{\frac{1}{2}-\delta}/n),$$

which in combination with the fact that $Z_n = \mathcal{O}_P(1)$ gives

$$(2.20) \quad U_{n-k_n,n} = 1 - \frac{k_n}{n} + \mathcal{O}_P(k_n^{\frac{1}{2}}/n).$$

Next, for fixed $0 < \delta < \frac{1}{4}$ and $0 < M < \infty$, set

$$A_n(M) = \{u_n(-M) \le U_{n-k_n,n} \le u_n(M)\}$$

and

$$B_{n,\delta}(M) = \{|U_{n-k_n,n} - u_n(-Z_n)| \le Mk_n^{\frac{1}{2}-\delta}/n\}.$$

Then (2.19) and (2.20) imply

$$(2.21) \quad \lim_{M \to \infty} \liminf_{n \to \infty} P\{A_n(M) \cap B_{n,\delta}(M)\} = 1.$$

On the event $D_n \cap A_n(M) \cap B_{n,\delta}(M) =: E_n$, we have for all large n,

$$(2.22) \quad u_n(-2M) \le u_n(-Z_n) \le u_n(2M)$$

and we get the estimate after integrating by parts

$$\left| nA_n^{-1} \int_{u_n(-Z_n)}^{U_{n-k_n,n}} K(u)dH_n(u)I(E_n) \right| = nA_n^{-1}I(E_n)|K(U_{n-k_n,n})\{H_n(U_{n-k_n,n})$$

$$- H_n(u_n(-Z_n))\}$$

$$+ \int_{u_n(-Z_n)}^{U_{n-k_n,n}} \{H_n(u_n(-Z_n)) - H_n(u)\}dK(u)|$$

$$\leq 2k_n^{\frac{1}{4}}A_n^{-1}K(u_n(2M))nk_n^{-\frac{1}{4}}\Delta_n(M),$$

where

$$\Delta_n(M) = \sup_{u_n(-M)\leq b\leq u_n(M)} \sup_{0\leq|h|\leq Mk_n^{\frac{1}{2}-\delta}/n} |H_n(b+h) - H_n(b)|.$$

In view of Lemma 4, (2.16) and (2.21), the proof of (2.18) and hence of (2.17) will be complete if we show that for all $0 < \delta < \frac{1}{4}$ and $0 < M < \infty$,

(2.23) $$\qquad\qquad nk_n^{-\frac{1}{4}}\Delta_n(M) \to_P 0 \text{ as } n \to \infty.$$

Consider a grid

$$u_n(-M) = b_0 < b_1 < \cdots < b_\ell < \cdots < b_L = u_n(M)$$

such that

$$b_\ell - b_{\ell-1} = Mk_n^{\frac{1}{2}-\delta}/n \text{ for } \ell = 1,\ldots,L-1$$

and

$$b_L - b_{L-1} \leq Mk_n^{\frac{1}{2}-\delta}/n.$$

For each $b \in [u_n(-M), u_n(M))$ there exists exactly one $\ell \in \{1,\ldots,L\}$ such that $b_{\ell-1} \leq b < b_\ell$. This fact gives the bounds

$$nk_n^{-\frac{1}{4}}\Delta_n(M) \leq 2nk_n^{-\frac{1}{4}} \max_{1\leq\ell\leq L} \sup_{0\leq|h|\leq 2Mk_n^{\frac{1}{2}-\delta}/n} |H_n(b_{\ell-1}+h) - H_n(b_{\ell-1})|$$

$$= 2k_n^{-\frac{1}{4}}n^{\frac{1}{2}} \max_{1\leq\ell\leq L} \omega_n(2Mk_n^{\frac{1}{2}-\delta}/n, b_{\ell-1}),$$

where

$$\omega_n(a, b) = \sup_{0 \le |h| \le a} n^{\frac{1}{2}} |H_n(b + h) - H_n(b)|.$$

To finish the proof of (2.23) it suffices to show that for each fixed $0 < M < \infty$ and $0 < \delta < \frac{1}{4}$ the right side of the last inequality converges to zero in probability. For this, fix $\epsilon > 0$ and note that

$$P\{n^{\frac{1}{2}}k_n^{-\frac{1}{4}} \max_{1 \le \ell \le L} \omega_n(2Mk_n^{\frac{1}{2}-\delta}/n, b_{\ell-1}) \ge \epsilon\} \le \sum_{\ell=1}^{L} P\{\omega_n(2Mk_n^{\frac{1}{2}-\delta}/n, b_{\ell-1}) \ge \epsilon k_n^{\frac{1}{4}} n^{-\frac{1}{2}}\},$$

which by Theorem 1.1 of Einmahl and Ruymgaart [3] is

$$\le CL \exp\{-\frac{A\epsilon^2}{2M} k_n^\delta \psi(\frac{B\epsilon}{4M} k_n^{\delta-\frac{1}{4}})\} =: I_n(\epsilon, M),$$

where A, B and C are universal positive constants and ψ is an increasing function on $(0, \infty)$ such that $\psi(z) \downarrow 1$ as $z \downarrow 0$. This when combined with the easy bound $L \le 2k_n^\delta + 1$ yields $I_n(\epsilon, M) \to 0$ as $n \to \infty$, assuming that $0 < \delta < \frac{1}{4}$. This completes the proof of (2.23) from which (2.17) follows. □

Finally, we require one more lemma.

LEMMA 6. *Under (1.3) we have*

$$nA_n^{-1} \int_0^{u_n(-Z_n)} K(u)dH_n(u)I(D_n)$$

$$= A_n^{-1} \left\{ -n^{\frac{1}{2}} \int_0^{u_n(-Z_n)} W(u)dK(u) + n^{\frac{1}{2}}K(u_n(-Z_n))W(u_n(-Z_n)) \right\} I(D_n) + o_P(1).$$

Proof. On the event D_n we obtain after integrating by parts that the left side

$$= nA_n^{-1}K(u_n(-Z_n))H_n(u_n(-Z_n)) - nA_n^{-1}\int_0^{u_n(-Z_n)} H_n(u)dK(u)$$

$$= -n^{\frac{1}{2}}A_n^{-1}\int_0^{u_n(-Z_n)} W(u)dK(u)$$

$$+ n^{\frac{1}{2}}A_n^{-1}\int_0^{u_n(-Z_n)} \{W(u) - n^{\frac{1}{2}}H_n(u)\}dK(u)$$

$$+ n^{\frac{1}{2}}A_n^{-1}K(u_n(-Z_n))W(u_n(-Z_n))$$

$$+ n^{\frac{1}{2}}A_n^{-1}K(u_n(-Z_n))\{n^{\frac{1}{2}}H(u_n(-Z_n)) - W(u_n(-Z_n))\},$$

so that to finish the proof we must verify that

$$(2.24) \qquad n^{\frac{1}{2}}A_n^{-1}\int_0^{u_n(-Z_n)} \{W(u) - n^{\frac{1}{2}}H_n(u)\}dK(u)I(D_n) \to_P 0 \text{ as } n \to \infty$$

and

$$(2.25)$$

$$n^{\frac{1}{2}}A_n^{-1}K(u_n(-Z_n))\{n^{\frac{1}{2}}H_n(u_n(-Z_n)) - W(u_n(-Z_n))\}I(D_n) \to_P 0 \text{ as } n \to \infty.$$

For the proof of (2.24) fix $0 < \nu < \frac{1}{2}, 0 < \delta < \frac{1}{4}$ and $0 < M < \infty$. In view of (2.22) we obtain on the event E_n

$$|n^{\frac{1}{2}}A_n^{-1}\int_0^{u_n(-Z_n)} \{W(u) - n^{\frac{1}{2}}H_n(u)\}dK(u)|$$

$$\leq \left\{ \sup_{0 \leq u \leq 1-n^{-1}} \frac{|W(u) - n^{\frac{1}{2}}H_n(u)|}{(1-u)^{\frac{1}{2}-\nu}} \right\} n^{\frac{1}{2}}A_n^{-1}\int_0^{u_n(2M)} (1-u)^{\frac{1}{2}-\nu}dK(u),$$

which by (2.3) is equal to

$$(2.26) \qquad \qquad \mathcal{O}_P(n^{-\nu})n^{\frac{1}{2}}A_n^{-1}\int_0^{u_n(2M)} (1-u)^{\frac{1}{2}-\nu}dK(u).$$

Now by integrating by parts and some elementary bounds,

$$n^{\frac{1}{2}}A_n^{-1}\int_0^{u_n(2M)} (1-u)^{\frac{1}{2}-\nu}dK(u) \leq n^{\frac{1}{2}}A_n^{-1}K(u_n(2M))(k_n/n)^{\frac{1}{2}-\nu}$$

$$+ (\frac{1}{2} - \nu)n^{\frac{1}{2}}A_n^{-1}\int_0^{u_n(2M)} K(u)(1-u)^{-\nu-\frac{1}{2}}du,$$

which, by the Cauchy-Schwarz inequality applied to the second term above and by (1.3) and (2.15),

$$\leq n^\nu k_n^{\frac{1}{4}-\nu}\{k_n^{\frac{1}{4}}A_n^{-1}K(u_n(2M))\} + \mathcal{O}(1)A_n^{-1}\tau_n(2M)(k_n/n)^{-\nu}$$

$$= \mathcal{O}(n^\nu k_n^{\frac{1}{4}-\nu}).$$

This bound when combined with (2.26), (2.16) and (2.21) yields (2.24) as long as $\frac{1}{4} < \nu < \frac{1}{2}$.

To establish (2.25), we fix $\frac{1}{4} < \nu < \frac{1}{2}$ and $0 < M < \infty$. On the event E_n (recall (2.22)),

$$|n^{\frac{1}{2}}A_n^{-1}K(u_n(-Z_n))\{n^{\frac{1}{2}}H_n(u_n(-Z_n)) - W(u_n(-Z_n))\}|$$

$$\leq \sup_{0 \leq u \leq 1-n^{-1}} \frac{|n^{\frac{1}{2}}H_n(u) - W(u)|}{(1-u)^{\frac{1}{2}-\nu}} n^{\frac{1}{2}}A_n^{-1}K(u_n(2M))(1 - u_n(-2M))^{\frac{1}{2}-\nu},$$

which by (2.3), (2.15) and $\frac{1}{4} < \nu < \frac{1}{2}$,

$$= \mathcal{O}_P(1)k_n^{\frac{1}{4}-\nu}k_n^{\frac{1}{4}}A_n^{-1}K(u_n(2M)) = o_P(1),$$

proving (2.25). This completes the proof of Lemma 6. $\square$

According to Lemmas 5 and 6 combined with (2.1), the assertion of the Theorem will follow if we show that

$$V_n := A_n^{-1}\{-n^{\frac{1}{2}}\int_0^{u_n(-Z_n)} W(u)dK(u) + n^{\frac{1}{2}}K(u_n(-Z_n))W(u_n(-Z_n)) - B_n\}I(D_n)$$

$$\to_D V \quad \text{as} \quad n \to \infty.$$

We prove this using characteristic functions. Since

$$|E\{\exp(itV_n)(1 - I(D_n))\}| \leq 1 - P(D_n) \to 0 \quad \text{as} \quad n \to \infty,$$

we have

$$E\{\exp(itV_n)\} = E\{\exp(itV_n)I(D_n)\} + o(1) \text{ as } n \to \infty.$$

Notice that for $x \geq -(1 - k_n/n)nk_n^{-\frac{1}{2}}$ the random variable

$$-n^{\frac{1}{2}} \int_0^{u_n(x)} W(u)dK(u) + n^{\frac{1}{2}}K(u_n(x))W(u_n(x))$$

is normally distributed with mean zero and variance $\tau_n^2(x)$. Therefore by the independence of Z_n and W, for any t,

$$E\{\exp(itV_n)I(D_n)\} = \exp(-itA_n^{-1}B_n) \int_{-(1-\frac{k_n}{n})\frac{n}{\sqrt{k_n}}}^{\infty} \exp(-\frac{v_n^2(x)t^2}{2})dP(-Z_n \leq x).$$

From

$$-Z_n \to_D N(0,1) \text{ and } -(1 - \frac{k_n}{n})\frac{n}{\sqrt{k_n}} \to -\infty \text{ as } n \to \infty$$

it follows by conditions (1.2) and (1.3) that for any t,

$$E\{\exp(itV_n)I(D_n)\} \to \exp(iCt) \int_{-\infty}^{\infty} \exp(-\frac{v^2(x)}{2}t)d\Phi(x),$$

completing the proof of the Theorem. $\square$

REFERENCES

[1] CSÖRGŐ, S., HAEUSLER, E. and MASON, D.M. (1988). A probabilistic approach to the asymptotic distribution of sums of independent, identically distributed random variables. *Adv. in Appl. Math.* **9**, 259-333.

[2] CSÖRGŐ, M., CSÖRGŐ, S., HORVÁTH, L. and MASON, D.M. (1986). Weighted empirical and quantile processes. *Ann. Probab.* **14**, 31-85.

[3] EINMAHL, J.H.J. and RUYMGAART, F.H. (1986). Some properties of weighted compound multivariate empirical processes. *Sankhyā, Series A* 48, 393-403.

[4] GRIFFIN, P.S. and PRUITT, W.E. (1987). The central limit theorem for trimmed sums. *Math. Proc. Camb. Phil. Soc.* 102, 329-349.

[5] MASON, D.M. (1990). A note on weighted empirical and quantile processes. This volume.

[6] MASON, D.M. and VAN ZWET, W.R. (1987). A refinement of the KMT inequality for the uniform empirical process. *Ann. Probab.* 15, 871-884.

[7] PRUITT, W.E. (1988). Sums of independent random variables with the extreme terms excluded. In: *Probability and Statistics. Essays in Honor of Franklin A. Graybill* (J.N. Srivastava, Ed.), pp. 201-216, Elsevier, Amsterdam.

Sándor Csörgő   Erich Haeusler   David M. Mason

Dept. of Statistics   University of Munich   Dept. of Mathematical Sciences

University of Michigan   Therresienstrasse 39   University of Delaware

Ann Arbor, Mich. 48109   8000 Munich 2   Newark, Del. 19716

                          West Germany

ON THE ASYMPTOTIC BEHAVIOR OF SUMS OF ORDER STATISTICS FROM A DISTRIBUTION WITH A SLOWLY VARYING UPPER TAIL

Erich Haeusler[1] and David M. Mason[2]

Let $(X_n)_{n \geq 1}$ be a sequence of independent non–negative random variables from a common distribution function F with a regularly varying upper tail. A number of results are presented on the stability, asymptotic distribution and law of the iterated logarithm for trimmed sums formed by deleting a number of the upper extreme values from the partial sum $X_1 + \cdots + X_n$ at each stage n. The methods of proof are entirely based on quantile function techniques. This paper should provide the reader with a good introduction to some of the possibilities and the scope of this methodology.

1. Introduction and statements of results. Let $(X_n)_{n \geq 1}$ be a sequence of independent non–negative random variables with common non–degenerate distribution function F. For integers $0 \leq k \leq n - 1$, $n \geq 2$, consider the trimmed sums

$$S_n(k) = \sum_{i=1}^{n-k} X_{i,n},$$

---

[1] Research partially supported by the Deutsche Forschungsgemeinschaft while the author was visiting the University of Delaware.

[2] Research partially supported by the Alexander von Humboldt Foundation while the author was visiting the University of Munich and NSF Grant DMS–8803209.

where $X_{1,n} \leq \cdots \leq X_{n,n}$ denote the order statistics based on $X_1, \ldots, X_n$. When $k = 0$ we write $S_n = S_n(0)$. In this paper we shall investigate the asymptotic behavior of these trimmed sums when it is assumed that F has a slowly varying upper tail, which means that $1 - F(x) = L(x)$, where L is a slowly varying function at infinity, i.e. for all $0 < \rho < \infty$, $L(x\rho)/L(x) \to 1$ as $x \to \infty$.

Let

$$Q(s) = \inf\{x : F(x) \geq s\} \quad \text{for} \quad 0 < s \leq 1$$

and $Q(0) = Q(0+)$ denote the inverse or quantile function of F. Frequent use in this paper will be made of the fact that $1 - F(x)$ is slowly varying at infinity if and only if $Q(1 - s)$ is rapidly varying at zero, i.e. for all $0 < \rho < 1$

$$Q(1 - \rho s)/Q(1 - s) \to \infty \quad \text{as} \quad s \downarrow 0,$$

cf. Corollary 1.2.1.5 in de Haan (1975).

It has been known since Darling (1952) that when F has a slowly varying upper tail the maximum term completely dominates the entire sum in the sense that for non–negative summands

$$S_n/X_{n,n} \xrightarrow{P} 1 \quad \text{as} \quad n \to \infty.$$

Recently, among many other results, Maller and Resnick (1984) and Pruitt (1987) obtained necessary and sufficient conditions under which one has, in addition,

$$\lim_{n\to\infty} S_n/X_{n,n} = 1 \quad \text{a.s.}$$

Our Theorem 1 shows, in particular, that for non–negative summands the maximum completely dominates the whole sum if and only if F has a slowly varying upper

tail. Though Theorem 1 is contained in the results of Mori (1981) and Maler and
Resnick (1984), for didactic reasons we provide an alternate proof here based on
quantile function techniques. We mention that Darling (1952), Mori (1981), Maller
and Resnick (1984) and Pruitt (1987) consider the more general situation of not
necessarily non–negative summands and compare the partial sums with the term of
maximum absolute value.

*In the statements of all of our results that follow it is assumed implicitly that
F has positive support.*

THEOREM 1. *For any fixed integer $k \geq 0$, the following three statements are
equivalent:*

(1.1) $\quad\quad 1 - F(x) = L(x), \quad where \quad L \quad is\ slowly\ varying\ at\ infinity;$

(1.2) $\quad\quad\quad\quad\quad\quad X_{n-k-1,n}/X_{n-k,n} \xrightarrow{P} 0 \quad as \quad n \to \infty;$

(1.3) $\quad\quad\quad\quad\quad\quad S_n(k)/X_{n-k,n} \xrightarrow{P} 1 \quad as \quad n \to \infty.$

Our next result is an extension of a stability result due to Teicher (1979). The
case $k = 0$ of our Theorem 2 is Theorem 2 of Teicher (1979). For further results on
the almost sure stability of partial sums of independent and identically distributed
random variables from a distribution with a slowly varying upper tail see Pruitt
(1981), Zhang (1986), Révész and Willekens (1987) and Mikosch (1988). Also the
interested reader is referred to Einmahl, Haeusler and Mason (1988) for a study of
the almost sure stability of trimmed sums of i.i.d. non–negative random variables
in the domain of attraction of a non–normal stable law.

THEOREM 2. *Let $(b_n)_{n \geq 1}$ be a non-decreasing sequence of positive constants. Whenever (1.1) holds, then for any fixed integer $k \geq 0$*

(1.4) $\limsup\limits_{n \to \infty} S_n(k)/b_n = 0 \quad or \quad \infty \quad$ a.s.

according as the following series is finite or infinite

(1.5) $\sum\limits_{n=1}^{\infty} n^k (1 - F(b_n))^{k+1}.$

A distribution with positive support and a slowly varying upper tail is a classic example of a distribution which is not in the domain of partial attraction of any non–degenerate law. This means that if $(X_n)_{n \geq 1}$ is a sequence of independent non-negative random variables from a distribution with a slowly varying upper tail, it is impossible to find a subsequence $(n_j)_{j \geq 1} \subset (n)_{n \geq 1}$ and sequences of norming and centering constants $(A_{n_j})_{j \geq 1}$ and $(B_{n_j})_{j \geq 1}$ such that $A_{n_j}^{-1}(S_{n_j} - B_{n_j})$ converges in distribution as $j \to \infty$ to a non–degenerate law. From Corollary 6 of S. Csörgő, Haeusler and Mason (1988a) it can be easily inferred that the same is true for the trimmed sums $S_n(k)$ for fixed integers $k \geq 1$. The question naturally arises as to whether the partial sums can be slightly trimmed as a function of the sample size n so that the remaining sum when properly normalized converges in distribution to a non–degenerate law. The answer to this question is yes and in fact the non–degenerate law can be $N(0,1)$. In order to describe how this can happen, let $(k_n)_{n \geq 1}$ denote a sequence of integers such that

(K) $0 \leq k_n \leq n - 1 \quad$ for $\quad n \geq 1, \ k_n \to \infty \quad$ and $\quad k_n/n \to 0.$

Furthermore, for any such sequence of integers $(k_n)_{n \geq 1}$ satisfying (K) set

$$a_n(k_n) = k_n^{1/2} Q(1 - k_n/n) \quad \text{and} \quad \mu_n(k_n) = n \int_0^{1 - k_n/n} Q(s)\,ds.$$

An obvious modification of the proof of the Corollary in S. Csörgő, Haeusler and Mason (1988b) yields the following result concerning asymptotic normality of $S_n(k_n)$.

THEOREM A. *Let F be any distribution function as in (1.1) and $(k_n)_{n\geq 1}$ be any sequence of integers satisfying (K). Then*

$$(i) \qquad a_n(k_n)^{-1}(S_n(k_n) - \mu_n(k_n)) \overset{\mathcal{D}}{\to} N(0,1) \quad as \quad n \to \infty$$

if and only if

$$(ii) \qquad X_{n-k_n,n}/Q(1 - k_n/n) \overset{P}{\to} 1 \quad as \quad n \to \infty$$

if and only if for all c

$$(iii) \qquad Q(1 - k_n/n + ck_n^{1/2}/n)/Q(1 - k_n/n) \to 1 \quad as \quad n \to \infty.$$

REMARK 1. If k_n is equal to the integer part of nb for some $0 < b < 1$, for $n \geq 1$, and $1 - b$ is a continuity point of Q then $S_n(k_n)$ is asympototically normally distributed when appropriately normalized. Refer to Stigler (1973).

It is straightforward to obtain from a slight adaptation of the proof of Theorem A the following result which describes what can happen when condition (iii) of Theorem A does not hold.

THEOREM B. *Let F be any distribution function as in (1.1) and $(k_n)_{n\geq 1}$ be any sequence of integers satisfying (K). Whenever for all $0 \leq c_1 < c_2 < \infty$*

$$(i) \quad Q(1 - k_n/n + c_2 k_n^{1/2}/n)/Q(1 - k_n/n + c_1 k_n^{1/2}/n) \to \infty \quad as \quad n \to \infty,$$

then there exist sequences $(A_n)_{n\geq 1}$ and $(B_n)_{n\geq 1}$ of norming and centering constant such that the sequence of random variables

$$(ii) \qquad A_n^{-1}(S_n(k_n) - B_n)$$

is bounded in probability if and only if

(iii) $$A_n^{-1}(S_n(k_n) - \mu_n(k_n)) \xrightarrow{P} 0 \quad \text{as} \quad n \to \infty$$

and

(iv) $$\sup_{n \geq 1} |A_n^{-1}\{B_n - \mu_n(k_n)\}| < \infty.$$

Moreover, (iii) holds with $A_n = k_n^{1/2}Q(1 - k_n/n + \ell_n k_n^{1/2}/n)$ *whenever* $\ell_n \to \infty$ *and* $\ell_n/k_n^{1/2} \to 0$ *as* $n \to \infty$.

Notice that Theorem B says that under condition (i) there do not exist sequences of norming and centering constants $(A_n)_{n \geq 1}$ and $(B_n)_{n \geq 1}$ such that the sequence in (ii) converges in distribution to a non-degenerate law. If both (iii) of Theorem A and (i) of Theorem B do not hold it may still be possible to find sequences of norming and centering constant $(A_n)_{n \geq 1}$ and $(B_n)_{n \geq 1}$ so that $A_n^{-1}(S_n(k_n) - B_n)$ converges in distribution to a non–degenerate random variable which is necessarily non–normal. See S. Csörgő, Haeusler and Mason (1988b).

Our final result provides a set of sufficient conditions for the law of the iterated logarithm to hold for $S_n(k_n)$. For our law of the iterated logarithm we need to impose some additional regularity conditions on the sequence $(k_n)_{n \geq 1}$. We require that

(K') $\quad 0 \leq k_n \leq n-1$, for $n \geq 1$, $k_n \sim \alpha_n \uparrow \infty$ and $k_n/n \sim \beta_n \downarrow 0$.

THEOREM 3. *Let F be any distribution function as in (1.1) and $(k_n)_{n \geq 1}$ be any sequence of integers satisfying (K'). Whenever*

(1.6) $$k_n/\log \log n \to \infty \quad \text{as} \quad n \to \infty$$

and

(1.7) $X_{n-k_n,n}/Q(1-k_n/n) \to 1 \quad as \quad n \to \infty \quad a.s.,$

then

(1.8) $\limsup_{n\to\infty} (S_n(k_n) - \mu_n(k_n))/(a_n(k_n)(2 \, log \, log \, n)^{1/2}) = 1 \quad a.s.,$

with corresponding lim inf equal to -1 a.s.

REMARK 2. It is shown in Section 2 that if $k_n/log \, log \, n \to d$ for some $0 \le d < \infty$ then (1.7) cannot hold for any F of the form given in (1.1). Moreover it is proven that if a sequence $(k_n)_{n \ge 1}$ satisfies (K'), (1.6) and

(1.9) $Q(1 - k_n/n \pm c(k_n log \, log \, n)^{1/2}/n)/Q(1 - k_n/n) \to 1 \quad as \quad n \to \infty$

for some $c > 2^{1/2}$ then (1.7) holds.

Using Remark 2 it is easy to verify the following examples for Theorem 3.

EXAMPLE 1. Set $Q(1 - s) = \exp(s^{-\alpha})$ for some $0 < \alpha < \infty$. For any sequence $(k_n)_{n \ge 1}$ satisfying (K') and

$$n(log \, log \, n)^{1/(2\alpha)}/k_n^{1+1/(2\alpha)} \to 0 \; as \quad n \to \infty,$$

we have (1.6) and (1.7).

EXAMPLE 2. Set $Q(1 - s) = \exp(\exp(1/s))$. In this case for any sequence $(k_n)_{n \ge 1}$ satisfying (K') and

$$n(log \, log \, n)^{1/2} \exp(n/k_n)/k_n^{3/2} \to 0 \quad as \quad n \to \infty,$$

we have (1.6) and (1.7).

EXAMPLE 3. Set $Q(1 - s) = \exp(log(1/s)/s)$. In this example for any sequence $(k_n)_{n\geq 1}$ satisfying (K') and

$$n(log\ log\ n)^{1/2} log(n/k_n)/k_n^{3/2} \to 0 \quad as \quad n \to \infty,$$

we have (1.6) and (1.7).

The proofs of Theorem 1, 2, and 3 are provided in Section 2.

2. Proofs. Let $(U_n)_{n\geq 1}$ be a sequence of independent uniform $(0,1)$ random variables. For any integer $n \geq 1$ let $U_{1,n} \leq \cdots \leq U_{n,n}$ denote the order statistics and G_n the right continuous empirical distribution function based on the first n of these random variables. The two sequences $(X_n)_{n\geq 1}$ and $(Q(U_n))_{n\geq 1}$ are equal in law and consequently the two processes $\{X_{i,n} : 1 \leq i \leq n,\ n \geq 1\}$ and $\{Q(U_{i,n}) : 1 \leq i \leq n,\ n \geq 1\}$ are equal in law too. Therefore without loss of generality we may assume $X_{i,n} = Q(U_{i,n})$ for all $1 \leq i \leq n$ and $n \geq 1$.

PROOF OF THEOREM 1. Let $\xi_1, \xi_2, \ldots,$ be a sequence of independent exponential random variables with mean one and for each integer $j \geq 1$ set

$$E_j = \xi_1 + \cdots + \xi_j$$

and for integers $0 \leq k \leq n - 1, n \geq 1$, let

$$1 - U_{n-k,n}^* = E_{k+1}/E_{n+1}.$$

It is well-known that for each integer $n \geq 1$

$$(1 - U_{n-j,n})_{j=0}^{n-1} \overset{\mathcal{D}}{=} (1 - U_{n-j,n}^*)_{j=0}^{n-1}.$$

Therefore to prove the equivalence of (1.1) and (1.2) it suffices to show that for each fixed integer $k \geq 0$

(2.1) $$Q(U^*_{n-k-1,n})/Q(U^*_{n-k,n}) \xrightarrow{P} 0 \quad as \quad n \to \infty$$

if and only if (1.1) holds.

First assume (1.1). Choose any $0 < \lambda < 1$ and set for $n \geq 1$

$$A_n(k,\lambda) = \{(1 - U^*_{n-k,n})/(1 - U^*_{n-k-1,n}) = E_{k+1}/E_{k+2} \leq \lambda\}$$

and

$$R_n(k,\lambda) = Q(1 - (1 - U^*_{n-k-1,n}))/Q(1 - \lambda(1 - U^*_{n-k-1,n}))1_{A_n(k,\lambda)}.$$

Obviously,

$$R_n(k,\lambda) \geq \{Q(U^*_{n-k-1,n})/Q(U^*_{n-k,n})\}1_{A_n(k,\lambda)}.$$

Since $1 - U^*_{n-k-1,n} \xrightarrow{P} 0$ and (1.1) holds we have for each $0 < \lambda < 1$

$$R_n(k,\lambda) \xrightarrow{P} 0 \quad as \quad n \to \infty.$$

Observing that

$$\lim_{\lambda \uparrow 1} P(A_n(k,\lambda)) = \lim_{\lambda \uparrow 1} P(E_{k+1}/E_{k+2} \leq \lambda) = 1$$

we conclude (1.2)

Next assume (1.2) and choose any $\lambda > 1$. Notice that

$$P(Q(U^*_{n-k-1,n})/Q(U^*_{n-k,n}) \geq Q(1 - \lambda n^{-1})/Q(1 - n^{-1})),$$
$$\geq P(n(1 - U^*_{n-k-1}) \leq \lambda \quad and \quad n(1 - U^*_{n-k,n}) \geq 1).$$

This last probability converges as $n \to \infty$ to

$$P(E_{k+2} \leq \lambda, \; E_{k+1} \geq 1),$$

which in turn is greater than or equal to

$$P(1 \leq E_{k+1} \leq (\lambda+1)/2, \xi_{k+2} \leq (\lambda-1)/2)$$

$$=P(1 \leq E_{k+1} \leq (\lambda+1)/2)P(\xi_{k+2} \leq (\lambda-1)/2) > 0.$$

Thus whenever (2.1) holds we must have for each $\lambda > 1$

$$Q(1 - \lambda n^{-1})/Q(1 - n^{-1}) \to 0 \quad \text{as} \quad n \to \infty.$$

By a routine argument this is equivalent to $Q(1 - s)$ being rapidly varying at zero.

That (1.3) implies (1.2) is obvious. Finally we prove (1.2) implies (1.3). It is enough to establish

$$S_n(k+1)/X_{n-k,n} \overset{P}{\to} 0 \quad \text{as} \quad n \to \infty.$$

Integrating by parts we obtain

$$S_n(k+1) = n \int_0^{U_{n-k-1,n}} Q(s) \, dG_n(s)$$

$$= -(k+1)Q(U_{n-k-1,n}) + n \int_0^{U_{n-k-1,n}} (1 - G_n(s)) \, dQ(s)$$

$$=: -(k+1)X_{n-k-1,n} + n \, I_n(k+1).$$

Clearly by (1.2) it suffices to prove that

$$n \, I_n(k+1)/X_{n-k,n} \overset{P}{\to} 0 \quad \text{as} \quad n \to \infty.$$

Since

$$\sup_{0 \leq s \leq 1} (1 - G_n(s))/(1 - s) = O_p(1),$$

eg. Shorack and Wellner (1986), page 345, we need only show that

$$(2.2) \qquad n \int_0^{U_{n-k-1,n}} (1-s) \, dQ(s)/X_{n-k,n} \xrightarrow{P} 0 \quad \text{as} \quad n \to \infty.$$

For this we require the following lemma.

LEMMA 2.1. *Let V be a non–increasing non–negative function which is rapidly varying at zero. Then for all real β*

$$(2.3) \qquad \int_s^1 u^{\beta-1} V(u) \, du/(s^\beta V(s)) \to 0 \quad \text{as} \quad s \downarrow 0,$$

$$(2.4) \qquad \int_s^1 u^\beta \, dV(u)/(s^\beta V(s)) \to 1 \quad \text{as} \quad s \downarrow 0$$

and

$$(2.5) \qquad s^\beta V(s) \to \infty \quad \text{as} \quad s \downarrow 0.$$

PROOF. Assertion (2.3) follows from Theorem 1.3.2 of de Haan (1975), (2.4) from (2.3) and integration by parts and (2.5) from the definition of a rapidly varying function at zero. ∎

Now by (1.2) implies (1.1), (2.3) and $U_{n-k-1,n} \xrightarrow{P} 1$

$$n \int_0^{U_{n-k-1,n}} (1-s) \, dQ(s) = Op(1)n(1 - U_{n-k-1,n}) \, Q(U_{n-k-1,n}).$$

Noting that

$$n(1 - U_{n-k-1,n}) = Op(1)$$

we conclude that the expression in (2.2) equals

$$Op(1)X_{n-k-1,n}/X_{n-k,n}$$

The convergence statement in (2.2) now follows from (1.2). ■

PROOF OF THEOREM 2. We shall require the following lemmas.

LEMMA 2.2. (Mori (1976)). *Let $(b_n)_{n\geq 1}$ be a non-decreasing sequence of positive constants, then for each fixed integer $k \geq 0$*

(2.6) $P(X_{n-k,n} > b_n \text{ i.o.}) = 0 \quad \text{or} \quad 1$

according as the series in (1.5) finite or infinite.

In particular, from Lemma 2.2 it is easy to infer that whenever (1.1) holds

(2.7) $\limsup_{n\to\infty} X_{n-k,n}/b_n = 0 \quad \text{or} \quad \infty \quad \text{a.s.}$

according as the series in (1.5) is finite or infinite. Also from Lemma 2.2 one obtains the following result.

LEMMA 2.3 *Let $(a_n)_{n\geq 1}$ be a non-decreasing sequence of positive constants then for each fixed integer $k \geq 0$*

$$\liminf_{n\to\infty} a_n(1 - U_{n-k,n}) = \infty \quad \text{or} \quad 0 \quad \text{a.s.}$$

according as the following series is finite or infinite

(2.8) $$\sum_{n=1}^{\infty} n^k a_n^{-(k+1)}.$$

For integers $0 \leq k \leq n - 1, n \geq 1$, let

$$M_n(k) = \sup_{0 < s \leq U_{n-k,n}} (1 - G_n(s))/(1 - s).$$

LEMMA 2.4. (Einmahl, Haeusler and Mason (1988)). *Let* $(c_n)_{n\geq 1}$ *be a sequence of positive constants such that* nc_n *is non-decreasing then for each fixed integer* $k \geq 0$

(2.9) $$\limsup_{n\to\infty} M_n(k)/c_n = 0 \quad \text{or} \quad \infty \quad \text{a.s.}$$

according as the following series is finite or infinite

(2.10) $$\sum_{n=1}^{\infty} n^{-1} c_n^{-(k+1)}.$$

Turning now to the proof of Theorem 2, first assume that the series in (1.5) is infinite. This implies that the lim sup in (2.7) is infinite almost surely. Therefore, obviously, the lim sup in (1.4) is equal to infinity with probability one.

Now assume that the series in (1.5) is finite. We can write

$$S_n(k) = n \int_0^{U_{n-k,n}} Q(s)\, dG_n(s),$$

which after integrating by parts equals

$$-kQ(U_{n-k,n}) + n \int_0^{U_{n-k,n}} (1 - G_n(s))\, dQ(s) =: -kX_{n-k,n} + n\, I_n(k).$$

Since the lim sup in (2.7) is equal to the zero almost surely when the series in (1.5) is finite, to finish the proof of Theorem 2 it suffices to show that

(2.11) $$\lim_{n\to\infty} n\, I_n(k)/b_n = 0 \quad \text{a.s.}$$

Observe that

$$n\, I_n(k) \leq M_n(k) n \int_0^{U_{n-k,n}} (1 - s)\, dQ(s)/b_n =: M_n(k)\, n\, J_n(k).$$

Since $U_{n-k,n} \nearrow 1$ almost surely as $n \to \infty$ we infer from (2.4) with $\beta = 1$ that

$$J_n(k)/\{(1 - U_{n-k,n})\,Q(U_{n-k,n})\} \to 1 \quad \text{a.s.} \qquad as \quad n \to \infty.$$

Therefore to establish (2.11) it is enough to prove that

(2.12) $$\lim_{n \to \infty} M_n(k)\,n(1 - U_{n-k,n})\,Q(U_{n-k,n})/b_n = 0 \quad \text{a.s.}$$

The function $Q(1-s)$ being rapidly varying at zero and non–increasing implies that there exists a non–increasing function ψ on $(0,1)$ such that

(2.13) $$s\,Q(1-s)/\psi(s) \to 1 \quad \text{as} \quad s \downarrow 0,$$

cf. Proposition 2.4.4 (iv), page 85, of Bingham, Goldie and Teugels (1987). Set

$$Q^*(1-s) = s^{-1}\psi(s),\ 0 < s < 1.$$

Noting that

$$Q^*(1-s)/Q(1-s) \to 1 \quad \text{as} \quad s \downarrow 0,$$

we see that to show (2.12), we must only establish

(2.14) $$\limsup_{n \to \infty} M_n(k)\,n(1 - U_{n-k,n})\,Q^*(U_{n-k,n})/b_n = 0 \quad \text{a.s.}$$

The series in (1.5) being finite implies by Lemma 2.3 that

(2.15) $$\lim_{n \to \infty} (1 - U_{n-k,n})/(1 - F(b_n)) = \infty \quad \text{a.s.}$$

This combined with (2.13) says that the lim sup in (2.14) is almost surely less than or equal to

(2.16) $$\limsup_{n \to \infty} M_n(k)\,n(1 - F(b_n))\,Q^*(F(b_n))/b_n.$$

Noticing that with $c_n^{-1} = n(1 - F(b_n))$ the series in (2.10) is finite we have by Lemma 2.4

$$(2.17) \qquad \lim_{n \to \infty} M_n(k) n(1 - F(b_n)) = 0 \quad \text{a.s.}$$

Since $Q(F(b_n)) \leq b_n$, we also have

$$(2.18) \qquad Q^*(F(b_n))/b_n = O(1).$$

Statements (2.17) and (2.18) imply that the lim sup in (2.16) is equal to zero almost surely. This completes the proof of Theorem 2. ■

PROOF OF THEOREM 3. The following computation will be crucial to the proof of Theorem 3. We can write

$$a_n(k_n)^{-1} \left\{ \sum_{i=1}^{n-k_n} X_{i,n} - \mu_n(k_n) \right\} = a_n(k_n)^{-1} \left\{ \sum_{i=1}^{n-k_n} Q(U_{i,n}) - \mu_n(k_n) \right\}$$

$$= a_n(k_n)^{-1} n \left\{ \int_0^{U_{n-k_n,n}} Q(s) \, dG_n(s) - \int_0^{1-k_n/n} Q(s) \, ds \right\},$$

which by two integrations by parts equals

$$a_n(k_n)^{-1} n \int_0^{1-k_n/n} (s - G_n(s)) \, dQ(s)$$

$$(2.19) \qquad + a_n(k_n)^{-1} n \int_{U_{n-k_n,n}}^{1-k_n/n} \left(G_n(s) - \frac{n - k_n}{n} \right) dQ(s) =: Y_n + \Delta_n.$$

We shall require the following technical lemma.

Set

$$\sigma^2(s) = \int_0^{1-s} \int_0^{1-s} (u \wedge v - uv) \, dQ(u) \, dQ(v), \quad 0 < s < 1,$$

where $u \wedge v = min(u, v)$.

LEMMA 2.5 *Whenever F satisfies* (1.1)

$$\sigma^2(s) \sim sQ^2(1-s) \quad as \quad s \downarrow 0.$$

PROOF. An elementary but somewhat lengthy computation gives

$$\sigma^2(s)/\{sQ^2(1-s)\}$$

$$=1 - s + \int_s^1 Q^2(1-u) \, du/\{sQ^2(1-s)\} - 2s \left(\int_s^1 Q(1-u) \, du/\{sQ(1-s)\}\right)$$

$$- s \left(\int_s^1 Q(1-u) \, du/\{sQ(1-s)\}\right)^2.$$

An application of (2.3) to the function $u \to Q^i(1-u)$ with $i = 1, 2$ and $\beta = 1$ yields the desired result. ∎

Turning now to the proof of Theorem 3 we see that for s in the closed interval formed by $U_{n-k_n,n}$ and $1 - k_n/n$ we have

$$|G_n(s) - (1 - k_n/n)| \le |G_n(1 - k_n/n) - (1 - k_n/n)|$$

so that

$$\frac{|\Delta_n|}{(2 \log \log n)^{1/2}} \le \frac{n|G_n(1 - k_n/n) - (1 - k_n/n)|}{(2 k_n \log \log n)^{1/2}} \left| \frac{Q(U_{n-k_n,n})}{Q(1 - k_n/n)} - 1 \right|.$$

The second factor on the right side converges to zero almost surely by assumption (1.7), and the first one is bounded with probability one because of assumption (1.6) as can be seen from Theorem 3.2 of Csáki (1977). Thus $|\Delta_n|/(2 \log \log n)^{1/2} \to 0$ a.s., and it remains to prove that

$$\limsup_{n \to \infty} Y_n/(2 \log \log n)^{1/2} = 1 \quad \text{a.s.}$$

with the corresponding liminf equal to -1 almost surely. We shall consider only the limsup statement here, the proof for the liminf statement being analogous. The proof of

$$\limsup_{n \to \infty} Y_n/(2 \log \log n)^{1/2} \geq 1 \quad \text{a.s.}$$

can be based on classical techniques and is similar to the proof of (2.11) in Haeusler and Mason (1987). The main difference is that instead of (2.4) from that paper one now has to use Lemma 2.5 of the present paper. All other modifications are straightforward so that we do not repeat the details here.

For the proof of

$$(2.20) \qquad\qquad \limsup_{n \to \infty} Y_n/(2 \log \log n)^{1/2} \leq 1 \quad \text{a.s.}$$

the classical blocking technique as used in the corresponding proof in Haeusler and Mason (1987) is not appropriate because of rapid variation of $Q(1-s)$. Instead, we shall use

LEMMA 2.6 *Whenever* $(k_n)_{n>1}$ *is a sequence satisfying* (K') *and (1.6), for any* $1 < \tau < \infty$

$$\limsup_{n \to \infty} \left(\frac{n}{2 \log \log n} \right)^{1/2} \sup_{1-\tau k_n/n \leq s \leq 1-k_n/n} \frac{|G_n(s) - s|}{(1-s)^{1/2}} \leq \tau^{1/2} \quad \text{a.s.}$$

PROOF. The lemma follows immediately from Theorem 2 of Einmahl and Mason (1988). Alternatively, a direct proof can be given using inequality (16) in Lemma 3 of Shorack (1980) combined with an exponential inequality as Inequality 1.1 of Shorack and Wellner (1982) or Lemma 2.3 of Stute (1982). Since the details of the computations are routine, they are omitted.

For the proof of (2.20) we fix $1 < \tau < \infty$ and write

$$
Y_n = a_n(k_n)^{-1}n \int_0^{1-\tau k_n/n} (s - G_n(s))\, dQ(s)
$$

(2.21)

$$
+ a_n(k_n)^{-1} n \int_{1-\tau k_n/n}^{1-k_n/n} (s - G_n(s))\, dQ(s) =: \Delta_n(\tau) + Y_n(\tau).
$$

For $\Delta_n(\tau)$ we find by (2.4) that

$|\Delta_n(\tau)|$

$$
\leq \frac{n}{(2k_n \log\log n)^{1/2}Q(1-k_n/n)} \int_0^{1-\tau k_n/n} (1-s)^{1/2}\, dQ(s) \sup_{0\leq s\leq 1-\tau k_n/n} \frac{|G_n(s)-s|}{(1-s)^{1/2}}
$$

$$
\sim \tau^{1/2}\frac{Q(1-\tau k_n/n)}{Q(1-k_n/n)} \left(\frac{n}{2\log\log n}\right)^{1/2} \sup_{0\leq s\leq 1-\tau k_n/n} \frac{|G_n(s)-s|}{(1-s)^{1/2}}.
$$

By rapid variation of $Q(1-s)$ we have $Q(1-\tau k_n/n)/Q(1-k_n/n) \to 0$, whereas Theorem 3.2 of Csáki (1977) shows that under assumption (1.6) the remaining factor on the right side of the last inequality is bounded with probability one. Consequently

(2.22) $\Delta_n(\tau) \to 0$ a.s.

Furthermore, by (2.4)

$|Y_n(\tau)|$

$$
\leq \frac{n}{(2k_n \log\log n)^{1/2}Q(1-k_n/n)} \int_0^{1-k_n/n} (1-s)^{1/2}dQ(s) \sup_{1-\tau k_n/n\leq s\leq 1-k_n/n} \frac{|G_n(s)-s|}{(1-s)^{1/2}}
$$

$$
\sim \left(\frac{n}{2\log\log n}\right)^{1/2} \sup_{1-\tau k_n/n\leq s\leq 1-k_n/n} \frac{|G_n(s)-s|}{(1-s)^{1/2}}.
$$

Applying Lemma 2.6 we get

$$
\limsup_{n\to\infty} |Y_n(\tau)| \leq \tau^{1/2} \quad \text{a.s.}
$$

Combining this with (2.21) and (2.22) gives

$$\limsup_{n\to\infty} Y_n \le \tau^{1/2} \quad \text{a.s.}$$

for each $1 < \tau < \infty$ which proves (2.20) and completes the proof of Theorem 3. ∎

PROOF OF REMARK 2. First consider the first part of the remark. First assume $0 < d < \infty$. Applying Theorem 6 of Kiefer (1971) we have when $k_n \sim d \, loglog \, n$

$$\limsup_{n\to\infty} n(1 - U_{n-k_n,n})/k_n = \lambda_d \quad \text{a.s.}$$

for some $1 < \lambda_d < \infty$. This says that for all $1 < \lambda < \lambda_d$

$$P(1 - \lambda k_n/n > U_{n-k_n,n} \text{ i.o}) = 1,$$

which by the assumption that $Q(1 - s)$ is a non–negative rapidly varying function at zero yields with probability one

$$\liminf_{n\to\infty} Q(U_{n-k_n,n})/Q(1 - k_n/n) \le \lim_{n\to\infty} Q(1 - \lambda k_n/n)/Q(1 - k_n/n) = 0,$$

showing that (1.1) cannot hold.

Next assume that $d = 0$. In this case by Theorem 2 of Kiefer (1972) for all $0 < \epsilon < 1$

$$P(n(1 - U_{n-k_n,n})) > \epsilon \log \log \, n \text{ i.o.}) = 1,$$

which by an argument similar to that just given yields

$$\liminf_{n\to\infty} Q(U_{n-k_n,n})/Q(1 - k_n/n) = 0 \quad \text{a.s.}$$

Thus (1.1) also cannot hold in this case.

The proof of the second part of the remark is a direct consequence of the fact that when $(k_n)_{n \geq 1}$ satisfies (K') and (1.6)

$$\limsup_{n \to \infty} \pm n(U_{n-k_n,n} - (1 - k_n/n))/(k_n \log \log n)^{1/2} \leq 2^{1/2} \quad \text{a.s.}$$

This fact is proved by a straightforward argument based on a combination of Theorem 2 of Einmahl and Mason (1988) and Theorem 4 of Wellner (1978). ∎

REFERENCES.

[1] BINGHAM, N.H., GOLDIE, C.M. and TEUGELS, J.L. (1987). *Regular Variation*. Cambridge University Press, Cambridge.

[2] CSÁKI, E. (1977). The law of the iterated logarithm for normalized empirical distribution function. *Z. Wahrsch. Verw. Gebiete* **38** 147-167.

[3] CSÖRGŐ, S., HAEUSLER, E. and MASON, D.M. (1988a). A probabilistic approach to the asymptotic distribution of sums of independent, identically distributed random variables. *Adv. in Appl. Math.* **9** 259-333.

[4] CSÖRGŐ, S., HAEUSLER, E. and MASON, D.M. (1988b). Asymptotic distribution of trimmed sums. *Ann. Probab.* **16** 672-699.

[5] DARLING, D.A. (1952). The influence of the maximum term in the addition of independent random variables. *Trans. Amer. Math. Soc.* **73** 95-107.

[6] EINMAHL, J.H.J. and MASON, D.M. (1988). Laws of the iterated logarithm in the tails for weighted uniform empirical processes. *Ann. Probab.* **16** 126-141.

[7] EINMAHL, J.H.J., HAEUSLER, E., and MASON, D.M. (1988). On the relationship between the almost sure stability of weighted empirical distributions and sums of order statistics. *Prob. Th. Rel. Fields* **79** 59-74.

[8] HAAN, DE, L. (1975). *On Regular Variation and Its Application to the Weak Convergence of Sample Extremes*. Mathematical Centre Tracts **32**, Amsterdam.

[9] HAEUSLER, E. and MASON, D.M. (1987). Laws of the iterated logarithm for the middle portion of the sample. *Math. Proc. Cambridge Philos. Soc.* **101** 301-312.

[10] KIEFER, J. (1972). Iterated logarithm analogues for sample quantities when $p_n \downarrow 0$. *Proc. Sixth Berkeley Symp. Math. Statist. Probab.* Vol. I, 227-244.

[11] MALLER, R.A. and RESNICK, S.I. (1984). Limiting behaviour of sums and the term of maximum modulus. *Proc. London Math. Soc.* **49** 385-422.

[12] MIKOSCH, T. (1988). Interated logarithm results for rapidly growing random walks. *Statistics* **19** 107-115.

[13] MORI, T. (1976). The strong law of large numbers when extreme terms are excluded from sums. *Z. Wahrsch. Verw. Gebiete* **36**, 189-194.

[14] MORI, T. (1981). The relation of sums and extremes of random variables. Session Summary Booklet, Invited Papers, Buenos Aires Session, Nov. 30-Dec. 11, (International Statistical Institute) 879-894.

[15] PRUITT, W.E. (1981). General one-sided laws of the iterated logarithm. *Ann. Probab.* **9** 1-48.

[16] PRUITT, W.E. (1987). The contribution to the sum of the summand of maximum modulus. *Ann. Probab.* **15** 885-896.

[17] RÉVÉSZ, P. and WILLEKENS, E. (1987). On the maximal distance between two renewal epochs. *Stoc. Proc.* **27** 21-41.

[18] SHORACK, G.R. (1980). Some laws of the iterated logarithm type results from the empirical process. *Austral. J. Statist.* **22** 50-59.

[19] SHORACK, G.R. and WELLNER, J.A. (1982). Limit theorems and inequalities for the uniform empirical process indexed by intervals. *Ann. Probab.* **10**, 639-652.

[20] SHORACK, G.R. and WELLNER, J.A. (1986). *Empirical Processes with Application to Statistics.* Wiley, New York.

[21] STIGLER, S.M. (1973). The asymptotic distribution of the trimmed mean. *Ann. Statist.* **1** 472-477.

[22] STUTE, W. (1982). The oscillation behavior of empirical processes. *Ann. Probab.* **10** 86-107.

[23] TEICHER, H. (1979). Rapidly growing random walks and a associated stopping time. *Ann. Probab.* **7** 1078-1081.

[24] WELLNER, J.A. (1978). Limit theorems for the ratio of the empirical distribution function to the true distribution function. *Z. Wahrsch. Verw. Gebiete* **45** 73-88.

[25] ZHANG, C.-H. (1986). The lower limit of a normalized random walk. *Ann. Probab.* **14** 560-581.

Eric Haeusler   David M. Mason
Mathematical Institute   Department of Mathematical Sciences
University of Munich   University of Delaware
Theresienstrasse 39   Newark, Delaware 19716
8000 Munich 2
West Germany

LIMIT RESULTS FOR LINEAR COMBINATIONS

Galen R. Shorack*

1. Asymptotic Normality of L-statistics.

Let $X_1, \ldots, X_n$ denote a random sample from the $df\, F$, and let $X_{1,n} \leq \cdots \leq X_{n,n}$ denote their order statistics. Suppose the statistician specifies a known function h and known constants $c_{n1}, \ldots, c_{nn}$. Our interest is in establishing the asymptotic normality of L-statistics of the form

$$(1.1) \qquad L_n \equiv \frac{1}{n} \sum_{i=1}^{n} c_{ni} h(X_{i,n})$$

under very mild regularity conditions. If X has $df\, F$ and U is a Uniform $(0,1)$ rv, then $F^{-1}(U)$ has $df\, F$. For this reason, there is no loss in assuming that

$$(1.2) \qquad L_n = \frac{1}{n} \sum_{i=1}^{n} c_{ni} g(U_{i,n})$$

where $g \equiv h(F^{-1})$ and $0 \leq U_{1,n} \leq \cdots \leq U_{n,n} \leq 1$ are the order statistics of independent Uniform $(0,1)$ rv's $U_1, \ldots, U_n$.

A reasonable centering constant to use for L_n is, with $a_n \equiv 1/(n+2)$,

$$(1.3) \qquad \mu_n \equiv \frac{(n+2)}{n} \int_{a_n}^{1-a_n} g(t) J_n(t) dt = \frac{n+2}{n} \int_{a_n}^{1-a_n} g(t) d\Gamma_n(t)$$

where we define the function J_n on $[0,1]$ by letting

$$(1.4) \qquad J_n(t) = c_{ni} \quad \text{for} \quad i/(n+2) < t \leq (i+1)/(n+2) \quad \text{and} \quad 1 \leq i \leq n$$

* Supported in part by National Science Foundation Grant DMS-8801083 and The Netherlands Organization of Scientific Research (ZWO).

with $J_n(t) \equiv c_{n1}$ for $0 \le t \le a_n$ and $J_n(t) \equiv c_{nn}$ for $1 - a_n \le t \le 1$, and then define

$$(1.5) \qquad\qquad \Gamma_n(t) \equiv \int_{1/2}^{t} J_n(s)ds \quad \text{for} \quad 0 \le t \le 1.$$

We find it convenient to assume initially that

$$(1.6) \qquad\qquad\qquad J_n \ge 0 \quad \text{on} \quad [0,1]$$

and

$$(1.7) \qquad\qquad g \text{ is } \nearrow \text{ and left continuous on } (0,1).$$

We will assume that J_n converges in some sense to a limiting continuous function J. Specifically, we suppose for all $n \ge 1$ and all $0 < \lambda < 1$ we can choose M_λ so large that

$$|J_n(s) - J(t)| \le M_\lambda[t(1-t)]^{-1}J(t)[|s - t| + n^{-1}] \text{ for all } 0 < s,t < 1$$

$$\text{for which} \quad \lambda/n \le t \le 1 - \lambda/n \text{ and } |s - t| \le 1/(\lambda\sqrt{n})$$

$$(1.8) \qquad \text{with } \lambda t \le s \le t/\lambda \text{ and } 1 - (1-t)/\lambda \le s \le 1 - \lambda(1-t).$$

Now define (if $1/2$ is not a continuity point of g, uses a "nearby" continuity point)

$$(1.9)$$

$$K(t) \equiv \int_{[1/2,t)} J(s)\,dg(s) \text{ with } K(1/2) \equiv 0; \text{ with } \int_{[1/2,t)} = -\int_{[t,1/2)} \quad \text{if } t < 1/2.$$

Because of (1.6), (1.7) and (1.8),

$$(1.10) \qquad K(t) \text{ is } \nearrow \text{ and left continuous on } (0,1) \text{ with } K(1/2) = 0.$$

Thus K is the inverse of some df, which we will label as H; that is,

$$(1.11) \qquad\qquad\qquad K = H^{-1} \quad \text{for some } df \ H.$$

We call K the *quantile function*, abbreviated as *qf*. It follows that the rv's

(1.12) $Y_i = K(U_i) = H^{-1}(U_i)$,   for $1 \le i \le n$,   are iid with *df H*.

Let K_+ denote the right continuous version of K, and set

(1.13) $Y_{ni} = K_n(U_i) = (\text{Winsorized } Y_i)$

where $K_n(t)$ equals $K(a_n), K(t), K_+(1 - a_n)$ according as $0 \le t \le a_n$, $a_n < t \le$
$1 - a_n$, $1 - a_n < t \le 1$. Let $R_n =_a S_n$ mean that $R_n - S_n \to_p 0$ and D(Normal)
denote the domain of attraction of the normal *df*.

THEOREM 1. Suppose J_n, J, g and K satisfy (1.6)-(1.9). Also assume that

(1.14)   K is in D (Normal).

Then

(1.15)
$$\sqrt{n}(L_n - \mu_n)/\sigma_n =_a \frac{1}{\sqrt{n}\sigma_n} \sum_{i=1}^{n}(Y_{ni} - EY_{ni}) =_a \frac{1}{\sqrt{n}\sigma_n} \sum_{i=1}^{n}(Y_i - EY_{ni}) \to_d N(0,1),$$

where

(1.16) $\sigma_n^2 \equiv \int_{[a_n,1-a_n]} \int_{[a_n,1-a_n]} (s \wedge t - st) \, dK(s) \, dK(t) = \text{Var}[Y_{ni}].$

REMARK 1.1. One could give conditions on J and g that allow $\mu \equiv \int_0^1 g(t)J(t) \, dt$
to replace μ_n. If there is symmetry of $F(\cdot - \theta)$ about 0 for some θ as well as
antisymmetry (except at the points $(i/(n + 2))$ of J_n and J about $t = 1/2$, then
$\mu_n = \mu = \theta$ in the case $g = F^{-1}$ whenever
$$\int_0^1 J(t)dt = \int_{a_n}^{1-a_n} J_n(t)dt = 1.$$

Barring such symmetry, it seems to the author that μ is less appropriate than μ_n as a measure of centrality. (For typical smooth J's, this replacement is a simple deterministic exercise treated in quite a number of papers on L-statistics.)

REMARK 1.2. Suppose $J_n = J_{1n} - J_{2n}$, $J = J_1 - J_2$ and $g = g_1 - g_2$ where each of the four possibilities J_{in}, J_i and g_j, with $1 \le i, j \le 2$, satisfies (1.6)-(1.9) with each $dK_{ij} = J_i dg_j$ in D(Normal). Then (1.15) still holds; just make four applications of the proof of Theorem 1 and add the results together.

REMARK 1.3. Condition (1.8) typically fails if J has any discontinuities. We now indicate how *discontinuous J functions* can be treated. We can add to the J above a finite number of terms of the form $a1_{[r,1)}(t)$, provided g is continuous at r; this supposes adding to J_n the corresponding term $a1_{[r_n,1)}(t)$ where $r_n \to r$. We prove this fact at the end of the proof of Theorem 1.

REMARK 1.4. Suppose J is of one of the following forms:

(i) J is bounded, continuous and non-zero on $[0,1]$, with a bounded derivative.

(ii) $J(t) = \Phi^{-1}(t)$ for the $N(0,1)$ *df* Φ.

(iii) $J(t) = [t(1-t)]^r$ for $0 < t < 1$, and any $r \le 0$.

(iv) $J(t) = [t(1-t)]^r$ for $0 < t < 1$, and any $r > 0$.

In cases (i)-(iv), or for any of the typical J's they suggest, virtually any "reasonable" way of defining J_n will satisfy (1.8). These examples should make it clear that we have established the following result: If K is in D(Normal) where $K(t) \equiv \int_{1/2}^{t} J(s)dg(s)$ with a "reasonably smooth" J, then for any "reasonable" J_n converging to J a.e. dg we have $\sqrt{n}(L_n - \mu_n)/\sigma_n \to_d N(0,1)$. (Note that in cases like (iii), the original X's need not have finite variance, or mean.) Though the theorem is really joint in J and $g = h(F^{-1})$, we prefer to think of the statistician as choosing "reasonable" J, J_n, and h and then using the theorem to determine distributions F (note Remark 1.6 below) for which L_n is asymptotically normal.

REMARK 1.5. The ordinary CLT is a corollary to Theorem 1. That is,

$$\sqrt{n}(\bar{X} - \mu)/\text{StDev}[X] \to_d N(0,1) \quad \text{whenever} \quad \text{StDev}[X] \in (0, \infty).$$

Just let $J_n(t) = J(t) = 1$ for all $0 \le t \le 1$ in Theorem 1, noting that $\sqrt{n}(\mu_n - EX) \to 0$ will be shown following the proof of Theorem 1 in Section 2. Most theorems on L-statistics in the literature that apply for unbounded J do not contain the ordinary CLT. (We can likewise replace EY_{ni} by EY_i in (1.15) when $0 < \text{Var}[Y] < \infty$; of course $\text{StDev}[Y]$ can also replace σ_n.)

Among the many interesting results in the literature on asymptotic normality of L-statistics are Chernoff, Gastwirth and Johns (1967), Shorack (1969 and 1972), Stigler (1969 and 1974), Ruymgaart and van Zuijlen (1977), Sen (1978), Boos (1979), Mason (1981), Singh (1981), and Helmers and Ruymgaart (1988).

REMARK 1.6 Suppose that (1.6), (1.7), (1.8) hold and that J is such that $J(t)$ and $J(1-t)$ are regularly varying at 0. Then Mason and Shorack (1991) show that

$$\sqrt{n}(L_n - B_n)/A_n \to_d N(0,1) \quad \text{for some } B_n \text{ and } A_n > 0$$

(1.17)   if and only if K is in D(Normal) .

When normality holds, then μ_n and σ_n are acceptable choices. The proof with necessity included is much harder.

2. Proofs.

Proof of Theorem 1.1. We will employ a slight variation $\tilde{\mathbf{G}}_n$ on the usual empirical *df* $\mathbf{G}_n$, defined by

$$(2.1) \quad \tilde{\mathbf{G}}_n(t) \equiv (i+1)/(n+2) \quad \text{for} \quad U_{i,n} \le t < U_{i+1,n} \quad \text{and} \quad 0 \le i \le n+1.$$

The left continuous inverse of $\tilde{\mathbf{G}}_n$ satisfies

$$(2.2) \quad \tilde{\mathbf{G}}_n^{-1}(t) = U_{i,n} \quad \text{for} \quad i/(n+2) < t \le (i+1)/(n+2) \quad \text{and} \quad 0 \le i \le n+1.$$

Of course $0 \equiv U_{0,n} < U_{1,n} < \cdots < U_{n,n} < U_{n+1,n} \equiv 1 < U_{n+2,n} \equiv \infty$ a.s.

Now given $\epsilon > 0$, there exist $0 < \lambda \equiv \lambda_\epsilon < 1$ and a set $A_{n\epsilon}$ having $P(A_{n\epsilon}) > 1-\epsilon$ on which:

$$\lambda i/(n+2) \le U_{i,n} \le (i/(n+2))/\lambda \text{ for } 1 \le i \le n,$$

$$1 - (i/(n+2))/\lambda \le U_{n-i+1,n} \le 1 - \lambda i/(n+2) \text{ for } 1 \le i \le n,$$

$$(2.3) \, \lambda t \le \tilde{\mathbf{G}}_n(t) \text{ for } 0 \le t \le 1, \; \tilde{\mathbf{G}}_n(t) \le 1 - \lambda(1-t) \text{ for } 0 \le t \le 1,$$

$$\tilde{\mathbf{G}}_n(t) \le t/\lambda \text{ for } \lambda/n \le t \le 1, \; 1 - (1-t)/\lambda \le \tilde{\mathbf{G}}_n(t) \text{ for } 0 \le t \le 1 - \lambda/n$$

and, for $n \geq$ some n_ϵ,

(2.4) $\qquad\qquad |\tilde{G}_n(t) - t| \leq 1/(\lambda\sqrt{n}) \quad \text{for all} \quad 0 \leq t \leq 1.$

Let $1_{n\epsilon}$ denote the indicator function of the set $A_{n\epsilon}$. (This is essentially the same as Inequality 1 of Shorack and Wellner (1986, p. 419)).

We define empirical processes by

(2.5) $\qquad \tilde{\alpha}_n(t) \equiv \sqrt{n}[\tilde{G}_n(t) - t] \quad \text{and} \quad \alpha_n(t) \equiv \sqrt{n}[G_n(t) - t] \quad \text{for} \quad 0 \leq t \leq 1.$

Note that

$$E\tilde{\alpha}_n^2(t) = \text{Variance} + \text{Bias}^2 = (n/(n+2))^2 t(1-t) + n((1-2t)/(n+2))^2$$

(2.6) $\qquad \leq 2t(1-t) \quad \text{for} \quad 1/(n+2) \leq t \leq 1 - 1/(n+2).$

Now observe that, with g_+ being the right continuous version of g,

(2.7) $$\frac{n}{n+2} L_n = \int_{[U_{1,n}, U_{n,n}]} g \, d\Gamma_n(\tilde{G}_n)$$

(2.8) $$\underset{\text{a.s.}}{=} g(U_{n,n})\Gamma_n(1 - a_n) - g(U_{1,n})\Gamma_n(a_n) - \int_{[U_{1,n}, U_{n,n}]} \int_{1/2}^{\tilde{G}_n(t)} J_n(s) ds \, dg(t)$$

using integration by parts and the fact that $U_{n,n}$ is a.s. not equal to one of the, at most countable, discontinuities of g. As a centering constant for L_n we are using μ_n where

$$-\frac{n}{n+2}\mu_n = -\int_{a_n}^{1-a_n} g \, d\Gamma_n$$

(2.9) $$= -g_+(1 - a_n)\Gamma_n(1 - a_n) + g(a_n)\Gamma_n(a_n) + \int_{[a_n, 1-a_n]} \int_{1/2}^{t} J_n(s) \, ds \, dg(t).$$

A bit of rearrangement provides the *basic identity for L- statistics*

(2.10) $T_n \equiv \left[\sqrt{n}(L_n - \mu_n)/\sigma_n\right] (n/(n+2))$

$$= -\int_{[a_n,1-a_n]} \sqrt{n} \int_t^{\tilde{G}_n(t)} J_n(s)\, ds\, dg(t)/\sigma_n$$

$$- \int_{[U_{1,n},a_n)} \sqrt{n} \int_{a_n}^{\tilde{G}_n(t)} J_n(s)\, ds\, dg(t)/\sigma_n$$

$$- \int_{(1-a_n,U_{n,n}]} \sqrt{n} \int_{1-a_n}^{\tilde{G}_n(t)} J_n(s)\, ds\, dg(t)/\sigma_n$$

(2.11) $\equiv \theta_n + d_n + d'_n.$

(Using $\tilde{G}_n$ instead of G_n gave us an identity in which none of the limits of integration is ever equal to 0 or 1. This prevents certain integrals from ever being infinite.)

The following two facts will be important. First, from S. Csörgő, Haeusler and Mason (1988)

$$K(r/n)/(\sqrt{n}\sigma_n) \to 0 \text{ and } K_+(1 - r/n)/(\sqrt{n}\sigma_n) \to 0$$

(2.12)   for all r, when K is in D(Normal).

Secondly, we will show at the end of this proof that

(2.13) $D_n \equiv \int_{[a_n,1-a_n]} n^{-\nu}[t(1-t)]^{\frac{1}{2}-\nu}\, dK(t)/\sigma_n \to 0$ for all K in D(Normal)

follows from (2.12).

From here on, we let $\int_a^b$ denote $\int_{[a,b]}$.

Consider θ_n. Now

$$\theta_n = -\frac{1}{\sigma_n}\int_{a_n}^{1-a_n} \alpha_n\, dK - \frac{1}{\sigma_n}\int_{a_n}^{1-a_n} (\tilde{\alpha}_n - \alpha_n)\, dK$$

$$- \frac{1}{\sigma_n} \int_{a_n}^{1-a_n} \sqrt{n} \int_t^{\hat{G}_n(t)} [J_n(s) - J(t)] \, ds \, dg(t)$$

$$(2.14) \qquad \equiv \theta_{0n} + \gamma_{1n} + \gamma_{2n}.$$

Next, for the special construction of $U_{i,n}$'s and a Brownian bridge B of Mason and van Zwet (1987), one can define

$$(2.15) \qquad \theta_{*n} \equiv -\frac{1}{\sigma_n} \int_{a_n}^{1-a_n} B \, dK \stackrel{d}{=} N(0,1)$$

and obtain that

$$|\theta_{0n} - \theta_{*n}| \le \left(n^\nu \sup\left\{ \frac{|\alpha_n(t) - B(t)|}{[t(1-t)]^{\frac{1}{2}-\nu}} : \frac{\lambda}{n+2} \le t \le 1 - \frac{\lambda}{n+2} \right\} \right).$$

$$\left(\frac{1}{\sigma_n} \int_{a_n}^{1-a_n} n^{-\nu} [t(1-t)]^{\frac{1}{2}-\nu} \, dK(t) \right)$$

$$(2.16) \qquad \equiv M_n D_n = O_p(1) o(1) = o_p(1);$$

use Mason and van Zwet (1987) with $0 < \nu < 1/2$ for $M_n = O_p(1)$ and (2.13) for $D_n \to 0$.

As an alternative to the previous paragraph use (2.12) twice for

$$(2.17)$$
$$\theta_{0n} = \frac{1}{\sigma_n} \int_0^1 K_n \, d\alpha_n = \frac{1}{\sqrt{n}\sigma_n} \sum_{i=1}^n \int_0^1 K_n d[1_{[U_i \le t]} - t] = \frac{1}{\sqrt{n}\sigma_n} \sum_{i=1}^n (Y_{ni} - EY_{ni})$$

$$(2.18) \qquad = \frac{1}{a} \frac{1}{\sqrt{n}\sigma_n} \sum_{i=1}^n \left[K(U_i) - \int_0^1 K_n(t) \, dt \right] = \frac{1}{\sqrt{n}\sigma_n} \sum_{i=1}^n (Y_i - EY_{ni}),$$

with normality coming from (2.17) and the Lindeberg-Feller theorem. Lindeberg-Feller is applicable since, bounding the rv by its maximum, (2.12) gives

$$\sigma_n^{-2} \int_{[|Y_{ni} - EY_{ni}| \ge \epsilon \sqrt{n}\sigma_n]} (Y_{ni} - EY_{ni})^2 \, dP$$

$$\leq \sigma_n^{-2}[K^2(a_n) + K_+^2(1 - a_n)]P(|Y_{ni} - EY_{ni}| \geq \epsilon\sqrt{n}\sigma_n)$$

(2.19)
$$\leq \epsilon^{-2}[K^2(a_n) + K_+^2(1 - a_n)]/(n\sigma_n^2)$$

$$\rightarrow 0.$$

Consider γ_{1n}. Elementary computations show that for any $0 < \nu < 1/2$ and $0 < \lambda < 1$ we have

(2.20)
$$\sup_{\frac{\lambda}{n} \leq t \leq 1 - \frac{\lambda}{n}} \frac{n^\nu|\tilde{\alpha}_n(t) - \alpha_n(t)|}{[t(1 - t)]^{\frac{1}{2} - \nu}} \leq \sup_{\frac{\lambda}{n} \leq t \leq 1 - \frac{\lambda}{n}} \frac{n^{\nu + \frac{1}{2}}|\tilde{G}_n(t) - G_n(t)|}{[t(1 - t)]^{\frac{1}{2} - \nu}} \leq \sqrt{2/\lambda}.$$

Thus the same type of inequality as used in (2.16), followed by (2.13), gives

(2.21)
$$|\gamma_{1n}| \leq 3D_n \rightarrow 0.$$

Consider γ_{2n}. Now on the set $A_{n\epsilon}$ of (2.3) and (2.4) we have

(2.22)
$$1_{n\epsilon}|\gamma_{2n}| \leq \frac{1}{\sigma_n} \int_{a_n}^{1-a_n} \sqrt{n} \left| \int_t^{\tilde{G}_n(t)} M_{\lambda_\epsilon}[t(1 - t)]^{-1} J(t)[|s - t| + 1/n] \, ds \right| dg(t)$$

(2.23)
$$\leq M_{\lambda_\epsilon} \frac{1}{\sigma_n} \int_{a_n}^{1-a_n} \left[n^{-\frac{1}{2}}\tilde{\alpha}_n^2(t) + n^{-1}|\tilde{\alpha}_n(t)| \right] [t(1 - t)]^{-1} J(t) \, dg(t).$$

Hence Fubini's theorem and (2.6) give

$$E1_{n\epsilon}|\gamma_{2n}| \leq M_{\lambda_\epsilon} \frac{1}{\sigma_n} \int_{a_n}^{1-a_n} \left[n^{-\frac{1}{2}}2t(1 - t) + n^{-1}\sqrt{2t(1 - t)} \right] [t(1 - t)]^{-1} \, dK(t)$$

$$\leq 5M_{\lambda_\epsilon} \int_{a_n}^{1-a_n} dK/(\sqrt{n}\sigma_n)$$

(2.24) $\rightarrow 0$

by (2.12). Thus

(2.25) $\gamma_{2n} \rightarrow_p 0.$

Consider d_n. Observe that

(2.26) $\sqrt{n}\left|\tilde{G}_n(t) - a_n\right| \leq |\tilde{\alpha}_n(a_n)|$ for all t between $U_{1,n}$ and a_n.

Now $\tilde{\alpha}_n(a_n) = O_p(n^{-1/2})$ by Chebyshev's inequality and (2.6). Proceeding as with γ_{2n} we thus have, using (2.5) on the limits,

$$1_{n\epsilon}|d_n| \leq 1_{n\epsilon}M_{\lambda_\epsilon} \int_{[U_{1,n},a_n]} \left\{|\tilde{\alpha}_n(a_n)| + [n^{-1/2}\tilde{\alpha}_n^2(a_n) + n^{-1}|\tilde{\alpha}_n(a_n)|]\right.$$

(2.27) $\left. \cdot [t(1-t)]^{-1}\right\} dK(t)/\sigma_n$

(2.28) $\leq M_{\lambda_\epsilon} \left|\int_{\lambda_\epsilon a_n}^{a_n/\lambda_\epsilon}\right| \quad \text{same} \quad \left|/\sigma_n\right.$

$$\leq O_p(1) \int_{\lambda_\epsilon a_n}^{a_n/\lambda_\epsilon} (n^{-1/2} + n^{-3/2}[t(1-t)]^{-1})\, dK(t)/\sigma_n$$

$$= O_p(1) \int_{\lambda_\epsilon a_n}^{a_n/\lambda_\epsilon} dK(t)/(\sqrt{n}\sigma_n)$$

(2.29) $\to_p 0$

from (2.12). Hence

(2.30) $d_n \to_p 0; \quad \text{and} \quad d_n' \to_p 0$

by the symmetry obtained from transforming from U to $1 - U$ in d_n'.

Thus $T_n = \theta_{0n} + \gamma_{1n} + \gamma_{2n} + d_n + d_n' = \theta_{0n} + o_p(1) \to_d N(0,1)$ using (2.15), (2.16), (2.21), (2.25) and (2.30). It remains only to prove (2.13).

Now to show that $D_n \to 0$ for any K in D (Normal) as claimed in (2.13). Let

(2.31) $\tilde{\mu}_n \equiv \frac{r}{n}K\left(\frac{r}{n}\right) + \int_{[r/n,1-r'/n]} K(t)dt + \frac{r'}{n}K_+\left(1 - \frac{r'}{n}\right),$

which is a Winsorized mean. Then

$$
\begin{aligned}
M_n(r, r') \ &\equiv n^{-\nu} \int_{r/n}^{1-r'/n} [t(1-t)]^{1/2-\nu} dK(t)/\sigma_n \\
&= n^{-\nu} \int_{r/n}^{1-r'/n} [t(1-t)]^{1/2-\nu} d(K(t) - \tilde{\mu}_n)/\sigma_n \\
&\leq n^{-\nu} \left\{ \left(\frac{r}{n}\right)^{1/2-\nu} \left| K\left(\frac{r}{n}\right) - \tilde{\mu}_n \right| + \left(\frac{r'}{n}\right)^{1/2-\nu} \left| K_+\left(1 - \frac{r'}{n}\right) - \tilde{\mu}_n \right| \right. \\
&\quad \left. + \int_{r/n}^{1-r'/n} |K(t) - \tilde{\mu}_n| \frac{(\frac{1}{2}-\nu)|1-2t|}{[t(1-t)]^{1/2+\nu}} \, dt \right\} /\sigma_n
\end{aligned}
$$

using integration by parts. Let us apply Cauchy-Schwarz to the last term, and then use $(|a| + |b| + |c|)^2 \leq (3\max(|a|, |b|, |c|))^2 \leq 9(a^2 + b^2 + c^2)$ to get

$$
\begin{aligned}
M_n(r, r') \ &\leq n^{-\nu} \left\{ \left(\frac{r}{n}\right)^{1/2-\nu} \left| K\left(\frac{r}{n}\right) - \tilde{\mu}_n \right| + \left(\frac{r'}{n}\right)^{1/2-\nu} \left| K_+\left(1 - \frac{r'}{n}\right) - \tilde{\mu}_n \right| \right. \\
&\quad \left. + \frac{1}{\sqrt{\nu}} \left(\frac{r \wedge r'}{n}\right)^{-\nu} \left(\int_{r/n}^{1-r'/n} [K(t) - \tilde{\mu}_n]^2 dt \right)^{1/2} \right\} /\sigma_n \\
&\leq \frac{3/\sqrt{\nu}}{(r \wedge r')^{\nu}} \left\{ \frac{r}{n} \left[K\left(\frac{r}{n}\right) - \tilde{\mu}_n \right]^2 + \frac{r'}{n} \left[K_+\left(1 - \frac{r'}{n}\right) - \tilde{\mu}_n \right]^2 \right. \\
&\quad \left. + \int_{r/n}^{1-r'/n} [K(t) - \tilde{\mu}_n]^2 dt \right\}^{1/2} /\sigma_n \\
&\leq (3/\sqrt{\nu})(r \wedge r')^{-\nu} \sigma[r/n, 1 - r'/n]/\sigma[a_n, 1 - a_n]
\end{aligned}
$$

(2.32) $\leq (3/\sqrt{\nu})(r \wedge r')^{-\nu}$ for any qf K, any $0 \leq \nu < 1/2$, and any $r, r' \geq 1$

where

(2.33)

$$
\sigma^2[a, 1 - a'] \equiv \int_{[a, 1-a']} \int_{[a, 1-a']} (s \wedge t - st) dK(s) dK(t) \quad \text{is } \nearrow \text{ as a } \searrow \text{ and } a' \searrow.
$$

Also

$$
n^{-\nu} \int_{1/(n+2)}^{r/n} [t(1-t)]^{\frac{1}{2}-\nu} dK(t)/\sigma_n
$$

(2.34)
$$\leq \sqrt{r} \frac{1}{\sqrt{n}\sigma_n} \int_{1/(n+2)}^{r/n} dK \leq \sqrt{r} \frac{1}{\sqrt{n}\sigma_n} \left| K\left(\frac{1}{n+2}\right) \right|,$$

with an analogous bound in the other tail. Thus

$$D_n \leq \sqrt{r} \left[\frac{1}{\sqrt{n}\sigma_n} \left| K\left(\frac{1}{n+2}\right) \right| + \frac{1}{\sqrt{n}\sigma_n} K_+\left(1 - \frac{1}{n+2}\right) \right] + \frac{3/\sqrt{\nu}}{r^\nu}$$

(2.35)   for all qf K, all $0 \leq \nu < 1/2$, all $r \geq 1$.

By choosing r large and then applying (2.12), we have that (2.13) holds.

REMARK 2.1. This same proof shows that T_n may summed from k to $n+1-k'$ (instead of 1 to n) if $k/(n+2)$ and $(n+1-k')/(n+2)$ replace $1/(n+2)$ and $(n+1)/(n+2)$ in the definitions of a_n, μ_n and σ_n.

Proof of Remark 1.3. The identity (2.11) that $S_n = \theta_n + d_n + d_n'$ is still correct. The proof that $d_n \to_p 0$ and $d_n' \to_p 0$ remains unchanged. The previous proof that $\theta_n \to_d N(0,1)$ requires only one easy observation. Suppose $J = 1_{[r,1)}$ and $J_n = 1_{[r_n,1)}$. Note that the contribution to θ_n in (2.11) made by

$$- \int_{r-\epsilon}^{r+\epsilon} \sqrt{n} \int_t^{\tilde{G}_n(t)} J_n(s) \, ds \, dg(t)/\sigma_n$$

is trivially made as small as we require by choosing ϵ small enough, since $\int_{r-\epsilon}^{r+\epsilon} dg \to 0$ as $\epsilon \to 0$ by continuity of g at r.

Proof of Remark 1.5. We now replace μ_n by EX in case $\text{Var}[X] < \infty$. For n sufficiently large we have that

(2.36)
$$\sqrt{n} \left| \int_0^{1/n} F^{-1}(t) dt \right| \leq \sqrt{n} \left\{ \sup_{0 < t \leq 1/n} \sqrt{t} |F^{-1}(t)| \right\} \int_0^{1/n} t^{-1/2} dt$$
$$\leq \sqrt{n} \{o(1)\} 2/\sqrt{n} \to 0,$$

since, in particular, if $F^{-1}(0) < 0$, for all small $t > 0$

$$t[F^{-1}(t)]^2 \leq \int_0^t [F^{-1}(s)]^2 \, ds \to 0 \text{ as } t \to 0 \text{ when } \text{Var}[F^{-1}(U)] < \infty.$$

The other tail is similar. Thus $\sqrt{n}(\mu_n - EX) \to 0$.

References.

[1] BOOS, D. (1979). A differential for L-statistics. *Ann. Statist.* **7** 955-959.

[2] CHERNOFF, H., GASTWIRTH, J. and JOHNS, M. (1967). Asymptotic distribution of linear combinations of functions of order statistics with applications to estimation. *Ann. Math. Statist.* **38** 52-72.

[3] CSÖRGŐ, S., HAEUSLER, E., and MASON, D. (1988). A probabilistic approach to the asymptotic distribution of sums of independent identically distributed random variables. *Adv. in Appl. Statist.*, **9** 259-333.

[4] HELMERS, R., and RUYMGAART, F. (1988). Asymptotic normality of generalized L-statistics with unbounded scores. *J. Statist. Plann. Inf.* **19** 43-53.

[5] MASON, D. (1981). Asymptotic normality of linear combinations of order statistics with a smooth score function. *Ann. Statist.* **9** 899-904.

[6] MASON, D., and VAN ZWET, W. (1987). A refinement of the KMT inequality for the uniform empirical process. *Ann. Probability* **15** 871-884.

[7] MASON, D and SHORACK, G. (1991). Necessary and sufficient conditions for asymptotic normality of L-statistics. To appear in *Ann. Probability*, 1991.

[8] RUYMGAART, F., and VAN ZUIJLEN, M. (1977). Asymptotic normality of linear combinations of order statistics in the non-iid case. *Proc. Koninklijke Nederl. Akad. Wetensch. Ser. A* **80** (5), 432-447.

[9] SEN, P. (1978). An invariance principle for linear combinations of order statistics. *Z. Wahrsch. verw. Geb.* **42** 327-340.

[10] SHORACK, G. (1969). Asymptotic normality of linear combinations of functions of order statistics. *Ann. Statist.* **40** 2041-2050.

[11] SHORACK, G. (1972). Functions of order statistics. *Ann. Math. Statist.* **43** 412-427.

[12] SHORACK, G. and WELLNER, J. (1986). *Empirical Processes with Applications to Statistics.* John Wiley & Sons, New York.

[13] SINGH, K. (1981). On asymptotic representation and approximation to normality of L-statistics. I. *Sankhya A* **43** 67-83.

[14] STIGLER, S. (1969). Linear functions of order statistics. *Ann. Math. Statist.* **40** 770-788.

[15] STIGLER, S. (1974). Linear functions of order statistics with smooth weight functions. **2** 676-693.

Galen R. Shorack
Department of Statistics
GN-22
University of Washington
Seattle, Washington 98195

NON-NORMALITY OF A CLASS OF RANDOM VARIABLES

David M. Mason[1] and Galen R. Shorack[2]

Let Φ be a non-decreasing, non-positive, left continuous function defined on $(0, \infty)$ for which

$$(1) \qquad \int_{\epsilon}^{\infty} \Phi^2(s)ds < \infty \quad \text{for all } \epsilon > 0.$$

Thus $b\Phi^2(b) \to 0$ as $b \to \infty$, since $(b/2)\Phi^2(b) \le \int_{b/2}^{b} \Phi^2(s)ds \to 0$ as $b \to \infty$. Let $N(t), 0 \le t < \infty$, be a right continuous Poisson process with rate 1 and corresponding jump times $S_1, S_2, \ldots$. Set for all $-\infty < \rho < \infty$ and $0 < u, v < \infty$

$$h_\rho(u,v) = \int_u^v w^\rho dw/v^\rho = \begin{cases} (v^{\rho+1} - u^{\rho+1})/((\rho+1)v^\rho), & \rho \ne -1 \\ v\log(v/u), & \rho = -1. \end{cases}$$

For any $-\infty < \rho < \infty$ and integer $k \ge 1$ consider the random variable

$$V_{\rho,k} = \int_{S_k}^{\infty} h_\rho(N(s), s)d\Phi(s) + \int_k^{S_k} h_\rho(k, s)d\Phi(s).$$

This class of rv arises as natural limits for L-statistics in Mason and Shorack (1991). In that paper necessary and sufficient conditions are derived for the asymptotic normality of a general class of L-statistics. Essential to the proof of the necessity part was the following technical result.

THEOREM 1. *The random variable $V_{\rho,k}$ is never a non-degenerate normal random variable for any choice of $-\infty < \rho < \infty$ and integer $k \ge 1$.*

---

[1] Supported in part by National Science Foundation Grant DMS-8803209.
[2] Supported in part by National Science Foundation Grant DMS-8801083.

Proof. Roughly, the idea of the proof is to show that if $\rho \leq -1/2$, then for any integer $k \geq 1, V_{\rho,k}$, has a lighter left tail than any non-degenerate normal random variable, whereas if $\rho > -1/2$ and Φ is not identically equal to zero, then $V_{\rho,k}$ has a heavier left tail than any non-degenerate normal random variable. This of course proves that $V_{\rho,k}$ cannot be non-degenerate normal for any choice of ρ and integer $k \geq 1$. For convenient reference later on we begin by collecting a number of facts essential to the main body of the proof.

For all $-\infty < \rho < \infty$ and $0 < \lambda < \infty$ let

$$\zeta_\rho(\lambda) = \begin{cases} (1 - \lambda^{\rho+1})/(\rho + 1), & \rho \neq -1 \\ -\log \lambda, & \rho = -1. \end{cases}$$

Observe that ζ_ρ is strictly decreasing,

$$(2) \qquad\qquad\qquad \zeta_\rho(u/v) = \nu^{-1} h_\rho(u, v),$$

and for any $-\infty < \rho < \infty$ and $0 < u, v, b < \infty$ we have trivially that

$$(3) \qquad h_\rho(u, v + b) = (v + b)\zeta_\rho(u/(v + b)) = b\zeta_\rho(u/(v + b)) + v\zeta_\rho(u/(v + b)).$$

Notice that for each fixed v,

$$(4) \qquad\qquad\qquad h_\rho(u, v) \text{ is a strictly decreasing function of } u,$$

and for $0 < u, v < \infty$ (using a two term Taylor expansion)

$$(5) \qquad\qquad\qquad h_\rho(u, v) \leq (\geq) h_0(u, v) \text{ when } \rho \geq 0 \text{ (when } \rho \leq 0).$$

Also since ζ_ρ is a strictly decreasing function of λ, we have from (3) that for any $-\infty < \rho < \infty$ and $0 < u, v, b < \infty$

$$(6) \qquad\qquad\qquad h_\rho(u, v + b) \geq b\zeta_\rho(u/(v + b)) + h_\rho(u, v).$$

We shall require the following three Poisson distribution inequalities.

INEQUALITY 1. *For every $\lambda > 0$ we have*

(7)
$$P(N(\lambda) \geq x\lambda) \begin{array}{l} \leq \exp(\lambda(x - 1 - x \log x)), \quad \text{if } x > 1. \\ \leq \exp(-(\lambda x \log x)/2), \qquad \text{if } x \geq e^2. \end{array}$$

Proof. Use elementary moment generating function techniques. See, for example, Shorack and Wellner (1986, p. 486). □

INEQUALITY 2. *For every $\lambda > 0$ there exists a constant $0 < K(\lambda) < \infty$ such that for all $z \geq 1$ we have*

(8)
$$P(N(\lambda) \geq z) \geq K(\lambda) z^{-1/2} \exp(-z(\log z - \log(\lambda e))).$$

Proof. The proof follows easily from Stirling's formula. □

INEQUALITY 3. *For all $0 < \lambda < 1$ and $z > 0$ we have*

(9)
$$P\left(\sup_{t \geq 0}\{N(t) - t/\lambda\} > z\right) \leq D(\lambda) \exp(z \log \lambda),$$

where $0 < D(\lambda) = (1 - \lambda e^{-\lambda+1})^{-1} < \infty$.

Proof. From Dwass (1974), e.g. Shorack and Wellner (1986, p. 392);

(10)
$$P\left(\sup_{t \geq 0}\{N(t) - t/\lambda\} > z\right) = \sum_{n > z} \frac{(n - z)^n}{n!}(\lambda e^{-\lambda})^n e^{zz}(1 - \lambda).$$

Now by using $n! > (n/e)^n$, $(1 - z/n)^n \leq e^{-z}$ for $0 < z < n$ and $\lambda \exp(1 - \lambda) < 1$ for $0 < \lambda < 1$, we see that the right side of (10) is bounded above by

$$\sum_{n > z} (\lambda e^{-\lambda+1})^n \exp((\lambda - 1)z)(1 - \lambda) \leq D(\lambda)\lambda. \quad \square$$

For any $-\infty < \rho < \infty$ and integer $k \geq 1$, we can write

$$
(11) \quad \begin{aligned}
V_{\rho,k} &= \int_k^\infty h_\rho(N(s) \vee k, s)d\Phi(s) + \int_{S_k}^k h_\rho(N(s) \vee k, k)(k/s)^\rho d\Phi(s) \\
&\equiv V_{\rho,k}^{(1)} + V_{\rho,k}^{(2)}.
\end{aligned}
$$

For any $x \geq 0$, set

$$
V_{\rho,k}(x) = \int_x^\infty h_\rho(N(s-x) + k, s)d\Phi(s) + \int_k^x h_\rho(k, s)d\phi(s).
$$

Note from the original representation for $V_{\rho,k}$ that

$$
(12) \qquad\qquad \text{conditioned on } S_k = x, \quad V_{\rho,k} \stackrel{d}{=} V_{\rho,k}(x).
$$

We shall first prove the theorem for the harder case $\rho \leq -1/2$. This will require a number of lemmas.

LEMMA 1. *For all* $x, b \geq 0$, $-\infty < \rho < \infty$ *and integers* $k \geq 1$ *we have*

$$
(13) \qquad\qquad P(V_{\rho,k}(x) > z) \leq P(V_{\rho,k}(x+b) > z).
$$

Proof. By (4) we have both

$$
\int_k^x h_\rho(k, s)d\Phi(s) + \int_x^{x+b} h_\rho(N(s-x) + k, s)d\Phi(s) \leq \int_k^{x+b} h_\rho(k, s)d\Phi(s)
$$

and conditioned on $N(b) = m$,

$$
\begin{aligned}
\int_{x+b}^\infty h_\rho(N(s-x) + k, s)d\Phi(s) &\stackrel{d}{=} \int_{x+b}^\infty h_\rho(N(s-x-b) + m + k, s)d\Phi(s) \\
&\leq \int_{x+b}^\infty h_\rho(N(s-x-b) + k, s)d\Phi(s).
\end{aligned}
$$

By the above two inequalities we have (13). □

LEMMA 2. *For all x and z*

(14) $$P(V_{\rho,k}^{(1)} > z | S_k = x) \leq P(V_{\rho,k}(x+k) > z).$$

Proof. Since $V_{\rho,k}^{(1)} = V_{\rho,k}$ when $S_k \geq k$, by Lemma 1 we need only consider the case $0 < x < k$.

Given $S_k = x$ with $0 < x < k$,

$$V_{\rho,k}^{(1)} = \int_k^\infty h_\rho(N(s), s) d\Phi(s);$$

and this conditional distribution is the same as the distribution of

$$\int_k^\infty h_\rho(N(s-x) + k, s) d\Phi(s).$$

By (4), the latter rv is

$$\leq \int_k^\infty h_\rho(N(s-k) + k, s) d\Phi(s) = V_{\rho,k}(k).$$

Thus for $0 < x < k$

$$P(V_{\rho,k}^{(1)} > z | S_k = x) \leq P(V_{\rho,k}(k) > z),$$

which by Lemma 1 is less than or equal to $P(V_{\rho,k}(k+x) > z)$. □

LEMMA 3. *For each $k \geq 1$ there exists a constant $0 < c_k < \infty$ such that for all $-\infty < \rho, z < \infty$*

(15) $$P(V_{\rho,k}^{(1)} > z) \leq c_k P(V_{\rho,k} > z).$$

Proof. Let f_k denote the density of S_k,

$$P(V_{\rho,k}^{(1)} > z) = \int_0^\infty P(V_{\rho,k}^{(1)} > z | S_k = x) f_k(x) dx,$$

which by Lemma 2 is

$$\leq \int_0^\infty P(V_{\rho,k}(x+k) > z) f_k(x) dx$$

$$\leq c_k \int_0^\infty P(V_{\rho,k}(x) > z) f_k(x) dx,$$

where

$$c_k = \sup_{x \geq 0} f_k(x)/f_k(x+k).$$

By (12) the right side of the last inequality equals $c_k P(V_{\rho,k} > z)$. □

For any $b > 0$, set $\Phi_b(s)$ equal $\Phi(b)$, $\Phi(s)$ according as $0 < s \leq b$, $b < s$, and let

$$U(b) = \int_{S_1}^\infty (s - N(s)) d\Phi_b(s) + \int_1^{S_1} s d\Phi_b(s) + \Phi(b).$$

LEMMA 4. *For all $b \geq 1$ and $-\infty < t < -\gamma/\Phi(b)$, where $\gamma > 0$,*

(16) $$E \exp(-tU(b)) \leq \exp\left(-t\Phi(b) + \frac{t^2}{2} f(b) e^\gamma\right),$$

where

$$f(b) = \int_0^\infty \Phi_b^2(s) ds = b\Phi^2(b) + \int_b^\infty \Phi^2(s) ds.$$

Proof. Arguing as in the proof of Theorem 3 of CsHM [1988], one obtains

$$\log E \exp(-tU(b)) = \int_0^\infty \left\{ \exp(-t\Phi_b(u)) - 1 + t \frac{\Phi_b(u)}{1 + \Phi_b^2(u)} \right\} du - t\gamma_b,$$

where

$$\gamma_b = \int_0^1 \frac{\Phi_b(u)}{1 + \Phi_b^2(u)} du - \int_1^\infty \frac{\Phi_b^3(u)}{1 + \Phi_b^2(u)} du = \frac{\Phi(b)}{1 + \Phi^2(b)} - \int_1^\infty \frac{\Phi_b^3(u)}{1 + \Phi_b^2(u)} du.$$

Using

$$\exp(v) - 1 \leq v + \frac{v^2}{2} e^\gamma, \quad -\infty < v \leq \gamma,$$

we get

$$\log E \exp(-tU(b)) \leq \int_0^\infty \frac{-t\Phi_b^3(u)}{1 + \Phi_b^2(u)} du + \frac{t^2}{2} e^\gamma f(b) - t\gamma_b$$

$$= -t\Phi(b) + \frac{t^2}{2} e^\gamma f(b). \square$$

LEMMA 5. *For all* $\tau > 0$ *and* $0 < \lambda \leq 1$

(17) $$e^{\tau z^2} P(-U(z^2) > \lambda z) \to 0 \text{ as } z \to \infty.$$

Proof. By Lemma 4, for all $0 < t < -\lambda/\Phi(z^2)$ with $0 < \lambda \leq 1$ and $z > 1$

(18) $$P(-U(z^2) > \lambda z) \leq \exp(-tz\lambda - t\Phi(z^2) + \frac{t^2}{2} e^\lambda f(z^2)).$$

Setting

$$t = \lambda(e^\lambda f(z^2)/z^2)^{-1/2} \leq -\lambda/\Phi(z^2)$$

into (18) we obtain

$$P(-U(z^2) > z\lambda) \leq \exp\{-\lambda^2 z^2/(e^\lambda f(z^2))^{1/2} + (1/2)\lambda^2 z^2 \\ - z\lambda\Phi(z^2)/(e^\lambda f(z^2))^{1/2}\}$$

$$\leq \exp\{-\lambda^2 z^2/(e^\lambda f(z^2))^{1/2} + (1/2)\lambda^2 z^2 + \lambda\}.$$

Since $f(z^2) \to 0$ as $z \to \infty$, we have (17). $\square$

For any $b > 0$, set

$$W_{\rho,k}(b) = \int_b^\infty h_\rho(N(s) \vee k, s) d\Phi(s),$$

and when $b > k$ let

$$\bar{W}_{\rho,k}(b) = \int_k^b h_\rho(N(s) \vee k, s) d\Phi(s).$$

Notice that when $b > k$ we have

$$
\begin{aligned}
W_{\rho,k}(b) &= \int_k^\infty h_\rho(N(s) \vee k, s) d\Phi_b(s) + \int_{S_k}^k h_\rho(N(s) \vee k, k)(k/s)^\rho d\Phi_b(s) \\
&= \int_{S_k}^\infty h_\rho(N(s), s) d\Phi_b(s) + \int_k^{S_k} h_\rho(k, s) d\Phi_b(s),
\end{aligned}
$$

and

$$
(19) \qquad\qquad V_{\rho,k}^{(1)} = \bar{W}_{\rho,k}(b) + W_{\rho,k}(b).
$$

LEMMA 6. *For every $\rho \le 0$, $k \ge 1$ and $\tau > 0$ there exists an $m > 0$ such that for all $0 < \lambda \le 1$ we have*

$$
(20) \qquad\qquad e^{\tau z^2} P(W_{\rho,k}((mz)^2) < -\lambda mz) \to 0 \text{ as } z \to \infty.
$$

Proof. First assume $\rho = 0$ and $k = 1$. In this case the assertion follows from the inequality

$$
W_{0,1}(z^2) = U(z^2) - \Phi_{z^2}(S_1) \ge U(z^2)
$$

and Lemma 5. For $\rho < 0$ and $k = 1$, (20) is immediate from the inequality (see (5))

$$
W_{\rho,1}(z^2) \ge W_{0,1}(z^2).
$$

Now assume $k > 1$ and $\rho \le 0$. Notice that if $S_k < (mz)^2$ and $k < (mz)^2$, then

$$
\begin{aligned}
W_{\rho,k}((mz)^2) &= \int_k^\infty h_\rho(N(s) \vee k, s) d\Phi_{(mz)^2}(s) \\
&= \int_1^\infty h_\rho(N(s) \vee 1, s) d\Phi_{(mz)^2}(s) = W_{\rho,1}((mz)^2).
\end{aligned}
$$

Thus, for any $m > 0$, as soon as z is large enough so that $(mz)^2 > k$ we have

$$P(W_{\rho,k}((mz)^2) < -m\lambda z) \leq P(W_{\rho,1}((mz)^2) < -m\lambda z) + P(S_k \geq (mz)^2).$$

By choosing m large enough so that

$$e^{\tau z^2} P(S_k \geq (mz)^2) \to 0 \text{ as } z \to \infty,$$

we see that the assertion follows. $\square$

Our next goal is to prove the following lemma.

LEMMA 7. For all $\rho \leq -1/2$, $k \geq 1$ and $0 < \tau < \infty$

$$(21) \qquad \liminf_{z \to \infty} e^{\tau z^2} P(\bar{W}_{\rho,k}(z^2) < -\lambda z) = 0$$

for all $0 < \lambda \leq 1$.

Proof. We must consider a number of cases separately. We will need a number of sub-lemmas, to be numbered 7.1 to 7.6

For any $b > 1$ and $-\infty < \rho < \infty$, set

$$\mu_\rho(b) = \int_1^b u^{-\rho} d\Phi(u).$$

LEMMA 7.1. For all $-1 \leq \rho < -1/2$

$$(22) \qquad \mu_\rho(b) = o(b^{-\rho-1/2}).$$

Proof. Choose any $x > 1$ and $b > x$, we see by integrating by parts that

$$\mu_\rho(b) = b^{-\rho}\Phi(b) - \Phi(1) + \rho \int_1^b u^{-\rho-1}\Phi(u)du$$

$$\leq -\Phi(1) + \rho \int_1^x u^{-\rho-1}\Phi(u)du - \rho \left(\int_x^b \Phi^2(u)du\right)^{1/2} \left(\int_x^b u^{-2\rho-2}du\right)^{1/2}$$

$$\leq -\Phi(1) + \rho \int_1^x u^{-\rho-1}\Phi(u)du - \frac{\rho}{(-2\rho-1)^{1/2}} \left(\int_x^\infty \Phi^2(u)du\right)^{1/2} b^{-\rho-1/2}.$$

Therefore

$$\limsup_{b\to\infty} \mu_\rho(b)b^{\rho+1/2} \leq \frac{-\rho}{(-2\rho-1)^{1/2}} \left(\int_x^\infty \Phi^2(u)du\right)^{1/2} \quad \text{for all } x > 1.$$

Letting $x \to \infty$ proves (22). □

We must now consider four cases separately.

Case 1. $\rho < -1$.

In this case $\bar{W}_{\rho,k}(z^2)$ is bounded below by

$$\mu_{-1}(z^2)/(1+\rho),$$

which by Lemma 7.1 is equal to $o(z)$. This shows that (21) is true with lim replacing liminf.

Case 2. $-1 < \rho < -1/2$.

Notice that for $z^2 > 1$

$$\bar{W}_{\rho,k}(z^2) \geq - \left[N(z^2) \vee k\right]^{\rho+1} \mu_\rho(z^2)/(\rho+1).$$

On the event $B_{z,l} = [N(z^2) \leq lz^2]$, with $l > e^2$ required below, we have

(23) $$-N(z^2)^{\rho+1}\mu_\rho(z^2)/(\rho+1) \geq -l^{\rho+1}z^{2\rho+2}\mu_\rho(z^2)/(\rho+1).$$

By Lemma 7.1 the right side of (23) is equal to $o(z)$. On the other hand, by Inequality 1

$$P(\bar{B}_{z,l}) \leq \exp(-z^2 l \log l/2).$$

Since l can be chosen arbitrarily large, we have (21) with lim.

Case 3. $\rho = -1$.

For this case we require a lemma.

LEMMA 7.2.

(24)
$$\lim_{b\to\infty} (\log b) b^{-1/2} \mu_{-1}(b/(\log b)^2) = 0$$

and

(25)
$$\liminf_{b\to\infty} (\log\log b) b^{-1/2} \{\mu_{-1}(b) - \mu_{-1}(b/(\log b)^2)\} = 0.$$

Proof. First, (24) follows from (22). Integrating by parts we see that for $b > e$ we have

$$(\log\log b) b^{-1/2} \{\mu_{-1}(b) - \mu_{-1}(b/(\log b)^2)\}$$

$$= b^{-1/2}(\log\log b) \int_{b/(\log b)^2}^{b} u \, d\Phi(u)$$

$$\leq -b^{1/2}(\log b)^{-2}(\log\log b)\Phi(b/(\log b)^2) - b^{-1/2}(\log\log b) \int_{b/(\log b)^2}^{b} \Phi(u) \, du$$

$$\leq o(1)(\log\log b)/\log b + \{(\log\log b)^2 \int_{b/(\log b)^2}^{b} \Phi^2(u) \, du\}^{1/2},$$

using $z\Phi^2(z) \to 0$ as $z \to \infty$ from (1). Hence to prove (25) it suffices to show that

(26)
$$\liminf_{b\to\infty} (\log\log b)^2 \int_{b/(\log b)^2}^{b} \Phi^2(u) \, du = 0.$$

Suppose the lim inf in (26) is greater than $c > 0$, then for all $k \geq k_0$, for some $k_0 > 0$, we have

$$\int_{2^{(k+1)^2}/((k+1)^2 \log 2)^2}^{2^{(k+1)^2}} \Phi^2(u)du \geq c/(\log((k+1)^2 \log 2))^2 \geq c/(\log(k+1))^2.$$

Since for all large enough k

$$2^{(k+1)^2-k^2} \geq ((k+1)^2 \log 2)^2,$$

this gives for all large k

$$\int_{2^{k^2}}^{2^{(k+1)^2}} \Phi^2(u)du \geq c/(\log(k+1))^2,$$

which contradicts (1). Therefore (26) must be true. □

We are now prepared to finish the proof of Case 3.

Choose any $\tau > 0$ and $k \geq 1$. Notice that on the event

$$C_{z,l} = \left\{ \sup_{s \geq 0} \{N(s) - ls\} \leq z^2 \right\}, \quad l \geq k,$$

we have

$$(N(s) \vee k)/s \leq l + z^2, \quad \text{when } k \leq s \leq z^2/(\log(z^2))^2$$

and

$$(N(s) \vee k)/s \leq l + (\log(z^2))^2, \quad \text{when } z^2/(\log(z^2))^2 \leq s \leq z^2.$$

Setting

$$
\begin{aligned}
-z(l) \;=\; & -\log(l + z^2)\mu_{-1}(z^2/(\log(z^2))^2) \\
& - \log(l + (\log(z^2))^2)\{\mu_{-1}(z^2) - \mu_{-1}(z^2/\log(z^2))^2)\},
\end{aligned}
$$

we see that

$$P(\bar{W}_{-l,k}(z^2) \leq -z(l)) \leq P(\bar{C}_{z,l}),$$

which by Inequality 3 is

$$\leq D(l^{-1}) \exp(-z^2 \log l).$$

Notice by Lemma 7.2 that for all $l \geq k$ we have

$$\liminf_{z \to \infty} z(l)/z = 0,$$

so if we select l large enough we obtain (21) for $\rho = -1$.

Finally we consider

Case 4. $\rho = -1/2.$

We shall need a number of lemmas.

LEMMA 7.3.

(27) $$\lim_{b \to \infty} \mu_{-1/2}(b)/(\log b)^{1/2} = 0.$$

Proof. Integrating by parts, we have for $1 < x < b$,

$$
\begin{aligned}
\mu_{-1/2}(b) &\leq -\Phi(1) - (1/2) \int_1^b u^{-1/2} \Phi(u) du \\
&\leq -\Phi(1) - (1/2) \int_1^x u^{-1/2} \Phi(u) du + \left(\int_x^\infty \Phi^2(u) du \right)^{1/2} (\log b)^{1/2}.
\end{aligned}
$$

Therefore for all $x > 1$

$$\limsup_{b \to \infty} \mu_{-1/2}(b)/(\log b)^{1/2} \leq \left(\int_x^\infty \Phi^2(u) du \right)^{1/2}.$$

Letting $x \to \infty$ proves (27). □

LEMMA 7.4.

$$(28) \qquad \liminf_{b \to \infty} \log \log b \int_b^{b^2} \Phi^2(u) du = 0.$$

Proof. Suppose that the lim inf in (28) is greater than $c > 0$. Then for all large enough k

$$\int_{2^{2^k}}^{2^{2^{k+1}}} \Phi^2(u) du > c/(\log(2^k \log 2)) \geq c/(2k).$$

This contradicts (1). □

Set
$$k(z) = [\log z / \log \log z] \text{ and}$$

$$z_k = z^2/(\log z)^k, k = 1, \ldots, k(z), z_{k(z)+1} = z.$$

Also let
$$S(z) = \sum_{k=1}^{k(z)} \int_{z_{k+1}}^{z_k} \frac{u^{1/2} d\Phi(u)}{(k \log \log z)^{1/2}} + \int_{z_1}^{z^2} u^{1/2} d\Phi(u).$$

LEMMA 7.5.

$$(29) \qquad \liminf_{z \to \infty} S(z) = 0.$$

Proof. Notice that

$$0 \leq S(z) \leq 2^{1/2} \int_z^{z^2/\log z} (\log(z^2/u))^{-1/2} u^{1/2} d\Phi(u) + \int_{z_1}^{z^2} u^{1/2} d\Phi(u).$$

Integrate this by parts to find that for all large z (use also that (1) implies $\lim_{b \to \infty} b\Phi^2(b) = 0$)

$$S(z) \leq 2^{1/2}(\log \log z)^{-1/2} z_1^{1/2} \Phi(z_1) - 2^{1/2}(\log z)^{-1/2} z^{1/2} \Phi(z)$$

$$- \int_z^{z_1} (\log(z^2/u))^{-1/2} u^{-1/2} \Phi(u) du + z\Phi(z^2) - z_1^{1/2} \Phi(z_1)$$

$$-1/2 \int_{z_1}^{z^2} u^{-1/2} \Phi(u) du$$

$$\leq o(1) + \left(\int_z^{z_1} \Phi^2(u) du \right)^{1/2} \left(\int_z^{z_1} (\log(z^2/u)u)^{-1} du \right)^{1/2}$$

$$+ \left(\int_{z_1}^{z^2} \Phi^2(u) du \right)^{1/2} (\log \log z)^{1/2}$$

$$\leq o(1) + 2 \left(\log \log z \int_z^{z^2} \Phi^2(u) du \right)^{1/2} .$$

Lemma 7.4 completes the proof. ◻

LEMMA 7.6. *Let* $l > e^2$. *Then*

(30) $$P(N(z^2) > lz^2) < \exp(-lz^2/4).$$

Also, for all z sufficiently large

(31) $$P(N(z) > lz^2/\log z) < \exp(-lz^2/4),$$

and

(32) $$P(N(z_k) > lz^2/(k \log \log z)) < \exp(-lz^2/4)$$

uniformly in $1 \leq k \leq k(z)$.

Proof. Obviously (30) follows from Inequality 1. For (31) and (32) we see by the same inequality that for all large z

$$P(N(z) > lz^2/\log z) < \exp\left(-\frac{lz^2}{2\log z} \log(lz/\log z) \right),$$

and uniformly in $1 \leq k \leq k(z)$

$$P(N(z_k) > lz^2/(k \log \log z)) \quad \lessdot \quad \exp\left(-\frac{lz^2}{2} + \frac{z^2 l}{2k \log \log z} \log(k \log \log z)\right)$$

$$\leq \quad \exp\left(-\frac{lz^2}{2} + \frac{z^2 l}{2}\left\{\frac{K}{\log \log z} + \frac{\log \log \log z}{\log \log z}\right\}\right),$$

where

$$K = \sup_{k \geq 1} \frac{\log k}{k}.$$

We easily see now that for all z sufficiently large (31) and (32) hold. □

We are now ready to complete the proof of Case 4: $\rho = -1/2$. Choose any $0 < \tau < \infty$ and $l > e^2$, and set

$$E_{z,l} = \{N(z^2) \leq lz^2, N(z) \leq lz^2/\log z, N(z_k) \leq lz^2/(k \log \log z) \text{ for } 1 \leq k \leq k(z)\}.$$

Notice that

$$\bar{W}_{-1/2,k}(z^2) \quad \geq \quad -(N(z) \vee k)^{1/2}\mu_{-1/2}(z) - \sum_{k=1}^{k(z)}(N(z_k) \vee k)^{1/2}\int_{z_{k+1}}^{z_k} u^{1/2}d\Phi(u)$$

$$- \quad (N(z^2 \vee k))^{1/2}\int_{z_1}^{z^2} u^{1/2}d\Phi(u),$$

which on the event $E_{z,l}$ is for all z sufficiently large

$$\geq -zl^{1/2}\{(\log z)^{-1/2}\mu_{-1/2}(z) + S(z)\} \equiv -z(l).$$

Observe that by Lemma 7.6, for all z large

$$P(\bar{E}_{z,l}) \leq 2\log z \exp(-lz^2/4),$$

and by Lemmas 7.3 and 7.5 for all $l > 2e^2$

$$\liminf_{z \to \infty} z(l)/z = 0.$$

Thus by choosing l large enough we obtain (21) for the case $\rho = -1/2$. This completes the proof of Lemma 7. □

Combining Lemmas 6 and 7 with (19) yields

LEMMA 8. *For all* $0 < \tau < \infty$, $\rho \leq -1/2$ *and* $k \geq 1$ *we have*

$$(33) \qquad \liminf_{z \to \infty} e^{\tau z^2} P(V_{\rho,k}^{(1)} < -z) = 0.$$

Next we study the $V_{\rho,k}^{(2)}$ term. Recall (11) above.

LEMMA 9. *Whenever there exists a* $0 < \tau < \infty$ *such that*

$$(34) \qquad e^{-\tau z^2} \geq P(V_{\rho,k}^{(2)} \leq -z)$$

for all large enough z *with* $k \geq 1$ *and* $\rho \leq 0$, *then there exists a constant* $0 < b < \infty$ *such that*

$$(35) \qquad \Phi^2(u) \leq -b \log u \text{ for all } u > 0 \text{ small enough.}$$

Proof. We can assume that $\Phi(u) \to -\infty$ as $u \to 0$, otherwise (35) is trivial. Note

$$
\begin{aligned}
P(V_{\rho,k}^{(2)} \leq -z) &= P\left(\int_{S_{k+1}}^{k} h_\rho(k \vee N(s), k)(k/s)^\rho d\Phi(s) < -z \right) \\
&\geq P(h_\rho(k+1,k)(\Phi(k) - \Phi(S_{k+1})) < -z) \\
&= P(-h_\rho(k+1,k)\Phi(S_{k+1}) < -z - h_\rho(k+1,k)\Phi(k)),
\end{aligned}
$$

which for large enough z is

$$\geq P(\Phi(s_{k+1}) \leq -\lambda_k z)$$

where

$$\lambda_k = -2/h_\rho(k+1, k) > 0.$$

Set

$$L(x) = \sup\{s : \Phi(s) \le x\}, \quad -\infty < x < 0.$$

Obviously

(36) $$L(\Phi(s)) \ge s, \quad 0 < s < \infty.$$

Now

$$
\begin{aligned}
P(\Phi(S_{k+1}) \le \lambda_k z) &= \int_0^{L(-\lambda_k z)} \frac{u^k}{k!} e^{-u} du \\
&\ge \frac{e^{-k}}{(k+1)!} (L(-\lambda_k z))^{k+1} \text{ for large enough } z
\end{aligned}
$$

since $L(x) \searrow 0$ as $x \to -\infty$. From (34) we conclude that there exists a $0 < b < \infty$ such that for all large enough z

$$e^{-z^2/b} \ge L(-z),$$

which by (36) implies that for all $u > 0$ small enough

$$\exp(-\Phi^2(u)/b) \ge u.$$

This last inequality gives (35). □

LEMMA 10. *Suppose (34) holds for some* $\tau > 0$, $\rho \le -1/2$ *and* $k \ge 1$. *Then*

(37) $$e^{\lambda z^2} P(V_{\rho,k}^{(2)} \le -z) \to 0 \text{ as } z \to \infty$$

for all $\lambda > 0$.

Proof. Note from definition (11) that

$$
\begin{aligned}
V_{\rho,k}^{(2)} &= \int_{S_k}^{k} h_\rho(k \vee N(k), k)(k/s)^\rho d\Phi(s) \\
&= \int_{S_k}^{k} s^{-\rho} k^{\rho+1} \zeta_\rho \left(\frac{k \vee N(k)}{k} \right) d\Phi(s).
\end{aligned}
$$

Integration by parts gives

$$
C_\rho \equiv k^{\rho+1} \int_0^{k} s^{-\rho} d\Phi(s) < \infty
$$

using (35).

When $\rho < -1$ we see that

$$
0 \geq \zeta_\rho \left(\frac{k \vee N(k)}{k} \right) \geq 1/(\rho+1),
$$

so that in this case we have

$$
V_{\rho,k}^{(2)} \geq C_\rho/(\rho+1), \text{ with } C_\rho/(\rho+1) < 0.
$$

Thus (37) holds trivially in this case.

Now assume $-1 \leq \rho \leq -1/2$. For $z > 1$, let

$$
F_{\rho,z} = \begin{cases} \{N(k) < kz^{1/(\rho+1)}\}, & \text{if } -1 < \rho \leq -1/2 \\ \{N(k) < ke^z\}, & \text{if } \rho = -1. \end{cases}
$$

On the event $F_{\rho,z}$, $-1 < \rho \leq -1/2$, we have

$$
V_{\rho,k}^{(2)} \geq -\frac{C_\rho}{\rho+1} z;
$$

and on the event $F_{-1,z}$ we have

$$
V_{-1,k}^{(2)} \geq -C_{-1} z.
$$

Applying Inequality 1, it is easy to check that for all $-1 \leq \rho \leq -1/2$ and $0 < \lambda < \infty$,

$$e^{\lambda z^2} P(\bar{F}_{\rho,z}) \to 0 \text{ as } z \to \infty.$$

This completes the proof of (37). □

We are now prepared to finish the proof of Theorem 1 for the case $\rho \leq -1/2$.

Assume that $V_{\rho,k}$ is non-degenerate normal for some $\rho \leq -1/2$ and $k \geq 1$. This implies that there exists a $0 < \tau < \infty$ such that

$$(38) \qquad \limsup_{z \to \infty} e^{\tau z^2} \{P(V_{\rho,k} > z) + P(V_{\rho,k} < -z)\} < \infty$$

which by Lemma 3 gives

$$(38) \qquad \limsup_{z \to \infty} e^{\tau z^2} P(V_{\rho,k}^{(1)} > z) < \infty$$

Now by (11)

$$P(V_{\rho,k} < -z/2) \geq P(V_{\rho,k}^{(2)} < -z) - P(V_{\rho,k}^{(1)} > z/2).$$

Therefore from (38) and (39)

$$(40) \qquad \limsup_{z \to \infty} e^{\tau z^2/4} P(V_{\rho,k}^{(2)} < -z) < \infty.$$

This in turn implies by Lemma 10 that for all $0 < \gamma < \infty$

$$(41) \qquad \lim_{z \to \infty} e^{\gamma z^2} P(V_{\rho,k}^{(2)} < -z) = 0.$$

Finally, for all $z > 0$, we have

$$P(V_{\rho,k} < -z) \leq P(V_{\rho,k}^{(1)} < -z/2) + P(V_{\rho,k}^{(2)} < -z/2);$$

and from (41) and Lemma 8 we infer that for all $0 < \gamma < \infty$

$$\liminf_{z \to \infty} e^{\gamma z^2} P(V_{\rho,k} < -z) = 0.$$

This contradicts the assumption that $V_{\rho,k}$ is non-degenerate normal, since if it were

$$\lim_{z \to \infty} e^{\gamma z^2} P(V_{\rho,k} < -z) = \infty$$

for all γ large enough. This completes the proof of the theorem for $\rho \le -1/2$.

For the proof for the case $\rho > -1/2$, we require a number of additional lemmas. Given $x, b > 0$, $-\infty < \rho < \infty$ and $k \ge 1$ set

$$U_{\rho,k}^b(x) = \int_x^{b+x} h_\rho(N(s - x) + k, s) d\Phi(s),$$

$$\bar{U}_{\rho,k}^b(x) = \int_{b+x}^\infty h_\rho(N(s - x) + k, s) d\Phi(s),$$

$$g_k(x) = \int_k^x h_\rho(k, s) d\Phi(s),$$

and let $U_{\rho,k}(x) = U_{\rho,k}^\infty(x)$. Notice that

(42) $$V_{\rho,k}(x) = U_{\rho,k}^b(x) + \bar{U}_{\rho,k}^b(x) + g_k(x) = U_{\rho,k}(x) + g_k(x).$$

Also note that conditioned on $N(b) = m$, $\bar{U}_{\rho,k}^b(x)$ is equal in distribution to

$$U_{\rho,k+m}(x + b).$$

LEMMA 11. *For all $0 < x$, $b < \infty$, $-\infty < \rho$, $z < \infty$ and $k \ge 1$*

(43) $$P(\bar{U}_{\rho,k}^b(x) > z) \le P(U_{\rho,k}(x + b) > z).$$

Proof. Using the above conditional statement for the second equality, we have

$$P(\bar{U}_{\rho,k}^b(x) > z) = \sum_{m=1}^\infty P(\bar{U}_{\rho,k}^b(x) > z | N(b) = m) \frac{e^{-b} b^m}{m!}$$

$$= \sum_{m=0}^\infty P(U_{\rho,k+m}(x + b) > z) \frac{e^{-b} b^m}{m!} \le \sum_{m=0}^\infty P(U_{\rho,k}(x + b) > z) \frac{e^{-b} b^m}{m!}.$$

Since $h\rho(u,v)$ being a decreasing function of u implies $U_{\rho,k}(x+b)$

$$\geq U_{\rho,k+m}(x+b), \; m=1,2,\ldots. \text{ From this, (43) is obvious.} \quad \square$$

LEMMA 12. *Whenever there exist $0 < 2c < b < \infty$ such that Φ is non-constant on $[2c,b]$, then for all $c/4 \leq x \leq c/2$, $\rho > -1/2$, $k \geq 1$ and $z \geq 1$ we have*

(44)
$$P(U^b_{\rho,k}(x) \leq B_\rho - zD_\rho) \geq$$
$$K(c)z^{-1/(2(1+\rho))} \exp(-z^{1/(1+\rho)} \log z^{1/(1+\rho)} - \log(ce)),$$

where $K(c)$ is as in Inequality 2,

$$B_\rho = (\rho+1)^{-1} \int_{c/4}^{(c/2)+b} s\,d\Phi(s) \text{ and } D_\rho = (\rho+1)^{-1} \int_{3c/2}^{b+c/4} s^{-\rho}d\Phi(s) > 0.$$

Proof. Observe that $[2c,b] \subset \left[\dfrac{3c}{2}, b+\dfrac{c}{4}\right]$ so that $D_\rho > 0$ and

$$U^b_{\rho,k}(x) \leq \int_x^{b+x} \frac{s}{\rho+1}d\Phi(s) - (N(c)+k)^{\rho+1}\int_{c+x}^{b+x} \frac{s^{-\rho}}{\rho+1}d\Phi(s),$$

which, since $c/4 \leq x \leq c/2$, is

$$\leq \int_{c/4}^{b+c/2} \frac{s}{\rho+1}d\Phi(s) - (N(c))^{\rho+1}\int_{3c/2}^{b+c/4} \frac{s^{-\rho}}{(\rho+1)}d\Phi(s) = B_\rho - N(c)^{\rho+1}D_\rho.$$

Now apply Inequality 2. $\square$

Our last lemma completes the proof for the case $\rho > -1/2$.

LEMMA 13. *Whenever there exist $0 < 2c < b < \infty$ such that Φ is non-constant on $[2c,b]$, then $V_{\rho,k}$ is never a non-degenerate normal random variable for any $\rho > -1/2$ and $k \geq 1$.*

Proof. Choose any $c/4 \leq x \leq c/2$ and $z > 1$. We have by (42)

$$P(V_{\rho,k}(x) \leq -zD_\rho + B_\rho) \geq$$

$$P(U_{\rho,k}^b(x) \leq B_\rho - 2zD_\rho, \bar{U}_{\rho,k}^b(x) + g_k(x) \leq zD_\rho) \geq$$

$$P(U_{\rho,k}^b(x) \leq B_\rho - 2zD_\rho) - P(\bar{U}_{\rho,k}^b(x) + g_k(x) > zD_\rho)$$

$$\equiv P(z) - Q(z).$$

Note that by Lemma 11, for $c/4 \leq x \leq c/2$

$$Q(z) \leq P(U_{\rho,k}(x + b) + g_k(x + b) \geq zD_\rho - A_\rho) = P(V_{\rho,k}(x + b) \geq zD_\rho - A_\rho),$$

where

$$A_\rho = \sup_{c/4 \leq x \leq c/2} \{g_k(x) - g_k(x + b)\} < 0.$$

Also, by an application of Lemma 12, for all $1/(1 + \rho) < \delta < 2$

$$\lim_{z \to \infty} e^{z^\delta} P(z) = \infty.$$

Thus for all $c/4 \leq x \leq c/2$ and $1/(1 + \rho) < \delta < 2$ we have

$$\lim_{z \to \infty} e^{z^\delta} \{P(V_{\rho,k}(x) \leq -zD_\rho + B_\rho) + P(V_{\rho,k}(x + b) \geq zD_\rho - A_\rho)\}$$

$$\geq \lim_{z \to \infty} e^{z^\delta} \{P(V_{\rho,k}(x) \leq -zD_\rho + B_\rho) + Q(z)\} \geq \lim_{z \to \infty} e^{z^\delta} P(z) = \infty.$$

Since (12) gives

$$P(V_{\rho,k} \leq -zD_\rho + B_\rho) + P(V_{\rho,k} \geq zD_\rho - A_\rho)$$

$$\geq \int_{c/4}^{c/2} \{P(V_{\rho,k}(x) \leq -zD_\rho + B_\rho)f_k(x) + P(V_{\rho,k}(x + b) \geq zD_\rho - A_\rho)f_k(x + b)\}dx,$$

we have from the above that for all $1/(1 + \rho) < \delta < 2$

$$\lim_{z \to \infty} e^{z^\delta} \{P(V_{\rho,k} \leq -zD_\rho + B_\rho) + P(V_{\rho,k} \geq zD_\rho - A_\rho\} = \infty.$$

This shows that $V_{\rho,k}$ cannot be non-degenerate normal. □

References.

[1] CSÖRGŐ, S., HAEUSLER, E. and MASON, D. (1988). A probabilistic approach to the asymptotic distribution of sums of independent, identically distributed random variables. *Adv. Appl. Math.* **9**, 259-333.

[2] DASS, M. (1974). Poisson processes and distribution free statistics. *Adv. Appl. Prob.* **6**, 359-375.

[3] MASON, D. and SHORACK, G. (1991). Necessary and sufficient conditions for asymptotic normality of L-statistics. *Ann. Prob.*, to appear.

[4] SHORACK, G. and WELLNER, J. (1986). *Empirical Processes with Application to Statistics.* John Wiley and Sons, New York.

David M. Mason   Galen R. Shorack
Department of Mathematical Sciences   Department of Statistics
501 Ewing Hall   GN-22
University of Delaware   University of Washington
Newark, Delaware 19716   Seattle, Washington 98195

Progress in Probability

Edited by:

Professor Thomas M. Liggett
Department of Mathematics
University of California
Los Angeles, CA 90024-1555

Professor Charles Newman
Courant Institute of
  Mathematical Sciences
251 Mercer Street
New York, NY 10012

Professor Loren Pitt
Department of Mathematics
University of Virginia
Charlottesville, VA 22903-3199

Progress in Probability includes all aspects of probability theory and stochastic processes, as well as their connections with and applications to other areas such as mathematical statistics and statistical physics. Each volume presents an in-depth look at a specific subject, concentrating on recent research developments. Some volumes are research monographs, while others will consist of collections of papers on a particular topic.

We encourage preparation of manuscripts in LaTeX or AMS TeX for delivery in camera-ready copy, which leads to rapid publication, or in electronic form for interfacing with laser printers or typesetters.

Proposals should be sent directly to the series editors or to Birkhäuser Boston, 675 Massachusetts Avenue, Cambridge, MA 02139.